LONDON MATHEMATICAL SOCIETY LECTURE NOTE S...

Managing Editor: Professor J.W.S. Cassels, Department of Pure ...
University of Cambridge, 16 Mill Lane, Cambridge CB2 1SB, E...

The titles below are available from booksellers, or, in case of difficulty, from Cambridge University Press.

London Mathematical Society Lecture Note Series. 246

p-Automorphisms
of Finite *p*-Groups

E. I. Khukhro
Institute of Mathematics, Novosibirsk

CAMBRIDGE
UNIVERSITY PRESS

PUBLISHED BY THE PRESS SYNDICATE OF THE UNIVERSITY OF CAMBRIDGE
The Pitt Building, Trumpington Street, Cambridge CB2 1RP, United Kingdom

CAMBRIDGE UNIVERSITY PRESS
The Edinburgh Building, Cambridge, CB2 2RU, United Kingdom
40 West 20th Street, New York, NY 10011-4211, USA
10 Stamford Road, Oakleigh, Melbourne 3166, Australia

First published 1997

Printed in the United Kingdom at the University Press, Cambridge

A catalogue record for this book is available from the British Library

Library of Congress Cataloging in Publication data

Khukhro, Evgenii I., 1956–.
p-automorphisms of finite p-groups / E.I. Khukhro.
  p. cm. – (London Mathematical Society lecture note series; 246)
Includes bibliographical references (p. –) and index
ISBN 0 521 59717 X pb
1. Automorphisms. 2. Finite groups. I. Title. II. Series
QA174.2.K58 1998
512'.2–dc21 97–28654 CIP

ISBN 0 521 59717 X paperback

Contents

Preface

This is a compilation of the lectures given in 1990–97 in the universities of Novosibirsk, Freiburg (in Breisgau), Trento and Cardiff. The book gives a concise account of several, mostly very recent, theorems on the structure of finite p-groups admitting p-automorphisms with few fixed points. The proofs, given in full detail, require various powerful methods of studying nilpotent p-groups; these methods are presented in the manner of a textbook, accessible for students with only a basic knowledge of linear algebra and group theory. Every chapter ends with exercises which vary from elementary checks to relevant results from research papers (but none of them is referred to in the proofs).

By the classical theorems of G. Higman, V. A. Kreknin and A. I. Kostrikin, a Lie ring is soluble (nilpotent) if it has a fixed-point-free automorphism of finite (prime) order. (These Lie ring theorems are also included along with all necessary preliminary material.) Prompted by and based on these Lie ring results, the main theorems of the book state that a finite p-group is close to being soluble (nilpotent) in terms of the order of a p-automorphism and the number of its fixed points. These results can be viewed as general structure theorems about finite p-groups. They are closely related to the theory of (pro-) p-groups of maximal class and given coclass and have natural extensions to locally finite p-groups.

Presenting linear (mostly Lie ring) methods in the theory of nilpotent groups is another main objective of the book. Of course, the methods are judged as tools yielding certain results; on the other hand, the results themselves can be viewed as an excuse for presenting the methods. The proofs of the main results involve viewing automorphisms as linear transformations, associated Lie rings, theory of powerful p-groups, the correspondences of A. I. Mal'cev and M. Lazard given by the Baker–Hausdorff Formula. Applications of the Baker–Hausdorff Formula are rare in the theory of finite p-groups; remarkably, the Mal'cev Correspondence is an essential ingredient in the proof of one of the main results, and the Lazard Correspondence is used to make easier reductions to Lie rings in the proofs of two others.

During the preparation of this book the author enjoyed the hospitality of the School of Mathematics of the University of Wales, Cardiff, being a visiting research professor there, and, at earlier stages, of the Department of Mathematics of Trento University, as a *Professore a contratto*. The author is grateful to his colleagues, whose attention and advice helped a lot to improve the presentation; in particular, the author thanks A. Caranti, O. H. Kegel, J. C. Lennox, N. Yu. Makarenko, V. D. Mazurov, Yu. A. Medvedev, A. Shalev, and J. Wiegold.

Introduction

Many problems in group theory arise from the fact that the group operation may not be commutative: in general $ab \neq ba$, for elements a, b of a group. It is natural to distinguish classes of groups with respect to how close they are to commutative (abelian) ones. To measure the deviation from commutativity the *commutator* of the elements a, b is defined to be $[a, b] = a^{-1}b^{-1}ab$. It is easy to see that $ab = ba$ if and only if $[a, b] = 1$ (we use 1 to denote the neutral element of a group). Now, a group G is abelian if and only if $[x, y] = 1$ for all $x, y \in G$ (in other words, if the law $[x, y] = 1$ holds on G).

Iterated commutators give rise to generalizations of abelian groups, other classes of groups that are less commutative, although, in a way, close to commutative ones. So $[...[[x_1, x_2], x_3], \ldots, x_k]$ is a *simple* (or *left-normed*) commutator of weight k in the elements $x_1, x_2, \ldots, x_k$. A group is said to be *nilpotent of nilpotency class* $\leq c$ if $[...[[x_1, x_2], x_3], \ldots, x_{c+1}] = 1$ for any elements $x_1, x_2, \ldots, x_{c+1}$. Another way of taking iterated commutators defines the class of soluble groups of derived length $\leq d$. These are groups satisfying the law $\delta_d(x_1, x_2, \ldots, x_{2^d}) = 1$, where, recursively, $\delta_1(x_1, x_2) = [x_1, x_2]$, and $\delta_{k+1}(x_1, x_2, \ldots, x_{2^{k+1}}) = [\delta_k(x_1, x_2, \ldots, x_{2^k}), \delta_k(x_{2^k+1}, x_{2^k+2}, \ldots, x_{2^{k+1}})]$. These generalizations (and many others) can also be defined via existence of normal series with commutative or central factors.

The more commutative is the group operation, the friendlier seems the group. For example, the finite abelian groups admit a well-known description. On the other hand, each area of mathematics has its own problems, so does the theory of abelian groups (even the theory of finite abelian groups contains, in a way, a large portion of number theory with its difficult problems). In fact, many areas of mathematics are studied modulo others, more transparent from some viewpoint. Having commutativity in mind, we can build up the following kind of series of classes of finite groups, in the order of decreasing commutativity of the group operation:

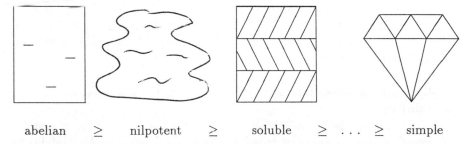

abelian $\geq$ nilpotent $\geq$ soluble $\geq \ldots \geq$ simple

(Extending this picture to arbitrary infinite groups, we could place polynilpotent and polycyclic between nilpotent and soluble, or insert other generalizations of solubility to the right of soluble; free groups, which do not satisfy any non-trivial laws, could be placed somewhere on the right, etc.)

We shall be dealing mostly with nilpotent groups in these lectures. The picture helps to describe a vague feeling about the place which the class of nilpotent groups occupies in finite group theory. Abelian groups seem to be the most transparent, admitting the well-known classification. The (non-abelian) simple groups are, probably, the most non-commutative finite groups: they have no proper normal subgroups, in contrast with abelian groups, all of whose subgroups are normal, or with nilpotent or soluble groups, which have plenty of normal subgroups. Nevertheless, the finite simple groups are also being classified, although this classification is very difficult, occupies thousands of pages of research articles, and is one of the main achievements of mathematics in this century. The finite simple groups seem to have very rigid structure, like crystals; they can be reconstructed from a small fragment (the centralizer of an involution, say). The above remark about studying modulo something is fully applied here: a typical result on non-soluble finite groups is a statement about the factor-group over the largest normal soluble subgroup, or a statement characterizing the (non-abelian) chief factors of a group. The structure of soluble groups appears in certain layers, nilpotent or abelian sections. A typical result about finite soluble groups is bounding the length of the shortest normal series with nilpotent factors, while the structure of these nilpotent factors remains unknown, but is considered good enough, so to say. The difference of soluble finite groups from simple ones seems similar to the difference of graphite from diamond.

The class of nilpotent groups is more like a marsh, a swamp, because their structure appears to be rather amorphous and because of their notorious diversity. There are actually special works showing that there are lots and lots of nilpotent groups and that a classification of nilpotent groups is, in a way, impossible. This is why it is important to have certain beacon lights, marks, showing the ways in this swamp, giving an idea where to work and what kind of results are good in the theory of nilpotent groups.

From the above "more or less commutativity" viewpoint, it is clear that a result in group theory is the better, the more commutativity it yields at the output. For example, one of the "beacon lights" is the Burnside Problem, which can be viewed as asking whether the identity $x^n = 1$ implies some kind of commutativity of the group operation.

Example. Suppose that $x^2 = 1$ for every element x in a group G. Then G is abelian.

Proof. For any two elements $a, b \in G$ we have $abab = 1$. Multiplying on the left by a and then by b we get $baabab = ba$. On the left-hand side we have $baabab = b(a^2)bab = b^2ab = ab$. As a result, $ab = ba$ for any a and b, as required.

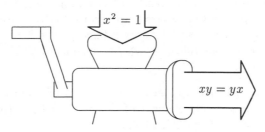

For sufficiently large exponents, there exist "bad" groups that are very far from being commutative (groups of S. I. Adyan and P. S. Novikov and of A. Yu. Ol'shanskii). However, if we restrict attention to finite groups only (which means considering the Restricted Burnside Problem), then the results are positive. Using the W. Magnus – A. N. Sanov reduction to Lie algebras, A. I. Kostrikin proved that a d-generator finite group of prime exponent p is nilpotent of class bounded by a function depending on p and d. (Yu. P. Razmyslov showed that the nilpotency class can increase unboundedly with the growth of d, for $p \geq 5$.) E. I. Zelmanov proved that a d-generator finite group of a prime-power exponent p^k is nilpotent of class bounded by a function of p^k and d. Together with the Reduction Theorem of P. Hall and G. Higman, this completes the positive solution of the Restricted Burnside Problem for all exponents in the class of soluble groups and, modulo the classification of the finite simple groups, for all finite groups: for every pair of natural numbers d and n, there exist only finitely many d-generator finite groups of exponent n.

Apart from groups of given exponent, there are other ways of choosing interesting classes of nilpotent groups. For example, N. Blackburn introduced p-groups of maximal class, that is, groups of order p^n and nilpotency class $n - 1$ (which is maximal possible for this order). These groups became a starting point for the theory of p-groups and pro-p-groups of given coclass (S. Donkin, C. R. Leedham-Green, A. Mann, S. McKay, M. Newman, W. Plesken, A. Shalev, E. I. Zelmanov and others). Another interesting generalization are the so-called thin p-groups (and pro-p-groups) where substantial progress was recently made by R. Brandl, A. Caranti, S. Mattarei, M. Newman and C. Scoppola.

In these lectures we shall consider nilpotent p-groups admitting certain p-automorphisms. A bijection φ of a group G onto itself is an *automorphism* if it preserves the group operation: $(xy)^\varphi = x^\varphi y^\varphi$ for all $x, y \in G$. For any $g \in G$, one can form the *inner automorphism* $\tau_g : x \to g^{-1}xg$. It is easy to see that τ_g is the identity mapping of G if and only if g commutes with all elements of G. So, modulo commutative groups, studying a group is equivalent to the study of its (inner) automorphisms. Conversely, all automorphisms of a group can be regarded as inner automorphisms of a larger group. Thus, studying automorphisms of groups is, in a way, equivalent to studying groups themselves; this approach sometimes provides certain advantages. For example, the theory of p-groups of maximal class is virtually equivalent to that of p-groups admitting an automorphism of order p with exactly p fixed points, and a large portion of the theory of p-groups of given coclass amounts to that of p-groups admitting an automorphism of order p^k with exactly p fixed points.

Again, results on automorphisms may be considered the better, the more commutativity they yield.

Example. Suppose that φ is an automorphism of a finite group G such that $\varphi^2 = 1$ and 1 is the only element of G left fixed by φ (in other words, $x^\varphi \neq x$ whenever $x \neq 1$). Then G is abelian.

Proof. We show first that $G = \{g^{-1}g^\varphi \mid g \in G\}$. Since G is finite, it is sufficient to show that the mapping $g \to g^{-1}g^\varphi$ is injective. If $g_1^{-1}g_1^\varphi = g_2^{-1}g_2^\varphi$, then $g_2 g_1^{-1} = g_2^\varphi (g_1^\varphi)^{-1} = (g_2 g_1^{-1})^\varphi$, whence $g_2 g_1^{-1} = 1$ by hypothesis, that is, $g_2 = g_1$. Now for each $h \in G$ we have $h = g^{-1}g^\varphi$ for some $g \in G$; then $hh^\varphi = g^{-1}g^\varphi(g^{-1}g^\varphi)^\varphi = g^{-1}g^\varphi(g^\varphi)^{-1}g^{\varphi^2} = g^{-1}g = 1$, since $\varphi^2 = 1$. In other words, $h^\varphi = h^{-1}$ for every $h \in G$. Finally, for any $a, b \in G$, we have $ba = (a^{-1}b^{-1})^\varphi = (a^{-1})^\varphi (b^{-1})^\varphi = ab$. $\qquad\qquad\square$

An automorphism that fixes only the identity element of a group is called *regular*. One can prove that if a finite group admits a regular automorphism of order 3, then it is nilpotent of class at most 2. By a theorem of J. G. Thompson, 1959, every finite group with a regular automorphism of prime order is nilpotent. G. Higman in 1957 proved that the nilpotency class of a nilpotent group with a regular automorphism of prime order p is bounded by a function $h(p)$, depending on p only. (V. A. Kreknin and A. I. Kostrikin in 1963 found a new proof giving an explicit upper bound for this Higman's function.) The theorem of J. G. Thompson is an example of a result modulo the theory of nilpotent groups, while the theorem of G. Higman deals with nilpotent groups from the outset.

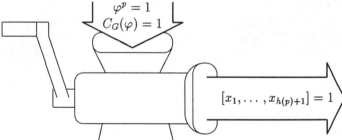

In fact, the works of G. Higman, V. A. Kreknin and A. I. Kostrikin are essentially about Lie rings with regular automorphisms: the group-theoretic corollaries are rather straightforward consequences. V. A. Kreknin in 1963 also proved that a Lie ring with a regular automorphism of arbitrary finite order n is soluble, of derived length bounded in terms of n. (However, it is still an open problem to obtain an analogous result for nilpotent groups.) These Lie ring results turned out to be useful in the study of p-automorphisms of finite p-groups with few fixed points ("almost regular" ones), although a p-automorphism of a finite p-group can never be regular.

The results that we are aiming at in these lectures are about the structure of a finite p-group P admitting a p-automorphism φ of order p^n with exactly p^m fixed points. First consider the case where $n = 1$, that is, $|\varphi| = p$.

- Then the derived length of P is (p, m)-bounded, that is, bounded in terms of p and m only [J. Alperin, 1962].

- Moreover, in a match to G. Higman's Theorem, P contains a subgroup of (p, m)-bounded index which is nilpotent of p-bounded class [E. I. Khukhro, 1985].

- In another direction, prompted by the results on p-groups of maximal class, P also contains a subgroup of (p, m)-bounded index which is nilpotent of m-bounded class [Yu. A. Medvedev, 1994a,b].

Now consider the general situation, with $|\varphi| = p^n$.

- Then the derived length of P is bounded in terms of p, n and m [A. Shalev, 1993a].

- Moreover, in a match to Kreknin's Theorem, P contains a subgroup of (p, n, m)-bounded index which is soluble of p^n-bounded derived length [E. I. Khukhro, 1993a].

- In the extreme case of $m = 1$, where the number of fixed points is p, minimal possible, P contains a subgroup of (p, n)-bounded index which is nilpotent of class 2 (for $|\varphi| = p$, [R. Shepherd, 1971], and [C. R. Leedham-Green and S. McKay, 1976]; for $|\varphi| = p^n$, [S. McKay, 1987], and [I. Kiming, 1988]).

(Most of the above results can be extended to certain classes of infinite groups, but we do not discuss these generalizations in the book.)

The "modular" case, where a p-automorphism acts on a p-group, turned out to be easier than the "ordinary" one, where the order of the automorphism is coprime to the order of the group. It is still an open problem to obtain an analogue of Kreknin's Theorem for nilpotent groups with regular automorphisms of finite order. The only known cases are those of prime order [G. Higman, 1957] and order four [L. G. Kovács, 1961]. There is also a generalization of Higman's Theorem for nilpotent groups with an almost regular automorphism of prime order ([E. I. Khukhro, 1990] and [Yu. Medvedev, 1994c]); see also the book [E. I. Khukhro, 1993b].

Applications of Lie rings and other linear tools are based on the fact that nilpotent groups are close to commutative ones. Abelian subgroups and sections are similar to vector spaces (or modules), and the action of automorphisms on such sections is similar to linear transformations. The group commutators can be used to define the structure of a Lie ring on the direct sum of the additively written factors of the lower central series of a group, the so-called associated Lie ring. Other ways of constructing a Lie ring from a nilpotent group, the correspondences of A. I. Mal'cev and M. Lazard, are based on the Baker–Hausdorff Formula. Under these correspondences, the Lie ring reflects the properties of the group in a much better way, but this technique cannot be applied to any nilpotent group.

The Lie ring method of solving group-theoretic problems consists of three major steps. First, the problem must be translated into a corresponding problem about Lie rings constructed from the groups. Then the Lie ring problem is solved. The results on Lie rings must then be translated back, into required conclusions about groups. The advantage lies in the fact that it is usually easier to deal with Lie rings as more linear objects; for example, one can extend the ground ring, which gives rise to the analogues of eigenspaces with respect to the automorphism, etc. On the other hand, both crossings over, from groups to Lie rings and back, may be quite non-trivial. For example, the number of fixed points of an automorphism may well be much greater on the associated Lie ring than on the group. Another example: if the Lie ring result gives a subring of small nilpotency class and small index, say, this does not immediately give the required subgroup in the group, since there is no good correspondence between subrings of the associated Lie ring and subgroups of the group (such kind of difficulty had to be overcome in the theorem on almost regular coprime automorphism in [E. I. Khukhro, 1990]). Thus, reductions to Lie rings and recovering information about the group from the Lie ring results may sometimes require even more effort than the Lie ring theorems themselves.

The difficulty in proving an analogue of Kreknin's Theorem for nilpotent groups with a regular automorphism of composite order lies in the fact that the derived length of the associated Lie ring may be smaller than that of the group. What makes the "modular" case that we are dealing with in this book much more friendly is the bounds for the ranks of all abelian sections that follow from the restrictions on the number of fixed points. This reduces the proofs to powerful p-groups whose nice linear properties make it easier to apply the Lie ring methods.

Lie rings are used in the book not only for proving the main results, but also for deriving many of the standard "linear" properties of nilpotent groups. Naturally, quite a lot of preliminary material on Lie rings is included.

We tried to make the book closer to a textbook, really accessible to students with only undergraduate knowledge in algebra and group theory; efforts were made to ensure that there are no stumbling-blocks disguised by the words "obvious" or "as is well-known". The relatively short chapters follow the pattern of the lecture course. The chapters on methods alternate with chapters on applications, so that the reader could see, as soon as possible, the results that can be achieved by these methods. Exercises included in every chapter vary from elementary ones to relevant results from research papers (but none of them is referred to in the proofs).

Chapter 1 contains preliminaries on groups, rings and modules, and on varieties of algebraic systems. Most of this material may well be not more than a reminder, but we note that varietal arguments, often in terms of groups or Lie rings with additional operations, are essential in the subsequent chapters.

Because of the outstanding role of automorphisms in the book, we devoted a special chapter, Chapter 2, to preliminary material on automorphisms, containing a few folklore elementary lemmas on fixed points.

Chapter 3 combines material on nilpotent and soluble groups. Besides definitions and basic properties, it contains some criteria for soluble groups to be nilpotent and a criterion for a variety to be soluble.

In Chapter 4 elementary properties of finite p-groups are proved, as well as a theorem of P. Hall on the orders of the lower central factors of a normal subgroup.

Chapter 5 introduces Lie rings. A section on soluble and nilpotent Lie rings collects the analogues of group-theoretic results from Chapter 3, including a criterion for a variety to be soluble, which is used in Chapter 7 for proving Kreknin's Theorem. Then free Lie rings are constructed within free associative algebras. The main use of this construction is in Chapters 9 and 10 on the Baker–Hausdorff Formula and the Mal'cev Correspondence; the only fact needed earlier is that free Lie rings are multihomogeneous with respect to free generators.

Chapter 6 introduces one of our main tools, the associated Lie rings; in § 6.3 they are used to derive several properties of nilpotent groups.

Theorems of G. Higman, V. A. Kreknin and A. I. Kostrikin on regular automorphisms of Lie rings are proved in Chapter 7 in generalized combinatorial form: first for $(\mathbb{Z}/n\mathbb{Z})$-graded Lie rings, then for free Lie rings, and finally for Lie rings with automorphisms. Although it is only these combinatorial results that are used later, we could not help deriving the consequences for Lie rings and finite nilpotent groups with regular automorphisms.

Technique accumulated to this point enables us to prove in Chapter 8 the first of the main results, an analogue of G. Higman's Theorem for p-groups with an almost regular automorphism of order p.

In Chapter 9 free nilpotent $\mathbb{Q}$-powered (torsion-free divisible) groups and nilpotent Lie $\mathbb{Q}$-algebras are constructed within associative algebras. We prove the basic property of the Baker–Hausdorff Formula, which links the group and the Lie ring operations.

In Chapter 10 the Baker–Hausdorff Formula is used to establish the Mal'cev Correspondence between nilpotent $\mathbb{Q}$-powered groups and nilpotent Lie $\mathbb{Q}$-algebras. There is also a similar correspondence of M. Lazard for nilpotent p-groups of class $\leq p - 1$. Applications of the Baker–Hausdorff Formula are rare in the theory of finite p-groups; remarkably, they are featured in the proofs of the rest of the main results. In Chapter 12 the Mal'cev Correspondence is substantially used to prove the analogue of Kreknin's Theorem for finite p-groups with an almost regular automorphism of order p^n, and in Chapters 13 and 14 the Lazard Correspondence makes it much easier to perform reductions to Lie rings.

Another important method used in Chapters 12, 13 and 14 is the Lubotzky–Mann theory of powerful p-groups, which is developed in Chapter 11. Every p-group of sectional rank r contains a powerful subgroup of (p, r)-bounded index, which is the "more linear part" of the group. So-called uniformly powerful p-groups enjoy even more linear properties similar to those of homocyclic abelian groups. Bounds for the number of fixed points of a p-automorphism imply bounds for the ranks; this is why powerful p-groups appear naturally in the theory of p-groups with almost regular p-automorphisms.

We already mentioned that Chapter 12 contains an analogue of Kreknin's Theorem for finite p-groups with an almost regular automorphism of order p^n, and that the Mal'cev Correspondence is used in the proof. Another ingredient of the proof is calculations in powerful p-groups, using an important Interchanging Lemma of A. Shalev. Kreknin's Theorem is used twice. First we apply it to the associated Lie ring of a powerful p-group, which already leads to a "weak" bound for the derived length depending on both the number of fixed points and the order of the automorphism. Then Kreknin's Theorem is applied via the Mal'cev Correspondence to a free nilpotent group with an automorphism of finite order. The general result obtained allows us to find a required subgroup of bounded index with a "strong" bound for its derived length, depending only on p^n, the order of the automorphism.

Chapter 13 deals with the extreme case where a finite p-group P admits a p-automorphism φ with just p fixed points, the least possible number. Then the result is extremely strong: P has a subgroup of bounded index which is nilpotent of class ≤ 2 (even abelian for $p = 2$). We give a proof which is different from the original proofs of C. R. Leedham-Green, S. McKay and R. Shepherd (for $|\varphi| = p$) and S. McKay and I. Kiming (for $|\varphi| = p^n$). Although with possibly worse bounds for the index of the subgroup, our proof is more Lie ring oriented, making use of Higman's and Kreknin's Theorems, theory of powerful p-groups, and the Lazard Correspondence. After reduction to Lie rings, we prove independently an analogous theorem on Lie rings which is interesting in its own right. There we adopt the approach of Yu. Medvedev, defining a new "lifted" Lie ring multiplication. Anticipated in the works of A. Shalev and E. I. Zelmanov on p-groups and pro-p-groups of given coclass, Yu. Medvedev's construction is remarkably transparent and elementary.

In Chapter 14 we prove that if a finite p-group P admits an automorphism of order p with p^m fixed points, then P has a subgroup of (p, m)-bounded index which is nilpotent of m-bounded class. Recall that in Chapter 8 we prove that P has a subgroup of bounded index which has p-bounded nilpotency class. Neither of these results follows from the other; which conclusion is better depends on which of the parameters p and m is "much less" than the other. The proof is quickly reduced to Lie rings via the Lazard Correspondence. This reduction to Lie rings based on the result of Chapter 8 is easier than in

Chapter 13, since here we are not constrained by the requirement to obtain such a strong bound for the nilpotency class as 2. In fact, the bulk of Chapter 14 is an independent proof of the analogous Lie ring theorem (of Yu. Medvedev). Much of the technique developed in Chapter 13 is used there, including the new lifted Lie products.

The book [E. I. Khukhro, 1993b] also contains the theorems from Chapters 8 and 12, but in a more condensed form. In the present book the proofs are rewritten (inflated) to make them more accessible for beginners, and all of the background material is included. (Among other topics on automorphisms of nilpotent groups in [E. I. Khukhro, 1993b] are splitting p-automorphisms of finite p-groups with applications to the Hughes problem, almost regular coprime automorphisms of nilpotent groups and Lie rings, and some generalizations of the Restricted Burnside Problem to varieties of operator groups.)

Besides the research papers mentioned in the book, we included several general references in the Bibliography; many of the textbooks may be indicated as our sources, especially for the preliminary chapters. The survey [A. Shalev, 1995] on finite p-groups reflects Lie ring methods, almost regular automorphisms, and applications in the theory of pro-p-groups. Almost regular automorphisms in a broader context of locally finite groups were also surveyed in [B. Hartley, 1987]. Linear methods in the theory of (residually) nilpotent groups and pro-p-groups are the subject of the survey [E. I. Zelmanov, 1995].

Chapter 1
Preliminaries

Here we record some basic definitions and elementary results about group, rings, and varieties of algebraic systems. We assume a basic knowledge in undergraduate algebra. In particular, in group theory, the reader is supposed to be familiar with definitions and basic properties of subgroups, cosets, cyclic subgroups, direct products, the structure of finite abelian groups, normal subgroups, the Homomorphism Theorems. The Sylow Theorems may be referred to occasionally, but they are not used in the proofs of the main results of the book, which are all about finite p-groups. Some familiarity with rings and modules is assumed, although many of the definitions are briefly reproduced.

We shall often use exponent notation for images under mappings, that is, a^φ or A^φ for $\varphi(a)$ or $\varphi(A)$ respectively, and sometimes also the right operator notation, $a\varphi$ for $\varphi(a)$, say. The identity mapping of a set M will be denoted by 1_M. Standard notation $\mathbb{N}$, $\mathbb{Z}$, $\mathbb{Q}$, $\mathbb{R}$, $\mathbb{C}$ is fixed for the sets of natural numbers, integers, rational, real and complex numbers, respectively. We shall say that a value is (a,b)-bounded, say, if there is a function depending only on a and b, such that the value does not exceed this function.

§ 1.1. Groups

Some basic definitions. The sets $\mathbb{Z}$, $\mathbb{Q}$, $\mathbb{R}$, $\mathbb{C}$ are groups with respect to addition. The set $\mathbb{Z}/n\mathbb{Z}$ of residues modulo n is a group with respect to addition mod n. Every vector space is a group with respect to addition. The set $\mathbb{S}_M$ of all bijections of a set M onto M is a group with respect to the composition of mappings, the identity mapping 1_M being the neutral element and the inverse mapping being the inverse element. When M is finite of order n, the notation $\mathbb{S}_n$ is often used; $\mathbb{S}_n$ is called the *symmetric group on n letters*.

The set of all bijective linear mappings of a vector space V over a field k is a subgroup of $\mathbb{S}_V$ denoted by $GL(V)$. The set $GL_n(k)$ of all non-degenerate $n \times n$ matrices over k is a group with respect to matrix multiplication. If n is the dimension of V, then fixing a basis of V, we associate a matrix to each linear transformation of V. This correspondence is an isomorphism of $GL(V)$ and $GL_n(k)$.

We use 1 to denote both the neutral element (identity) of a group and the *trivial subgroup* consisting of the neutral element only. To signify that H is a subgroup of G, we write $H \leq G$ or $G \geq H$; strict inequality $H < G$ means that $H \leq G$ and $H \neq G$, that is, H is a *proper* subgroup of G. Saying that a subset or a subgroup is *minimal* or *maximal* with respect to some property, we shall mean minimal or maximal with respect to inclusion.

A (sub)group generated by a subset M is the minimal subgroup containing M, denoted by $\langle M \rangle$; it is unique since it equals the intersection of all subgroups containing M. We know that $\langle M \rangle$ consists of all products $m_1^{\varepsilon_1} \cdots m_s^{\varepsilon_s}$, where $m_i \in M$ and $\varepsilon_i = \pm 1$. Note that if $M \subseteq H \leq G$, then $\langle M \rangle \leq H$. Usually, braces are omitted within the angle brackets, so that $\langle a \mid P(a) \rangle = \langle \{a \mid P(a)\} \rangle$; in a similar way, $\langle A_1, A_2, \ldots \rangle = \langle A_1 \cup A_2 \cup \ldots \rangle$.

The cardinality of a set M is denoted by $|M|$. The index of a subgroup $H \leq G$ is denoted by $|G : H|$. Recall a useful inequality

$$|G : H \cap K| \leq |G : H| \cdot |G : K| \tag{1.1}$$

for the indices of subgroups. It follows that the intersection of n subgroups of index at most m has (m, n)-bounded index (less than or equal to m^n).

An element $x^g = g^{-1}xg$ is the *conjugate* of x under g. Note that $(a^b)^c = a^{bc}$ and $(ab)^c = a^c b^c$ for any elements a, b, c in a group. For any subset M, a similar notation is used: $M^g = g^{-1}Mg = \{g^{-1}mg \mid m \in M\}$. A subgroup $H \leq G$ is *normal*, denoted by $H \trianglelefteq G$, if $H^g = H \Leftrightarrow Hg = gH$ for every $g \in G$. The *factor-group* G/N of a group G by a normal subgroup N is the set of cosets $\{Ng \mid g \in G\}$ with multiplication $Ng \cdot Nh = Ngh$; this operation is well-defined precisely because $N \trianglelefteq G$. The *normal closure* $\langle S^G \rangle$ of a subset S in a group G is the minimal normal subgroup of G containing S, the intersection of all normal subgroups containing S; clearly, $\langle S^G \rangle = \langle s^g \mid s \in S, \ g \in G \rangle$. For any subset S of a group G and a subgroup $H \leq G$, the set $N_H(S) = \{g \in H \mid S^g = S\}$ is a subgroup called the *normalizer* of S in H. The subset S is said to be K-*invariant* if $K \leq N_G(S)$.

Lemma 1.2. *A subgroup $H \leq G$ is contained in the normalizer $N_G(S)$ of a subset $S \subseteq G$ if and only if $S^h \subseteq S$ for every $h \in H$.*

Proof. Indeed, $S^{h^{-1}} \subseteq S \Rightarrow S = (S^{h^{-1}})^h \subseteq S^h$ for every $h \in H$, plus $S^h \subseteq S$ by the hypothesis; hence $S^h = S$ for every $h \in H$. (If S is finite, the result follows immediately from the fact that $|S| = |S^h|$.) $\qquad\square$

For a subgroup $H \leq G$ and a subset $S \subseteq G$, the set $C_H(S) = \{g \in H \mid sg = gs \text{ for all } s \in S\}$ is a subgroup called the *centralizer* of S in H. For a one-element $S = \{s\}$ we simply write $C_G(s)$.

Lemma 1.3. *For subgroups K, H, L, M, N in a group G*
  (a) $K \leq C_G(H)$ *if and only if* $H \leq C_G(K)$,
  (b) *if both L and M are N-invariant, then $C_L(M)$ and $N_L(M)$ are N-invariant too.*

Proof. (a) $K \leq C_G(H) \Leftrightarrow kh = hk$ for all $h \in H, k \in K \Leftrightarrow H \leq C_G(K)$.
  (b) For any $c \in C_L(M)$, $m \in M$ and $n \in N$ we have $c^n m = (cm^{n^{-1}})^n = (m^{n^{-1}}c)^n = mc^n$ since $m^{n^{-1}} \in M$, so $C_L(M)^n \subseteq C_L(M)$; hence $C_L(M)$ is

N-invariant by Lemma 1.2. For any $v \in N_L(M)$ and $n \in N$ we have $M^{v^n} = (M^n)^{v^n} = (M^v)^n = M^n = M$, so $N_L(M)^n \subseteq N_L(M)$; hence $C_L(M)$ is N-invariant by Lemma 1.2. □

The *centre* of a group G is $Z(G) = C_G(G) = \{z \in G \mid zg = gz$ for all $g \in G\}$. Clearly, $Z(G)$ is the intersection of the centralizers of all elements of G.

For any two subsets M and N of a group the *product* MN is the set $\{mn \mid m \in M, n \in N\}$. If M, say, consists of one element, $M = \{m\}$, then we simply write mN, which agrees with the usual notation for cosets if N is a subgroup. If H is a subgroup and N is a normal subgroup of a group G, then the set $HN = NH$ is also a subgroup of G, so that $\langle H, N \rangle = HN$; if both H and N are normal subgroups, then HN is a normal subgroup too.

The (external) *direct product* $A \times B$ of two groups A, B is the set of pairs (a, b), $a \in A$, $b \in B$, with the component-wise operation $(a_1, b_1) \cdot (a_2, b_2) = (a_1 a_2, b_1 b_2)$, so that $(1, 1)$ is the neutral element of $A \times B$ and $(a, b)^{-1} = (a^{-1}, b^{-1})$. Then A and B can be identified with the normal subgroups $\widehat{A} = \{(a, 1) \mid a \in A\}$ and $\widehat{B} = \{(1, b) \mid b \in B\}$ respectively, for which $\widehat{A} \cap \widehat{B} = 1$ and $\widehat{A}\widehat{B} = A \times B$. Conversely, if a group $G = MN$ is a product of two normal subgroups M, N with trivial intersection $M \cap N = 1$ (internal direct product), then G is isomorphic to the direct product $M \times N$, the isomorphism given by $mn \to (m, n)$. Thus, the notions of internal and external direct products are equivalent.

Example 1.4. Suppose that M and N are normal subgroups of a group G. Then $MN/(M \cap N) \cong \overline{M} \times \overline{N}$, where $\overline{M} = M/(M \cap N)$ and $\overline{N} = N/(M \cap N)$. Indeed, $MN/(M \cap N) = \overline{M}\overline{N}$, both $\overline{M}$ and $\overline{N}$ are normal subgroups, and $\overline{M} \cap \overline{N} = 1$ since $m(M \cap N) = n(M \cap N)$ for $m \in M$, $n \in N$ implies that $m \in N$, say.

The *Cartesian product* $\mathrm{Cr}_{i \in I} A_i$ of a family of groups $\{A_i \mid i \in I\}$ is the set of all functions $f : I \to \bigcup_{i \in I} A_i$ such that $f(i) \in A_i$; this is a group with respect to the coordinate-wise operation $(fg)(i) = f(i)g(i)$. The set of those functions f that have non-trivial values only on a finite subset of I (depending on f) is the *direct product* $\mathrm{Dr}_{i \in I} A_i$, a subgroup of $\mathrm{Cr}_{i \in I} A_i$. For a finite set of groups both definitions coincide and the resulting group is isomorphic to $A_1 \times \cdots \times A_n$ (with arbitrary order of parentheses). The equivalent definition of the internal direct product is that $G = \langle A_i \mid i \in I \rangle$, $A_i \trianglelefteq G$ for all $i \in I$, and $A_j \cap \langle A_i \mid i \neq j \rangle = 1$.

The *order* of an element a of a group denoted by $|a|$ is the least positive integer k such that $a^k = 1$, or ∞ if no such k exists. It is easy to see that $|a| = |\langle a \rangle|$. The *exponent* of a group G is the least $n \in \mathbb{N}$ such that $x^n = 1$ for all $x \in G$ (or ∞ if such n does not exist). Clearly, the exponent is the least common multiple of the orders of all elements of G. We denote by G^n the subgroup of a group G generated by all nth powers of the elements of G;

clearly, G^n is the smallest normal subgroup such that the factor-group is of exponent dividing n. A group is *torsion-free* if it has no non-trivial elements of finite order. Let p be a prime number. An element of a group is a *p-element*, if its order is a power of p. A group G is a *p-group*, if it consists of p-elements. For a p-group G, we define the subgroups $\Omega_i(G) = \langle x \in G \mid x^{p^i} = 1 \rangle$. A maximal (with respect to inclusion) p-subgroup is a *Sylow p-subgroup*. The Sylow Theorems state that in a finite group of order $p^k n$, where $p \nmid n$, all Sylow p-subgroups have order p^k (so that every p-subgroup is contained in a subgroup of order p^k), and all Sylow p-subgroups are conjugate.

A *section M/N* of a group G is a factor-group of any of its subgroups: $M \leq G$ and $N \trianglelefteq M$; a section is said to be normal if both M and N are normal in G. Any chain of nested subgroups

$$1 = G_0 \leq G_1 \leq G_2 \leq \ldots \leq G_n = G$$

is a *series* of G of length n. If all of the G_i are normal in G, the series is *normal*; if each G_i is a normal subgroup of G_{i+1}, the series is *subnormal*. The sections G_{i+1}/G_i of a subnormal series are the *factors* of the series. (Of course, one can number the terms of a series in any other way.)

Abelian groups. Recall that a group is *abelian* or *commutative* if $ab = ba$ for any two elements a and b. Abelian groups are often written additively, using $+$ for the operation and 0 for the neutral element (and for the trivial subgroup); but we still denote factor-groups as M/N, rather than $M - N$. For example, the additive group of integers $\mathbb{Z}$ is generated by 1, which is *not* the neutral element here. We denote by ka the kth power of an element a, speak about direct sums of abelian groups instead of direct products, etc.

If A is an abelian group and $n \in \mathbb{N}$, then $nA = \{na \mid a \in A\}$ is a subgroup of A. If n is coprime to the exponent of a finite abelian group A, then $a \to na$ is an injective mapping, since $na = nb \Rightarrow n(a - b) = 0 \Rightarrow a - b = 0$, and hence $nA = A$.

Suppose that $p_1, \ldots, p_n$ are all distinct prime divisors of the order of a finite abelian group A. It is clear that all p_i-elements in A form a subgroup A_{p_i} which is the unique Sylow p_i-subgroup of A. Suppose that, for an element $a \in A$, we have $|a| = m = p_1^{k_1} \cdots p_n^{k_n}$, $k_i \geq 0$ (only the p_i are involved by Lagrange's Theorem). Put $m_i = m/p_i^{k_i}$; the greatest common divisor of all of the m_i is 1, and hence there exist integers u_i such that $1 = u_1 m_1 + \cdots + u_n m_n$. It follows that $a = 1a = u_1 m_1 a + \cdots + u_n m_n a$, where $p_i^{k_i} m_i a = ma = 0$ so that $m_i a \in A_{p_i}$ for each i. Therefore, $A = A_{p_1} + \cdots + A_{p_n}$. Since $A_{p_i} \cap (A_{p_1} + \cdots + \widehat{A}_{p_i} + \cdots + A_{p_n}) = 0$ for each i (the order of an element in the intersection can be only 1), the sum is direct: $A = A_{p_1} \oplus \cdots \oplus A_{p_n}$.

The above decomposition is a part of the Structure Theorem for finite (finitely generated) abelian groups. This theorem takes especially simple form for finite abelian p-groups: every such a group is a direct sum of cyclic groups of orders $p^{k_1}, \ldots, p^{k_s} \geq p$ and the set $\{k_1, \ldots, k_s\}$ is an invariant, that is, does not depend on the choice of the summands (which are in no way unique).

A finite abelian group has *rank n*, if it is a direct sum of n cyclic groups, and n is minimal possible. For a finite p-group P, the rank r is equal to the number of (non-trivial) cyclic subgroups in the direct sum $P = \langle a_1 \rangle \oplus \cdots \oplus \langle a_r \rangle$ and does not depend on such decomposition, as follows from the Structure Theorem.

An abelian group E of prime exponent p is called *elementary abelian*. In the additive notation, E can be viewed as a vector space over $\mathbb{F}_p$, the field of p elements (of residues mod p, say). The addition of vectors is the group operation, and multiplying a vector g by a residue i is taking the ith power ig of the element g (this is well-defined since $pg = 0$); the axioms of a vector space over $\mathbb{F}_p$ are easily checked: for example, we have $i(x + y) = ix + iy$ for $x, y \in E$, $i \in \mathbb{F}_p$, since the group is abelian. One can say that the theory of elementary abelian groups of exponent p is "categorically equivalent" to the theory of vector spaces over $\mathbb{F}_p$: every statement about E as a group of exponent p can be translated into the language of a vector space over $\mathbb{F}_p$ and vice versa. The rank of E as a group is exactly the dimension of E as a space. The automorphism group $\mathrm{Aut}\, E$ coincides with the group of non-degenerate linear transformations over $\mathbb{F}_p$, and so on.

Suppose that $P = \langle a_1 \rangle \oplus \cdots \oplus \langle a_r \rangle$ is an abelian p-group with $|a_i| = p^{k_i} \geq p$; then $\Omega_1(P) = \langle p^{k_1-1} a_1 \rangle \oplus \cdots \oplus \langle p^{k_r-1} a_r \rangle$. The number of cyclic summands, r, is equal to the rank of $\Omega_1(P)$, which is the dimension of $\Omega_1(P)$, an invariant. Thus, the number of cyclic summands for P is always equal to the rank of P and to the rank of $\Omega_1(P)$. Considering the ranks of Ω_{i+1}/Ω_i is a way to prove the uniqueness part of the Structure Theorem (Exercise 1.18). We also have $pP = \langle pa_1 \rangle \oplus \cdots \oplus \langle pa_r \rangle$ and P/pP is an elementary abelian p-group of rank r.

A direct sum of several cyclic groups of equal order p^n is a *homocyclic* group of exponent p^n. Such groups have quite a homogeneous structure.

Lemma 1.5. *Suppose that A is a homocyclic group of exponent p^n and rank r. Then*

(a) *$p^i A = \Omega_{n-i}(A)$ for all $i \leq n$;*

(b) *if $p^i a \in p^j A$ for $i \leq j \leq n$, then $a \in p^{j-i} A$;*

(c) *the mapping $x \to px$ induces isomorphisms of the sections $p^i A / p^{i+1} A$, $i = 0, 1, \ldots, n-1$;*

(d) *if $p^i B \geq p^j A$ for a subgroup $B \leq A$, then $B \geq p^{j-i} A$ unless $j \geq n$;*

(e) *if $p^k A \geq B$, then $\Omega_k(A/B)$ is a homocyclic group of exponent p^k and rank r;*

(f) *if $A \geq B \geq p^k A$, then $B/p^{n-k} B$ is a homocyclic group of exponent p^{n-k} and rank r.*

Proof. Let $A = \langle a_1 \rangle \oplus \cdots \oplus \langle a_r \rangle$ where $|a_i| = p^n$ for all i. Every element $a \in A$ has a unique representation $a = \sum_{i=1}^r u_i a_i$ where $a_i \in \mathbb{Z}/p^n \mathbb{Z}$. It follows that $a \in p^s A \Leftrightarrow p^s | u_i$ for all i, and $p^s | |a| \Leftrightarrow p^{n-s} | u_i$ for all i. Hence (a) and (b) follow. It also follows that $|p^i A| = p^{r(n-i)}$ and hence $|p^i A / p^{i+1} A| = p^r$ for all $i \leq n - 1$. Since the mapping $x \to px$ obviously induces a homomorphism of $p^{i-1} A / p^i A$ onto $p^i A / p^{i+1} A$, this is actually an isomorphism.

To prove (d), choose k such that $B \geq p^k A$, but $B \not\geq p^{k-1} A$. The mapping $x \to px$ induces an isomorphism of $p^{k-1} A / p^k A$ onto $p^k A / p^{k+1} A$ which maps $B \cap p^{k-1} A$ onto $pB \cap p^k A$; hence $p^i B \not\geq p^{k-1+i} A$ as long as $k - 1 + i < n$. If $p^i B \geq p^j A$ for $j \leq n$, it follows that $j \geq k + i$ so that $B \geq p^k A \geq p^{i-j} A$.

(e) The factor-group $A/B = \langle g_1 \rangle \oplus \cdots \oplus \langle g_r \rangle$ is a direct sum of r cyclic p-groups of orders $|g_i| = p^{k_i} \geq p^k$, since $A/p^k A$ is a homomorphic image of A/B. Hence

$$\Omega_k(A/B) = \left\langle p^{k_1 - k} g_1 \right\rangle \oplus \cdots \oplus \left\langle p^{k_r - k} g_r \right\rangle$$

is a homocyclic group of exponent p^k and rank r.

(f) We have $B \geq \left\langle p^k a_1 \right\rangle \oplus \cdots \oplus \left\langle p^k a_r \right\rangle$; and hence $B = \langle g_1 \rangle \oplus \cdots \oplus \langle g_r \rangle$ is a direct sum of r cyclic groups of order $\geq p^{n-k}$ each. Then $B/p^{n-k} B = \langle g_1 \rangle / \left\langle p^{n-k} g_1 \right\rangle \oplus \cdots \oplus \langle g_r \rangle / \left\langle p^{n-k} g_r \right\rangle$ is a homocyclic group of exponent p^{n-k} and rank r. $\square$

Lemma 1.6. (a) *Suppose that B is a homocyclic subgroup of exponent p^n of an abelian group A of the same exponent p^n. Then $A = B \oplus C$ for some $C \leq A$.*

(b) *Suppose that U is a homocyclic abelian group of exponent p^n and U/B is a homocyclic group of the same exponent p^n for some $B \leq U$. Then $U = B \oplus C$ for some $C \leq U$.*

(c) *Suppose that U is a homocyclic abelian group of exponent p^n and $U = V \oplus W$; then V and $U/V \cong W$ are homocyclic groups of exponent p^n.*

Proof. (a) Induction on $|A|$. Suppose that $\Omega_1(A) \not\leq B$; then there is $c \in A \setminus B$ such that $pc = 0$ and hence $B \cap \langle c \rangle = 0$. Then the image $\overline{B}$ of B in $\overline{A} = A/\langle c \rangle$ is isomorphic to B. By the induction hypothesis $\overline{A} = \overline{B} \oplus \overline{C}$ for some $\overline{C} \leq \overline{A}$. Let C be the full preimage of $\overline{C}$; we claim that $A = B \oplus C$. Indeed, $A = B + C$ and $B \cap C \leq B \cap \langle c \rangle = 0$. It remains to prove that if $\Omega_1(A) \leq B$, then $B = A$. We use induction on k to prove that if $p^k a = 0$, then $a \in B$; for $k = 1$ this is true by the assumption $\Omega_1(A) \leq B$. For $k > 1$ we have $p^{k-1} a \in \Omega_1(A) \leq B$; since B is homocyclic of the same exponent as A, there is $b \in B$ such that $p^{k-1} b = p^{k-1} a$, whence $p^{k-1}(b - a) = 0$. By the induction hypothesis $b - a \in B$, whence $a \in B$.

(b) Let $U/B = \langle \bar{a}_1 \rangle \oplus \cdots \oplus \langle \bar{a}_k \rangle$ with $|\bar{a}_i| = p^n$ for all i. Choose some preimages a_i of the $\bar{a}_i$; then $|a_i| = p^n$ for all i since p^n is the exponent of U. We claim that $U = B \oplus C$ where $C = \langle a_1 \rangle + \cdots + \langle a_k \rangle$. Since $U = B + C$, we need only to show that $B \cap C = 0$. If $\sum_{i=1}^k k_i a_i \in B$, then $\sum_{i=1}^k k_i \bar{a}_i = 0$, whence $p^n | k_i$ for all i, since U/B is homocyclic of exponent p^n. But then

$\sum_{i=1}^{k} k_i a_i = 0$ too.

(c) Both V and W are direct sums of cyclic p-groups; together, these decompositions form a decomposition of $U = V \oplus W$ into the direct sum of cyclic groups. Since the set of the orders of the summands is unique by the Structure Theorem, both V and W must be homocyclic of exponent p^n. $\square$

The assertion (a) can be used to prove the existence part of the Structure Theorem for abelian p-groups.

Homomorphisms and automorphisms. Recall that a mapping φ of a group G into another group is a homomorphism if φ preserves the group operation: $(ab)^\varphi = a^\varphi b^\varphi$ (it follows that $(a^{-1})^\varphi = (a^\varphi)^{-1}$ and $1^\varphi = 1$). The set $\mathrm{Ker}\,\varphi = \{g \in G \mid g^\varphi = 1\}$ is always a normal subgroup of G, the *kernel* of φ. If $N \trianglelefteq G$, then $g \to gN$ is the so-called *natural homomorphism* of G onto the factor-group G/N. Congruences mod N denote equalities of the images in G/N of elements or subsets:

$$a \equiv b \,(\mathrm{mod}\,N) \quad \Leftrightarrow \quad aN = bN; \qquad A \equiv B \,(\mathrm{mod}\,N) \quad \Leftrightarrow \quad AN = BN.$$

Let φ be a homomorphism of a group G with kernel $N = \mathrm{Ker}\,\varphi$. The Homomorphism Theorems state that

- φ is a composition of the natural homomorphism of G onto G/N and some isomorphism of G/N onto G^φ;

- $H \to H^\varphi$ is a bijection of the set of all subgroups containing N and the set of subgroups of G^φ, and $N \le H \trianglelefteq G \Leftrightarrow H^\varphi \trianglelefteq G^\varphi$;

- for any subgroup $K \le G$ the full preimage of K^φ is KN; if $L \trianglelefteq K$, then $L^\varphi \trianglelefteq K^\varphi$ and $K^\varphi/L^\varphi \cong K/L(N \cap K)$; in particular, $K^\varphi \cong K/(K \cap N)$, and if $N \le M \trianglelefteq L$, then $L^\varphi/M^\varphi \cong L/M$.

We now prove a very useful though simple lemma.

Lemma 1.7. *If φ is a homomorphism of a group G, then $\langle M \rangle^\varphi = \langle M^\varphi \rangle$ for any subset $M \subseteq G$.*

Proof. We know that $\langle M \rangle = \{m_1^{\varepsilon_1} \cdots m_s^{\varepsilon_s} \mid m_i \in M, \ \varepsilon_i = \pm 1\}$. Since φ is a homomorphism, $(m_1^{\varepsilon_1} \cdots m_s^{\varepsilon_s})^\varphi = (m_1^\varphi)^{\varepsilon_1} \cdots (m_s^\varphi)^{\varepsilon_s}$ for every such a product. Hence $\langle M \rangle^\varphi$ coincides with the set of the products $(m_1^\varphi)^{\varepsilon_1} \cdots (m_s^\varphi)^{\varepsilon_s}$ which is exactly $\langle M^\varphi \rangle$. $\square$

A homomorphism of a group into itself is an *endomorphism*. A subgroup $H \le G$ is said to be *fully invariant*, if $H^\psi \le H$ for every endomorphism ψ of G. Lemma 1.7 implies that $\Omega_1(P)$ is a fully invariant subgroup in any p-group P, since elements of order p are mapped to elements of order dividing p by any endomorphism. Similarly, G^n is a fully invariant subgroup for any $n \in \mathbb{N}$. If a Sylow p-subgroup is unique, then it is fully invariant; this happens, for example, in abelian groups.

An isomorphism of a group onto itself is an *automorphism*; all automorphisms of a group G form the group $\operatorname{Aut} G$ as a subgroup of $\mathbb{S}_G$ (that is, with respect to composition of mappings). For any given $g \in G$, the mapping $\tau_g : x \to g^{-1}xg = x^g$ is the *inner automorphism* induced by g.

A subgroup H of a group G is said to be *characteristic* in G, if it is invariant under all automorphisms of G, that is, if $H^\varphi = H$ for every $\varphi \in \operatorname{Aut} G$. Similarly to Lemma 1.2, it is sufficient to require $H^\varphi \leq H$ for all $\varphi \in \operatorname{Aut} G$. Clearly, a characteristic subgroup is normal, since it is in particular invariant under all inner automorphisms of G. A section M/N is called characteristic, if both M and N are characteristic subgroups. Every fully invariant subgroup is, of course, characteristic. (So the examples above, $\Omega_1(P)$, G^n, a unique Sylow subgroup, are all characteristic subgroups.) The converse may not be true, for example, $Z(G)$ is always a characteristic subgroup, but need not be fully invariant (Exercise 1.3).

In general, a normal subgroup of a normal subgroup may not be normal in the whole group; for example, the subgroup $A = \langle (12)(34) \rangle$ of order 2 in $\mathbb{S}_4$ is normal in the subgroup $B = \langle (12)(34), (13)(24) \rangle$ of order 4 which, in turn, is normal in $\mathbb{S}_4$, but $A \ntrianglelefteq \mathbb{S}_4$.

Lemma 1.8. (a) *If C is a characteristic subgroup of N and N is a normal subgroup of a group G, then C is also normal in G.*

(b) *If, in addition, N is characteristic in G, then C is characteristic in G.*

(c) *If C is fully invariant in N and N is fully invariant in G, then C is fully invariant in G.*

Proof. (a) For every $g \in G$ the restriction $\tau_g|_N$ of the inner automorphism τ_g to N is an automorphism of N since $N^{\tau_g} = N$, and $\tau_g|_N$ preserves the operations since $\tau_g \in \operatorname{Aut} G$. Then $C^g = C^{\tau_g} = C$, since C is characteristic in N.

(b) If N is characteristic in G, then $\alpha|_N \in \operatorname{Aut} N$ for every $\alpha \in \operatorname{Aut} G$ for similar reasons, and hence $C^\alpha = C$ since C is characteristic in N.

(c) For any endomorphism φ of G its restriction to N is an endomorphism of N; hence C is φ-invariant. $\square$

Lemma 1.9. *For any subgroup $H \leq G$, $C_G(H)$ is a normal subgroup of $N_G(H)$ and $N_G(H)/C_G(H)$ is isomorphic to a subgroup of $\operatorname{Aut} H$.*

Proof. For every $g \in N_G(H)$, the restriction $\tau_g|_H$ is an automorphism of H since $N^g = N$. The mapping $\vartheta : g \to \tau_g|_H$ is a homomorphism of $N_G(H)$ into $\operatorname{Aut} H$ since $(x^{g_1})^{g_2} = x^{g_1 g_2}$ for any $x \in H$ so that $(g_1 g_2)^\vartheta = g_1^\vartheta g_2^\vartheta$. An element $g \in N_G(H)$ belongs to the kernel of ϑ if and only if $\tau_g|_H = 1_H \Leftrightarrow x^g = x$

for every $x \in H$, that is, $g \in C_G(H)$. By the Homomorphism Theorems, $N_G(H)/\mathrm{Ker}\,\vartheta = N_G(H)/C_G(H) \cong N_G(H)^\vartheta \le \mathrm{Aut}\,H$. $\qquad \square$

Lemma 1.10. *If H is a normal (or characteristic) subgroup of G, then both $N_G(H)$ and $C_G(H)$ are normal (or characteristic) subgroups of G.*

Proof. The proof is left as an exercise for the reader. $\qquad \square$

Commutators and commutator subgroups. We define recursively *commutators of weight* 1, 2, ... in variables $x_1, x_2, \ldots$ as formal bracket expressions. The letters $x_1, x_2, \ldots$ are commutators of weight 1; if c_1 and c_2 are commutators of weight w_1 and w_2, then $[c_1, c_2]$ is a commutator of weight $w_1 + w_2$. The *multiweight* of a commutator is the collection of the partial weights in particular variables, which are defined recursively in an obvious way. For example, $[[x_1, x_2], x_1]$ is a commutator of weight 3, and of weight 1 in x_2 and of weight 2 in x_1. A commutator $[\ldots[[a_1, a_2], a_3], \ldots, a_k]$ is called *simple* (or *left-normed*) and is denoted by $[a_1, a_2, \ldots, a_k]$.

The *commutator* $[a, b]$ of two elements in a group is defined to be $[a, b] = a^{-1}b^{-1}ab$; to avoid confusion, we had better say that $[a, b]$ is the value of the (formal) commutator $[x_1, x_2]$ on a, b. Then commutators of greater weight in the elements of a group are defined as the values of formal commutators on these elements. We may also use the same notions of weight and multiweight for these values. Commutators may be different as formal bracket structures, but their values, commutators in the group elements, may well be equal; for example, any commutator of weight ≥ 2 in the elements of an abelian group is trivial.

Lemma 1.11. *The following commutator formulae hold for any elements a, b, c in any group:*

   (a) $a^b = a[a, b]$,

   (b) $[ab, c] = [a, c]^b[b, c] = [a, c][a, c, b][b, c]$,

   (c) $[a, bc] = [a, c][a, b]^c = [a, c][a, b][a, b, c]$,

   (d) $[a, b]^{-1} = [b, a]$.

Proof. A direct verification, by expanding the commutators by definition; for example, in (b) we have $[ab, c] = (ab)^{-1}c^{-1}abc = b^{-1}a^{-1}c^{-1}abc$ on the left, $[a, c]^b[b, c] = b^{-1}[a, c]b[b, c] = b^{-1}a^{-1}c^{-1}acbb^{-1}c^{-1}bc = b^{-1}a^{-1}c^{-1}abc$ in the middle, and on the right $[a, c][a, c, b][b, c] = a^{-1}c^{-1}ac[a, c]^{-1}b^{-1}[a, c]bb^{-1}c^{-1}bc = a^{-1}c^{-1}acc^{-1}a^{-1}cab^{-1}a^{-1}c^{-1}acbb^{-1}c^{-1}bc = b^{-1}a^{-1}c^{-1}abc$, and all three results are the same. $\qquad \square$

For two subsets M and N in a group G, we define the *mutual commutator subgroup* to be $[M, N] = \langle [m, n] \mid m \in M, n \in N \rangle$. Note that $[M, N] = [N, M]$, since $[n, m] = [m, n]^{-1}$ and the inverses of the elements generate the same subgroup. If N, say, consists of one element n, we write simply $[M, n]$.

Lemma 1.12. *If M and N are subgroups of a group G, then $[M, N]$ is a normal subgroup of the group $\langle M, N \rangle$.*

Proof. It is clear that $[M, N] \leq \langle M, N \rangle$, so it suffices to prove that $\langle M, N \rangle \leq N_G([M, N])$; since $N_G([M, N])$ is a subgroup, it suffices to prove that both M and N are contained in $N_G([M, N])$. By symmetry, we need only show that $M \leq N_G([M, N])$. By Lemma 1.2, it suffices to show that $[M, N]^g \leq [M, N]$ for any $g \in M$. By Lemma 1.7, we need only prove that $[m, n]^g \in [M, N]$ for any $m \in M$, $n \in N$ (since $[M, N]$ is generated by the $[m, n]$). By Lemma 1.11(b), $[m, n]^g = [mg, n][g, n]^{-1} \in [M, N]$ because $mg \in M$ since M is a subgroup. $\square$

Lemma 1.13. *If M is a subgroup of G, then $[M, g] = [M, \langle g \rangle]$ for any $g \in G$.*

Proof. Clearly, $[M, g] \leq [M, \langle g \rangle]$. We denote by a bar the image in $M/[M, g]$. For any $\overline{m} \in \overline{M}$, we have $[\overline{m}, \overline{g}] = 1$ which implies $\overline{g} \in C_{\overline{M}}(\overline{m})$. Since $C_{\overline{M}}(\overline{m})$ is a subgroup, $\overline{g}^k \in C_{\overline{M}}(\overline{m})$ for every $k \in \mathbb{Z}$. Hence $[m, g^k] \in [M, g]$ for all $m \in M$, $k \in \mathbb{Z}$, and therefore $[M, \langle g \rangle] \leq [M, g]$ since $[M, g]$ is a subgroup. $\square$

For any homomorphism φ of a group G and any $a, b \in G$, we have

$$[a, b]^\varphi = [a^\varphi, b^\varphi], \qquad \text{whence} \quad [M, N]^\varphi = [M^\varphi, N^\varphi] \qquad (1.14)$$

by Lemma 1.7. It follows that if both M and N are characteristic (or fully invariant, or normal) subgroups, then $[M, N]$ is also a characteristic (respectively, fully invariant, normal) subgroup.

The *derived subgroup* of a group G is defined to be $[G, G]$ (often denoted by G'). The derived subgroup is fully invariant, since G is. It is easy to see that $[G, G]$ is the smallest normal subgroup such that the factor-group is abelian.

Lemma 1.15. *For a normal subgroup N of a group G, the factor-group G/N is abelian if and only if $[G, G] \leq N$.*

Proof. For $x, y \in G$, let $\bar{x}$ and $\bar{y}$ denote their images in G/N (under the natural homomorphism). By (1.14), $\overline{[x, y]} = [\bar{x}, \bar{y}]$. Then G/N is abelian $\Leftrightarrow \overline{[x, y]} = [\bar{x}, \bar{y}] = 1$ for all $\bar{x}, \bar{y} \in G/N \Leftrightarrow [x, y] \in N$ for all $x, y \in G \Leftrightarrow [G, G] \leq N$. $\square$

In fact, any subgroup containing $[G, G]$ is normal in G (Exercise 1.2).

We conclude this subsection with remarks on commutator subgroups. The same simple commutator notation is used for subgroups:

$$[A, B, C, \dots, Z] = [\dots[[A, B], C], \dots, Z].$$

Lemma 1.16. *Let H be a subgroup of a group G. For $g \in G$ and $F \leq G$,*

   (a) $g \in N_G(H) \Leftrightarrow [H,g] \leq H$;

   (b) $F \leq N_G(H) \Leftrightarrow [H,F] \leq H$;

   (c) $H \unlhd G \Leftrightarrow [H,G] \leq H$.

Proof. We have (a) $\Rightarrow$ (b), since $N_G(H)$ is a subgroup and $[H,F] \leq H$ $\Leftrightarrow [h,g] \in H$ for all $h \in H$, $g \in F \Leftrightarrow [H,g] \leq H$ for all $g \in F$. Clearly, we have (b) $\Rightarrow$ (c). So we need to prove (a) only. If $g \in N_G(H)$, then for any $h \in H$ we have $[h,g] = h^{-1}g^{-1}hg = h^{-1}h^g \in H$ and hence $[H,g]$, the subgroup generated by the $[h,g]$, lies in H. Conversely, if $[H,g] \leq H$, then $H \ni [h,g] = h^{-1}g^{-1}hg = h^{-1}h^g$ for any $h \in H$, whence $h^g \in H$ for any $h \in H$, that is, $g \in N_G(H)$. $\qquad\square$

The commutator formula $[ab,c] = [a,c]^b[b,c]$ from Lemma 1.11 can sometimes be used to compute the same for subgroups.

Lemma 1.17. *Suppose that $[A,C]$ is a normal subgroup in $\langle A,B,C \rangle$; then $[AB,C] = [A,C][B,C]$.*

Proof. It is clear that $[AB,C] \geq [A,C][B,C]$. For $a \in A$, $b \in B$, $c \in C$, for every generator of the left-hand side, $[ab,c] = [a,c]^b[b,c] \in [A,C][B,C]$ since $[A,C]$ is normal; hence $[AB,C] \leq [A,C][B,C]$ since $[A,C][B,C]$ is a subgroup. $\qquad\square$

The condition of Lemma 1.17 is satisfied, for example, if $B = C$ (by Lemma 1.12), or if both A and C are normal subgroups (by (1.14)).

Groups acting on sets. A group G is said to be *acting on a set M* if a bijection $\pi_g : M \to M$ corresponds to every element g of G in such a way that π_{ab} is the composition of π_a and π_b for any $a,b \in G$. (It follows at once that $\pi_1 = 1_M$ and $\pi_{g^{-1}} = (\pi_g)^{-1}$.) In other words, $g \to \pi_g$ is a homomorphism of G into $\mathbb{S}_M$, the group of all bijections of M. The action is *faithful* if the kernel of this homomorphism is 1; then G can be regarded as a subgroup of $\mathbb{S}_M$.

Let G be a group acting on a set M. The bijection corresponding to g is often denoted by the same letter, so that mg denotes the image of $m \in M$ under π_g, $g \in G$. The above requirement then takes the form $m(ab) = (ma)b$, for all $m \in M$ and for all $a,b \in G$. The elements of M are often called *points*. For a fixed $m \in M$, the set $mG = \{mg \mid g \in G\}$ is called a *G-orbit*. The terminology of permutation groups is usually applied to G with respect to the image of G in $\mathbb{S}_M$: for example, G is *transitive* on M, if M is the only orbit. The subset $G_m = \{g \in G \mid mg = m\}$ is a subgroup of G, the *stabilizer* of the point m.

Lemma 1.18. *If a group G acts on a set M, then M is the union of disjoint G-orbits.*

Proof. Since $m1 = m$, clearly M is the union of the orbits. If $m_1g_1 = m_2g_2$, then $m_1 = m_2g_2g_1^{-1}$, whence $m_1h = m_2g_2g_1^{-1}h = m_2(g_2g_1^{-1}h) \in m_2G$ for all $h \in G$, so that $m_1G \subseteq m_2G$; by symmetry, $m_2G \subseteq m_1G$ too.   □

Lemma 1.19. *If a finite group G acts on a set M, then $|mG| \cdot |G_m| = |G|$ for every $m \in M$.*

Proof. The idea is that the element mg is determined only by the coset of G_m which g belongs to. Indeed, if $g = hg_1$ for $h \in G_m$, then $mg = m(hg_1) = (mh)g_1 = mg_1$. If $G_mg_1 \neq G_mg_2$, then $mg_1 \neq mg_2$, for otherwise, if $mg_1 = mg_2$, then $mg_1g_2^{-1} = m$ so that $g_1g_2^{-1} \in G_m$, whence $G_mg_1 = G_mg_2$, a contradiction. Thus the number of elements in mG is equal to the number of the cosets of G_m, and the result follows by Lagrange's Theorem.   □

Corollary 1.20. *If a finite group G acts on a set M, then the cardinality of every orbit $|mG|$ divides the order of G.*   □

We now consider several important examples of actions. Every group G acts on itself by right multiplication: the action π_g of an element $g \in G$ is defined as $\pi_g : x \to xg$ for $x \in G$. Indeed, $(xg)h = x(gh)$ by the associative law. This action is faithful ($1\pi_g = 1g = 1 \Rightarrow g = 1$); all stabilizers are trivial ($ag = a \Rightarrow g = 1$); the whole set G is one orbit ($a\pi_x = b \Leftarrow x = a^{-1}b$).

Every group G acts on itself by conjugation, the action π_g of $g \in G$ being $\pi_g : x \to g^{-1}xg = x^g$ for $x \in G$. Indeed, $(x^g)^h = h^{-1}g^{-1}xgh = x^{gh}$. The kernel of this action is clearly $Z(G)$; an orbit is the set $a^G = \{g^{-1}xg \mid g \in G\}$, the *conjugacy class* containing a; the stabilizer of a point a is the centralizer $C_G(a)$. Therefore G is the union of disjoint conjugacy classes by Lemma 1.18.

Corollary 1.21. *The number $|a^G|$ of elements conjugate to an element $a \in G$ is equal to the index of the centralizer $|G : C_G(a)|$.*

Proof. Apply Lemma 1.19 and Lagrange's Theorem.   □

More generally, any subgroup $H \leq G$ acts by conjugation on any normal subgroup $N \trianglelefteq G$; the kernel is $C_H(N)$.

A group G is said to *act as a group of automorphisms* on another group F if the action is a homomorphism of G into $\text{Aut } F$. The action by conjugation is an example of action as automorphisms (inner ones). Action as automorphisms need not be faithful. For example, suppose that $\varphi \in \text{Aut } G$ and H is a characteristic subgroup of G. Then $\langle \varphi \rangle$ acts on H as the restriction $\langle \varphi \rangle|_H$; the kernel of this action is $C_{\langle \varphi \rangle}(H) = \{\psi \in \langle \varphi \rangle \mid \psi|_H = 1\}$, so that $\langle \varphi \rangle / C_{\langle \varphi \rangle}(H)$ is a subgroup of $\text{Aut } H$. If the order of φ is n, then the order of $\varphi|_H$ is a divisor of n (but it is often said that φ acts on H as an automorphism of order n).

Of course, any subgroup of $\mathbb{S}_M$ for any set M can be viewed as acting on M, so the group $GL(V)$ of all non-degenerate linear transformations of a vector space V acts on V, the automorphism group $\operatorname{Aut} G$ acts on G, and so on.

The following fact is a good illustration of the use of the notion of action.

Poincaré's Theorem 1.22. *If H is a subgroup of finite index n in a group G, then H contains a normal subgroup of G of finite index at most $n!$ in G.*

Proof. Consider the action of G on the set of right cosets of H by right multiplication, the action π_g of $g \in G$ being $\pi_g : Hx \to Hxg$. This really is an action since $Hx(g_1 g_2) = (Hxg_1)g_2$ by the associative law. The kernel of this action N is then the required normal subgroup. Indeed, N is normal in G as the kernel of the homomorphism of G into the group $\mathbb{S}_M$ of all bijections of the set $M = \{Hx \mid x \in G\}$. Next, $N \leq H$ since $Hg = H$ implies $g \in H$. Finally, $|G : N| = |G/N| \leq |\mathbb{S}_M| = n!$, since $|M| = n$ by the hypothesis. $\quad\square$

§ 1.2. Rings and modules

Rings. In general, the multiplication in a ring is only supposed to be left and right distributive with respect to addition. If a ring K has the neutral element for multiplication (unity), denoted by 1, then K is said to have 1; if the multiplication in K is associative, then K is *associative*; if the multiplication is commutative, then K is *commutative*.

Examples. 1.23. The sets $\mathbb{Z}$, $\mathbb{Q}$, $\mathbb{R}$, $\mathbb{C}$ with respect to the usual addition and multiplication are associative commutative rings with 1.

1.24. Let π be a set of prime numbers, then the set

$$\mathbb{Q}_\pi = \{m/n \mid m \in \mathbb{Z};\ n \text{ is a product of powers of primes in } \pi\}$$

is a subring of $\mathbb{Q}$.

1.25. The set $k[x]$ of polynomials in one variable x over a field k is also a commutative, associative ring with 1.

1.26. The set $k[x, y]$ of polynomials in two non-commuting variables is an associative ring with 1, but is non-commutative (a free associative k-algebra on free generators x and y). The same notation $k[x, y]$ is more often used for the polynomial ring in two commuting variables, which is a commutative associative ring with 1.

1.27. Vectors of the three-dimensional real vector space, with respect to the usual addition of vectors and the so-called vector multiplication $u \times v$, form a non-associative and non-commutative ring without 1 (this is an example of a Lie ring).

1.28. Let G be a group and K an associative commutative ring with 1. The *group ring KG* is the set of all (formal) finite sums $\sum_{g \in G} k_g g$, $k_g \in K$, equipped with addition subject to collecting terms, and with multiplication induced by multiplication in K and G, that is, $k_1 g_1 \cdot k_2 g_2 = (k_1 k_2)(g_1 g_2)$ for $k_1, k_2 \in K$, $g_1, g_2 \in G$, and extended by the distributive laws. If K is a field, then KG is a vector space with basis $\{g \mid g \in G\}$ (with multiplication as above).

1.29. The set $\operatorname{Hom} A = \operatorname{Hom}_{\mathbb{Z}} A$ of all endomorphisms of an abelian group A is a ring with respect to addition defined for $\varphi, \psi \in \operatorname{Hom} A$ as $x(\varphi + \psi) = x\varphi + x\psi$, and multiplication as composition of mappings. This is an associative ring with 1 (the identity mapping), which is not commutative in general.

A *subring* $H \leq K$ of a ring K is a subset closed under all operations of the ring K. *Direct (Cartesian) sums* of rings are direct (respectively, Cartesian) sums of their additive groups with the coordinate-wise multiplication. A (two-sided) *ideal* $I \trianglelefteq K$ of a ring K is an additive subgroup such that $ab \in I$ and $ba \in I$ for any $a \in I$ and $b \in K$; in particular, $I \leq K$. (If only one of these inclusions is required, we speak about a *right* or *left ideal*.) Ideals are exactly the kernels of homomorphisms of rings. The Homomorphism Theorems hold for rings, similar to those for groups. The *factor-ring* K/I of a ring K by an ideal I is the additive factor-group K/I with induced multiplication $(a + I)(b + I) = ab + I$, which is well-defined precisely because I is an ideal.

The factor-ring $\mathbb{Z}/n\mathbb{Z}$ of integers modulo n is a finite ring of order n. If p is a prime number, then $\mathbb{Z}/p\mathbb{Z}$ is a field of p elements, denoted also by $\mathbb{F}_p$.

We denote by $_+\langle M \rangle$ the *span* of a subset M in a ring, the additive subgroup generated by M; it consists of $\mathbb{Z}$-linear combinations of elements of M. The (sub)ring generated by a subset M is denoted by $\langle M \rangle$. A ring $R = \langle M \rangle$ is *homogeneous* with respect to the generating set M, if $R = \bigoplus_{i \in \mathbb{N}} R_i$, where R_i is the *homogeneous component of degree i*, the span of all monomials of degree i in the elements of M. The *multidegree* of a monomial is a collection of the partial degrees in particular elements; the ring $R = \langle M \rangle$ is *multihomogeneous* with respect to M if R is the direct sum of *multihomogeneous components*. The *ideal* $_{\text{id}}\langle M \rangle$ *generated by a subset* M of a ring K is the minimal ideal containing M, the intersection of all ideals containing M. The additive subgroup of $_{\text{id}}\langle M \rangle$ is generated (spanned) by all monomials involving at least one element of M.

Lie rings are considered in more detail in a separate chapter, Chapter 5.

Modules. Let K be an associative (but not necessarily commutative) ring with 1. An additive group M with unary operations $m \to mk \in M$, $m \in M$, $k \in K$, is a (right) K-module if $m1 = m$, $(m \pm m')k = mk \pm m'k$, $m(k \pm k') = mk \pm mk'$, and $(mk)k' = m(kk')$ for any $m, m' \in M$, $k, k' \in K$. A *K-submodule* $H \leq M$ of a K-module M is a subset closed under all module operations: H is an additive subgroup of M and $hk \in H$ for any $h \in H$, $k \in K$. The K-submodule generated by a subset M is denoted by $_+\langle M \rangle$ and consists of all K-linear combinations of elements of M. A mapping $\varphi : M \to N$ is a *K-homomorphism* of two K-modules M and N if φ preserves all module operations, that is, φ is a homomorphism of the additive groups M and N and $(mk)^{\varphi} = m^{\varphi}k$ for any $m \in M$, $k \in K$. A bijective homomorphism is an *isomorphism*. K-Submodules are kernels of homomorphisms of K-modules;

the Homomorphism Theorems hold for K-modules with natural adjustments. Direct (Cartesian) sums of K-modules are direct (Cartesian) sums of their additive groups with the coordinate-wise multiplication by elements of K.

Examples. 1.30. Vector spaces over a field k are exactly k-modules. Subspaces are k-submodules; factor-spaces are factor-modules; if $\{e_i \mid i \in I\}$ is a basis of a k-space V, then V is the direct sum of the $_+\langle c_i \rangle$. Every abelian group A is a $\mathbb{Z}$-module, with an being just the nth power of $a \in A$. In a natural way A is also a $(\mathrm{Hom}_{\mathbb{Z}} A)$-module (see 1.29).

1.31. Every ring K is a natural K-module; K-submodules of K are exactly (right) ideals of K.

1.32. Suppose that $G \leq \mathrm{Aut}\, A$ for an abelian group A. Then A is a $\mathbb{Z}G$-module, with

$$a \left(\sum_{g \in G} k_g g \right) = \sum_{g \in G} k_g a^g,$$

for $a \in A$ and $k_g \in \mathbb{Z}$. Submodules are precisely the G-invariant subgroups. For $\mathbb{Z}G$-modules, the module homomorphisms (isomorphisms) are often called G-homomorphisms (G-isomorphisms). An example of a G-isomorphism: $a \to ka$ for a fixed integer k.

Suppose that G is a group of linear transformations of a vector space V over a field k. Then V is a kG-module, with

$$a \left(\sum_{g \in G} k_g g \right) = \sum_{g \in G} k_g (ag),$$

for $a \in A$ and $k_g \in k$.

1.33. Let K be an associative ring with 1. A *free K-module F* on free generators $\{x_i \mid i \in I\}$ is the direct sum of the isomorphic copies $x_i K = \{x_i k \mid k \in K\}$ of the additive group of K with $(\sum_i x_i k_i)k = \sum_i x_i (k_i k)$. For any elements $m_i \in M$ in any other K-module M the mapping $x_i 1 \to m_i$ extends to a homomorphism of F into M.

Definition 1.34. Let K be an associative commutative ring with 1. The *tensor product $A \otimes B$* of two K-modules A and B is the factor-module of the free module $A \times B$ on the free generators (a, b), $a \in A$, $b \in B$, by the submodule generated by all elements of the forms

$$(ak, b) - (a, bk), \qquad (ak, b) - (a, b)k,$$

$$(a + a', b) - (a, b) - (a', b),$$

$$(a, b + b') - (a, b) - (a, b'),$$

for all $a, a' \in A$, $b, b' \in B$, $k \in K$. The image of $(a, b) \in A \times B$ in $A \otimes B$ is denoted by $a \otimes b$. In other words, $A \otimes B$ can be viewed as the set of all finite formal sums $\sum a_i \otimes b_i$, $a_i \in A$, $b_i \in B$, with the following identifications:

$$ak \otimes b = a \otimes bk = (a \otimes b)k,$$

$$(a + a') \otimes b = a \otimes b + a' \otimes b, \qquad a \otimes (b + b') = a \otimes b + a \otimes b',$$

for all $a, a' \in A$, $b, b' \in B$, $k \in K$.

Note that this definition is suitable only for commutative K. To indicate the ground ring, the notation $A \otimes_K B$ is used.

Lemma 1.35. *Let K be a commutative and associative ring with 1, and suppose that $A = \bigoplus_i A_i$ and $B = \bigoplus_j B_j$ for K-modules A_i, B_j. Then $A \otimes B$ is isomorphic to $\bigoplus_{i,j} A_i \otimes B_j$ under the isomorphism $\sum_i a_i \otimes \sum_j b_j \to \sum_{i,j} a_i \otimes b_j$, for finite sums of elements $a_i \in A_i$, $b_j \in B_j$.*

Proof. We consider only the simplest case of $A = A_1$, $B = B_1 \oplus B_2$, the proof in the general case being only an extension of the same argument. We adopt the notation of Definition 1.34. The mapping $(a, b_1 + b_2) \to (a, b_1) + (a, b_2)$ of $A \times (B_1 \oplus B_2)$ into $A \times B_1 \oplus A \times B_2$ is well-defined since the sum $B_1 \oplus B_2$ is direct. This mapping extends to a homomorphism ϑ of the free K-module $A \times (B_1 \oplus B_2)$ onto $A \times B_1 \oplus A \times B_2$. A routine check shows that ϑ maps the submodule of $A \times (B_1 \oplus B_2)$ that defines $A \otimes (B_1 \oplus B_2)$ as the factor-module by Definition 1.34 exactly onto the submodule of $A \times B_1 \oplus A \times B_2$ that defines $A \otimes B_1 \oplus A \otimes B_2$ as the factor-module. $\qquad\square$

Tensor products are used for *extending the ground ring*. Let $K \le L$ be commutative associative rings with 1, and let M be a K-module. The ring L is also a K-module with respect to multiplication by elements of K. The tensor product $M \otimes_K L$ can be regarded as an L-module, with respect to multiplication by elements of L defined as $(m \otimes a)b = m \otimes ab$, for $m \in M$, $a, b \in L$.

Now suppose that in this situation, M is also a KG-module, for a group G. Then $M \otimes_K L$ can be regarded as an LG-module: put $(m \otimes a)g = (mg \otimes a)$, for any $m \in M$, $a \in L$, $g \in G$, and extend by linearity. (Another way of defining the same thing is to consider L as a trivial *left* KG-module; then $M \otimes_{KG} L$ can be defined naturally.)

Example 1.36. Let φ be a linear transformation of a two-dimensional vector space V over $\mathbb{R}$, given in some basis e_1, e_2 by the matrix $\begin{pmatrix} 0 & 1 \\ -1 & 0 \end{pmatrix}$. Then V is an $\mathbb{R}\langle \varphi \rangle$-module. Extend the ground field forming $V \otimes_{\mathbb{R}} \mathbb{C}$. It is clear that the matrix of φ in the $\mathbb{C}$-basis $e_1 \otimes 1$, $e_2 \otimes 1$ is the same. But now it is possible to choose another basis consisting of eigenvectors, with respect to which the matrix of φ will be $\begin{pmatrix} i & 0 \\ 0 & -i \end{pmatrix}$, with eigenvalues on the diagonal.

Algebras. Suppose that R is a ring whose additive group is also a K-module for some commutative associative ring with 1 such that both structures agree in a natural way: $(r_1 r_2)k = (r_1 k)r_2 = r_1(r_2 k)$ for all $r_1, r_2 \in R$, $k \in K$. Then R is a K-*algebra*. (Of course, every ring is a $\mathbb{Z}$-algebra.) The definitions of subalgebras, ideals, factor-algebras, direct and Cartesian sums are similar to those of subrings, etc., but taking K-spans instead of $\mathbb{Z}$-spans.

If $K \le L$, then we can form the tensor product $R \otimes_K L$, which can be regarded as an L-algebra under multiplication defined as $(r_1 \otimes l_1)(r_2 \otimes l_2) = r_1 r_2 \otimes l_1 l_2$ on the generators of the additive group.

Suppose that R is a K-algebra; the set of all K-homomorphisms $\operatorname{Hom}_K R$ is itself a K-algebra with multiplication by composition of mappings, addition as $x(\varphi + \psi) = x\varphi + x\psi$, and multiplication by elements of K as $x(k\varphi) = k(x\varphi) = (kx)\varphi$, for $\varphi, \psi \in \operatorname{Hom}_K R$, $k \in K$, $x \in R$.

§ 1.3. Algebraic systems, varieties and free objects

The concept of an algebraic system comprehends virtually all objects in algebra: groups, rings, modules, vector spaces, etc., all are algebraic systems. Given a set A, a mapping $f : \underbrace{A \times \cdots \times A}_{n} \to A$ is an n-*ary algebraic operation* on A. The title "algebraic" simply means that the domain of f is the whole of $\underbrace{A \times \cdots \times A}_{n}$ and the values of f are all in A itself. The special case of a 0-ary (nullary) operation means taking a fixed element (a constant) in A. An *algebraic system* is a set with some algebraic operations on it, $f_j^{n_j}$, $j \in J$, of possibly different arities n_j; there can be infinitely many operations. To stress that the algebraic system is defined by both the set and the operations, more complex notation is used sometimes: $\mathfrak{A} = \langle A \mid f_j^{n_j}, \ j \in J \rangle$; here A is the underlying set of $\mathfrak{A}$ and $\{f_j^{n_j} \mid j \in J\}$ is the *signature* of $\mathfrak{A}$.

Subsystems are subsets that are closed under all operations. The subsystem $\langle M \rangle$ *generated* by a subset M is the smallest (with respect to inclusion) (sub)system containing M, the intersection of all subsystems containing M. It is clear that $\langle M \rangle$ consists of all possible expressions in the elements of M under arbitrarily repeated operations (finitely many times).

Homomorphisms of algebraic systems are mappings that preserve all operations; a correspondence between sets of operations themselves is assumed (in other words, homomorphisms make sense only for algebraic systems of the same signature). A homomorphism of a system into itself is an *endomorphism*. Bijective homomorphisms are *isomorphisms*; an isomorphism of a system onto itself is an *automorphism*. Subsystem that is invariant under any endomorphism is *fully invariant*; under any automorphism, *characteristic*. Under certain conditions (which will be satisfied for those algebraic systems that we shall use), the analogues of the Homomorphism Theorems hold for algebraic systems, and kernels of homomorphisms are defined naturally (*normal subsystems* or *ideals*, like normal subgroups in groups, or ideals in rings).

Examples. 1.37. Groups are algebraic systems $\langle G \mid f^2, g^1, h^0 \rangle$ with three operations satisfying certain laws. The binary operation is the group multiplication, usually denoted by dot which is often omitted, $f^2(a, b) = a \cdot b = ab$; the unary one is taking the inverses, denoted by $g^1(a) = a^{-1}$; the nullary one is taking the identity element, denoted by $h^0 = 1$. The laws are the group axioms: $a(bc) = (ab)c$, $aa^{-1} = a^{-1}a = 1$, $a1 = 1a = a$, for all $a, b, c \in G$.

1.38. A K-algebra is an algebraic system with two binary operations, multiplication and addition, the unary operation of taking the opposite element $-a$, the nullary operation of taking 0, plus $|K|$ unary operations of multiplying by scalars from K; of course, the axioms of algebras must hold.

1.39. Let a group Ω be acting as automorphisms on a group G (not necessarily faithfully). Then G can be regarded as an algebraic system which is a group with additional unary operations denoted by the elements of Ω that satisfy the laws $g^{ab} = (g^a)^b$ and $(gh)^a = g^a h^a$ for all $g, h \in G$ and $a, b \in \Omega$. With Ω fixed, such algebraic systems are called *operator groups* with operator domain Ω, or simply Ω-*groups*. Subsystems are the Ω-invariant subgroups called Ω-*subgroups*. A homomorphism φ of an Ω-group G (an Ω-*homomorphism*) must be a group homomorphism that commutes with the action of Ω, that is, $(g^a)^\varphi = (g^\varphi)^a$ for all $g \in G$, $a \in \Omega$. The kernels of Ω-homomorphisms are precisely the normal Ω-invariant subgroups.

1.40. Suppose that, for every positive integer k, a group G has a unique kth root of every element $a \in G$, an element $b \in G$ such that $b^k = a$. Then G can be viewed as an algebraic system with additional unary operations of taking kth roots, $k \in \mathbb{N}$, denoted by $a^{\frac{1}{k}}$, and satisfying the laws $(a^k)^{\frac{1}{k}} = (a^{\frac{1}{k}})^k = a$. Note that these laws imply the uniqueness of the kth root in G; in particular, G is necessarily torsion-free. Such groups are called $\mathbb{Q}$-*powered groups*, since $\mathbb{Q}$-powers can be defined as $a^{\frac{m}{n}} = (a^{\frac{1}{n}})^m$. (Sometimes such groups are also called $\mathbb{Q}$-groups, but we reserve the term Ω-groups for groups with operators.) The simplest example of a $\mathbb{Q}$-powered group is the additive group of $\mathbb{Q}$, which is generated as a $\mathbb{Q}$-powered group by one element 1, although it cannot be generated by finitely many elements as an abstract group ("abstract" means with respect only to the usual group operations). The subsystems are only $\mathbb{Q}$-powered subgroups, that is, closed under taking roots (*divisible*). Let G be a $\mathbb{Q}$-powered group and let φ be an isomorphism of G as an abstract group. Then φ is automatically an isomorphism of $\mathbb{Q}$-powered groups. Indeed, for any $a \in G$ and $\frac{m}{n} \in \mathbb{Q}$ we have $((a^{\frac{m}{n}})^\varphi)^n = ((a^{\frac{m}{n}})^n)^\varphi = (a^m)^\varphi = (a^\varphi)^m$ whence $(a^{\frac{m}{n}})^\varphi = (a^\varphi)^{\frac{m}{n}}$ since the nth root of $(a^\varphi)^m$ is unique. However, a homomorphism of G as an abstract group need not be a homomorphism of G as a $\mathbb{Q}$-powered group.

More generally, one can similarly define K-powered groups, for a commutative associative ring K with 1.

For a fixed signature $\{f_j^{n_j} \mid j \in J\}$, an algebraic system $F = \langle X \rangle$ is *free* on the set of free generators $X = \{x_i \mid i \in I\}$ if, for any elements g_i, $i \in I$, in any algebraic system G of the same signature, the mapping $x_i \to g_i$ extends to a homomorphism of F into G; this homomorphism is unique since $F = \langle X \rangle$. For a given cardinality $|I|$ of the set of free generators, a free system is unique up to isomorphism: if F' is also free on the free generators x_i', $i \in I$, then the composition of the homomorphisms extending the mappings $x_i \to x_i'$ and $x_i' \to x_i$ is identical on the x_i and hence identical on F, so that both homomorphisms are bijections. A free system F can be constructed as the set of all formal expressions (*terms*) in the x_i under arbitrarily repeated applications of the $f_j^{n_j}$. Indeed, the operations $f_j^{n_j}$ are defined on this set F, since $f_j^{n_j}(t_1, \dots, t_{n_j})$ is a term if the t_s are terms. If G is an arbitrary system of the same signature and the g_i are arbitrary elements of G, then the mapping $x_i \to g_i$ obviously extends to a homomorphism ϑ of F onto G: this homomorphism is simply the substitution of the g_i in place of the x_i in all terms. The elements of F are also called *words* of given signature; each word depends only on a finite subset of free generators. It is convenient to denote by $w(g_1, \dots, g_m) = w(x_1, \dots, x_m)^\vartheta$ the image of $w = w(x_1, \dots, x_m) \in F$ and call it the *value* of w on the g_i.

A *variety* is the class of all algebraic systems of a fixed signature defined by a given set of *laws* that are equalities of terms. More precisely, for a set of pairs of words $W \subseteq F \times F$ (where F is the free system), a system A belongs to the variety $\mathfrak{V}_W$ defined by W, if $u(a_1, \dots, a_m) = v(a_1, \dots, a_m)$ for all $a_1, \dots, a_m \in A$ and for all $(u, v) \in W$ (u and v can be assumed to depend on the same finite set $x_1, \dots, x_m$ of free generators). For example, the class of all groups is the variety defined by the group laws; in fact, all classes of algebraic systems in Examples 1.37–1.40 are varieties. Clearly, additional laws define a subvariety of a given variety. For example, abelian groups of exponent 2 form a subvariety of all groups defined by the additional laws $x_1 x_2 = x_2 x_1$ and $x_1^2 = 1$. Subvarieties of the variety of groups are called *varieties of groups*; the subvarieties of (associative, or Lie) rings are called *varieties of (associative, or Lie) rings*.

The *Cartesian product* $\mathrm{Cr}_{i \in I} A_i$ of a family of systems $\{A_i \mid i \in I\}$ of given signature is the set of all functions $g : I \to \bigcup_{i \in I} A_i$ such that $g(i) \in A_i$, with the coordinate-wise operations: $f_j^{n_j}(g_1, \dots, g_{n_j})(i) = f_j^{n_j}(g_1(i), \dots, g_{n_j}(i))$. The following is a straightforward consequence of the definitions.

Lemma 1.41. *Any variety is closed under taking subsystems, homomorphic images and Cartesian products.* □

The converse is also true (G. Birkhoff's Theorem), but we shall not need this fact.

Every variety of algebraic systems has its own (relatively) free objects: an algebraic system $F = \langle X \rangle$ is *free in the variety* $\mathfrak{V}_W$ on the set of free generators $X = \{x_i \mid i \in I\}$ if, for any elements g_i, $i \in I$, in any algebraic system $G \in \mathfrak{V}_W$, the mapping $x_i \to g_i$ extends to a (unique) homomorphism of F into G. Similarly to "absolutely free" systems above, it is easy to see that, for a given cardinality $|X|$, a free system in $\mathfrak{V}_W$ is unique up to isomorphism. A free system of the variety $\mathfrak{V}_W$ can be constructed as the factor-system of an "absolutely" free system of given signature by the congruence defined by the laws from W. However, such a straightforward way of constructing free objects usually does not give much information on their properties.

Examples. 1.42. A free group F on free generators $x_1, x_2, \ldots$ can be viewed as the collection of all group words in the x_i, with multiplication by juxtaposition, with cancellation by the group axioms: $v1w = vw$, $vx_i^{-1}x_iw = vx_ix_i^{-1}w = vw$. It still requires some effort to check that the multiplication is well-defined and associative; the empty word is the neutral element of F. This description does not say much about the structure of the free group F. In Chapter 9 we discuss another way of constructing free groups, which gives more information, like existence of an infinite descending series with abelian factors and trivial intersection (which may be rather surprising, since F has all nasty groups as its images; for example, simple ones).

1.43. As a rare exception, the structure of a free associative $\mathbb{Q}$-algebra A on free generators $x_1, x_2, \ldots$ seems to be quite transparent: A is the set of all (finite) linear combinations of monomials in the x_i subject to collection of terms, the monomials are multiplied by juxtaposition (no cancellation is allowed), and arbitrary linear combinations are multiplied by the distributive laws. Thus, A is homogeneous (even multihomogeneous) with respect to the generators x_i, so that $A = \bigoplus_{k \in \mathbb{N}} A_k$, where A_k is the $\mathbb{Q}$-span of all monomials of degree k in the x_i. If the number of free generators is finite, n, say, then $\dim A_k = n^k$.

1.44. Suppose that Ω is a fixed group; consider the class of all Ω-groups (in the sense of Example 1.39). A free Ω-group with free generators x_i, $i \in I$, can be constructed as an abstract free group F on free generators x_i^a, $i \in I$, $a \in \Omega$, admitting Ω as a group of automorphisms induced by permutations of the free generators: $(x_i^a)^b = x_i^{ab}$, where $b \in \Omega$, while x_i^a, x_i^{ab} are formal letters. We claim that F is a free Ω-group on the free generators x_i^1. Clearly, F is generated by the x_i^1 as an Ω-group. If g_i, $i \in I$, are some elements of an arbitrary Ω-group G, then the mapping $x_i^a \to g_i^a$ extends to a homomorphism φ of the abstract group F into G. This homomorphism is also a homomorphism of Ω-groups extending the mapping $x_i^1 \to g_i$. Indeed, we need to show that $(w^b)^\varphi = (w^\varphi)^b$ for any $w \in F$ and $b \in \Omega$; since both b and φ are homomorphisms, it suffices to prove that they commute on the generators: $((x_i^a)^b)^\varphi = ((x_i^a)^\varphi)^b$ for any letter x_i^a. By definition, on the left we have $((x_i^a)^b)^\varphi = (x_i^{ab})^\varphi = g_i^{ab}$, and on the right

$((x_i^a)^{\varphi})^b = (g_i^a)^b = g_i^{ab}$, the same.

1.45. Although we may think of free $\mathbb{Q}$-powered groups in terms of the general construction, it is not even clear if there are any non-abelian $\mathbb{Q}$-powered groups. We shall construct free nilpotent (non-abelian) $\mathbb{Q}$-powered groups in Chapter 9.

From now on we shall be dealing exclusively with algebraic systems that are groups with respect to one of their operations, and $f^n(1,\dots,1) = 1$ for all operations, where 1 is the neutral element with respect to the chosen group operation. For such systems any law $u = v$ can be rewritten in the form $w = 1$ with $w = uv^{-1}$ (or $w = 0$, if the group operation is written additively, as in rings, say). The *kernel* of a homomorphism is then defined as the full inverse image of 1 (of 0); kernels of homomorphisms are also called *normal subsystems* or *ideals*. Note that the neutral element with respect to the chosen group operation forms a normal subsystem as the kernel of the identity mapping. The Homomorphism Theorems hold for such systems, with formulations quite analogous to those for groups.

Let $\mathfrak{V}$ denote some variety of such systems. Changing notation, let $F = \langle X \rangle$ be a free system in the variety $\mathfrak{V}$ on the set of free generators X; elements of F are called $\mathfrak{V}$-*words* (for example, group words, if $\mathfrak{V}$ is the variety of all groups). For a set of $\mathfrak{V}$-words $W \subseteq F$, we define the (sub)variety $\mathfrak{V}_W$ as the class of all systems in $\mathfrak{V}$ satisfying the laws $w = 1$ for all $w \in W$ (or $w = 0$, in additive notation). Free systems in the subvariety $\mathfrak{V}_W$ then take the form of factor-systems of free systems in $\mathfrak{V}$ by the so-called verbal subsystems.

Definition 1.46. For a system $G \in \mathfrak{V}$ and a set of $\mathfrak{V}$-words $W \subseteq F$, the *verbal subsystem* $W(G)$ is the smallest (with respect to inclusion) normal subsystem containing the values of all words $w \in W$ on all elements of G.

Thus, by definition $G \in \mathfrak{V}_W$ if and only if $W(G) = 1$.

Lemma 1.47. *For any homomorphism φ of G, we have $W(G)^{\varphi} = W(G^{\varphi})$.*

Proof. For any $w \in W$ we have $w(g_1,\dots,g_s)^{\varphi} = w(g_1^{\varphi},\dots,g_s^{\varphi})$. By the Homomorphism Theorems, $W(G)^{\varphi}$ is the smallest normal subsystem that contains all $w(g_1,\dots,g_s)^{\varphi}$, and $W(G^{\varphi})$ is the smalest normal subsystem that conatins all $w(g_1^{\varphi},\dots,g_s^{\varphi})$. Hence they coincide. □

In particular, $W(G)$ is a fully invariant subsystem of G. The converse is not true in general, but is true for the free systems.

Lemma 1.48. *Every fully invariant subsystem H of a free system F is a verbal subsystem; in fact, $H = H(F)$.*

Proof. Obviously, $H \le H(F)$. Let $h = h(x_1,\dots,x_n)$ be any element of H

as the set of $\mathfrak{V}$-words, and let $f_1, \ldots, f_n$ be arbitrary elements in F. Then the mapping $x_i \rightarrow f_i$ extends to a homomorphism ϑ of F into F (with other free generators mapped to 1, say). Since H is fully invariant, $h(f_1, \ldots, f_n) = h(x_1, \ldots, x_n)^\vartheta \in H^\vartheta \subseteq H$. Thus, $H(F) \leq H$. $\square$

The verbal subsystem $W(F)$ is itself a set of words; by Lemma 1.48 this set of words defines the same verbal subsystem: $W(F)(F) = W(F)$. If $V \subseteq F$ is another set of $\mathfrak{V}$-words, then $V(W(F))$ is fully invariant in F by Lemma 1.8(c) and hence is a verbal subsystem of F by Lemma 1.48.

Lemma 1.49. *For any system $G \in \mathfrak{V}$, the subsystem $V(W(G))$ is verbal in G.*

Proof. There is a homomorphism ϑ of a suitable free system F onto G. Then, by Lemma 1.47, $V(W(G)) = V(W(F^\vartheta)) = V(W(F)^\vartheta) = V(W(F))^\vartheta$, and this is a verbal subsystem of G, again by Lemma 1.47, as the image of the verbal subsystem $V(W(F))$ of F under ϑ. $\square$

The factor-system $F/W(F)$ is the free system in $\mathfrak{V}_W$ on free generators y_i, $i \in I$, the images of the x_i. Indeed, for any elements $g_i \in G$ of a system $G \in \mathfrak{V}_W$, there is a homomorphism ϑ of F into G extending the mapping $x_i \rightarrow g_i$. Since all words in W are trivial on the g_i, the kernel of ϑ contains $W(F)$ (by Lemma 1.47), so that ϑ induces the required homomorphism of $F/W(F)$ into G extending the mapping $y_i \rightarrow g_i$.

Another important though simple observation: every possible law depends only on finitely many variables. Therefore, defining (sub)varieties, there is no loss of generality in assuming that the set of $\mathfrak{V}$-words is a subset of a countably generated free system in $\mathfrak{V}$. The same idea leads to the following lemma.

Lemma 1.50. *All systems in the variety $\mathfrak{V}_W$ satisfy all laws from a set of $\mathfrak{V}$-words V if and only if the free countably generated system U in $\mathfrak{V}_W$ satisfies all laws from V: that is, $\mathfrak{V}_W \subseteq \mathfrak{V}_V \Leftrightarrow V(U) = 1 \Leftrightarrow U \in \mathfrak{V}_V$.*

Proof. We need only to prove that $V(U) = 1 \Rightarrow \mathfrak{V}_W \subseteq \mathfrak{V}_V$. Every $v \in V$ depends only on finitely many variables: $v = v(x_1, \ldots, x_n)$. Let $u_1, \ldots, u_n$ be (a subset of) free generators of U; then $v(u_1, \ldots, u_n) = 1$ by the hypothesis. If $g_1, \ldots, g_n$ are any elements in any system $G \in \mathfrak{V}_W$, the mapping $u_i \rightarrow g_i$ extends to a homomorphism ϑ of U into G. Then $v(g_1, \ldots, g_n) = v(u_1^\vartheta, \ldots, u_n^\vartheta) = v(u_1, \ldots, u_n)^\vartheta = 1$. $\square$

Now we consider some specific features of verbal subgroups in the variety of groups. Let G be a group and let W be a set of group words. Then the verbal subgroup $W(G)$ is just the subgroup generated by all values of words from W on G:

$$W(G) = \left\langle w(g_1, \ldots, g_{n(w)}) \mid g_i \in G, \ w \in W \right\rangle.$$

This subgroup is always normal, since $w(g_1, \ldots, g_{n(w)})^h = w(g_1^h, \ldots, g_{n(w)}^h)$ for any $h \in G$. For example, the derived subgroup $[G, G]$ is the verbal subgroup of G with respect to the word $[x, y] = x^{-1}y^{-1}xy$, and G^n is the verbal subgroup with respect to the word x^n.

Lemma 1.51. *If $M = V(G)$ and $N = W(G)$ are verbal subgroups of a group G, then $[M, N]$ is also a verbal subgroup.*

Proof. Let F be a free group. Since $V(F)$ and $W(F)$ are fully invariant, $[V(F), W(F)]$ is a fully invariant subgroup of F by (1.14). Hence $[V(F), W(F)]$ is a verbal subgroup of F by Lemma 1.48. For an arbitrary group G, there is a homomorphism ϑ of a free group F onto G; then by Lemma 1.47 and (1.14)

$$[M, N] = [V(F)^\vartheta, W(F)^\vartheta] = [V(F), W(F)]^\vartheta$$

is a verbal subgroup of G as a homomorphic image of a verbal subgroup of F. □

Many results in various areas of algebra can be expressed in terms of varieties. For example, the negative solution of the Burnside Problem for groups of exponent $n \geq 667$ by S. I. Adyan and P. S. Novikov means that the variety of groups of exponent n is not locally soluble. A. I. Kostrikin's positive solution of the Restricted Burnside Problem for groups of prime exponent p means that all locally finite groups of exponent p form a variety; this fact follows from A. I. Kostrikin's Theorem stating that the variety of $(p-1)$-Engel Lie algebras of characteristic p is locally nilpotent. There are also specific varietal arguments, which can sometimes be very useful.

Example 1.52. Suppose that we succeeded in proving that all groups in a variety $\mathfrak{V}_W$ are periodic, that is, have no elements of infinite order. Then, in fact, there is a bound $n = n(\mathfrak{V}_W)$ for the orders of elements in $\mathfrak{V}_W$. Indeed, a free one-generator group $F = \langle x_1 \rangle$ in $\mathfrak{V}_W$ is periodic: $x_1^n = 1$ for some $n \in \mathbb{N}$. Then $g^n = 1$ for all $g \in G$ for any $G \in \mathfrak{V}_W$, since the mapping $x_1 \to g$ extends to a homomorphism of F into G.

Similar arguments can be applied to get a bound for the derived length of all groups in a variety all of whose groups are soluble, or for the nilpotency class, if all groups in the variety are nilpotent. In Chapter 3 we prove a less obvious varietal criterion for a variety to be soluble.

Exercises 1

1. Prove that $|HK| = |H| \cdot |H|/|H \cap K|$ for any two subgroups K, H of a finite group (although HK need not be a subgroup).

2. Prove that any subgroup containing $[G, G]$ is normal in G. [*Hint:* Use the correspondence for normal subgroups by the Homomorphism Theorems and the fact that all subgroups of an abelian group are normal.]

3. Prove that $Z(G)$ is always a characteristic subgroup of G. Show that the centre of the direct product $\mathbb{S}_3 \times \langle a \rangle$ of the symmetric group on three symbols and a cyclic group of order $|a| = 2$ is not fully invariant.

4. Prove that $|G : H \cap K| \leq |G : H| \cdot |G : K|$ for any subgroups $H, K \leq G$. Deduce that the intersection of n subgroups of index at most m has index at most m^n.

5. Prove that if H is a characteristic subgroup of a group G, then both $N_G(H)$ and $C_G(H)$ are characteristic subgroups. Is the same true for fully invariant instead of characteristic?

6. Show that the mapping $g \to \tau_g$, where τ_g is the inner automorphism of a group G induced by $g \in G$, is a homomorphism of G into $\operatorname{Aut} G$ with kernel $Z(G)$.

7. Prove that if, for a given prime p, there is only one Sylow p-subgroup P in a group G, then P is fully invariant (and hence characteristic) in G.

8. Suppose that N is a normal subgroup of a finite group G such that $(|N|, |G : N|) = 1$. Prove that N is characteristic in G.

9. Prove R. Remak's Theorem: if N and M are normal subgroups of a group G, then $G/(M \cap N)$ is isomorphic to a subgroup of the direct product $G/M \times G/N$. [*Hint:* Consider the mapping $g \to (gM, gN)$.] Deduce that if both G/M and G/N are abelian, then $G/(M \cap N)$ is abelian; if both are of exponent dividing n, then so is $G/(M \cap N)$; if both are torsion-free, then so is $G/(M \cap N)$. More generally, if $\{M_i \mid i \in I\}$, is a family of normal subgroups of a group G, then $G/\bigcap_{i \in I} M_i$ is isomorphic to a subgroup of the Cartesian product $\operatorname{Cr}_{i \in I} G/M_i$.

10. Suppose that both a group G and its normal subgroup N are subgroups of the direct product $H_1 \times H_2$ such that both the projection of G and the projection of N on H_i coincide with H_i, for each $i = 1, 2$. Prove that G/N is abelian.

11. Prove that the set $\mathbb{Q}_2 = \{m/2^n \mid m \in \mathbb{Z}, \ n \in \mathbb{N}\}$ is a subgroup of the additive group of $\mathbb{Q}$. Prove that if $\mathbb{Q}_2 = \langle M \rangle$, then $\mathbb{Q}_2 = \langle M \setminus \{m\} \rangle$ for any $m \in M$.

12. Suppose that a group $G = \langle M \rangle$ is generated by a subset M. Prove that $Z(G) = \bigcap_{m \in M} C_G(m)$.

13. Suppose that a group $G = \langle M \rangle$ is generated by a subset M. If $m_1 m_2 = m_2 m_1$ for any $m_1, m_2 \in M$, then G is abelian.

14. Prove that if $G/Z(G)$ is cyclic, then G is abelian.

15. Let p be a prime; suppose that a group G of order p^n acts on a set M of order coprime to p. Prove that there is at least one fixed point for G on M.

16. Show that a group of order greater than 6 with a subgroup of index 3 must have a proper non-trivial normal subgroup. [*Hint:* Use 1.22.]

17. Suppose that H is a subgroup of finite index in a torsion-free group G. Prove that $K \cap H \neq 1$ for every non-trivial subgroup $K \neq 1$.

18. For a finite abelian p-group P, use the dimensions of the $\Omega_{i+1}(P)/\Omega_i(P)$ to prove the uniqueness of the cyclic decomposition of P. [*Hint:* Show that $\dim(\Omega_{i+1}(P)/\Omega_i(P))$ is equal to the number of the cyclic summands of order $\geq p^{i+1}$.]

19. Prove that any group G acts by conjugation on the set of its subgroups. What is the stabilizer of a point H?

20. Prove that if H is a proper subgroup of a finite group G, then $G \neq \bigcup_{g \in G} H^g$.

21. Suppose that A and B are abelian groups of exponents m and n, respectively, such that $(m, n) = 1$. Regarding A and B as $\mathbb{Z}$-modules, prove that $A \otimes_{\mathbb{Z}} B = 0$.

22. Prove that the tensor product $V \otimes U$ of two vector spaces over a field k of dimensions s and t is a vector space of dimension st.

23. Suppose that $[G, G] \leq Z(G)$ in a group G. Let the abelian groups $A = G/[G, G]$ and $[G, G]$ be regarded as $\mathbb{Z}$-modules. Prove that $a \otimes a' \to [a, a']$ is a homomorphism of $A \otimes A$ onto $[G, G]$.

24. Suppose that every group G in some variety of groups $\mathfrak{V}$ satisfies the law $[x, \underbrace{y, \ldots, y}_{n}] = 1$ for some $n = n(G)$. Prove that there exists n_0 such that all groups in $\mathfrak{V}$ satisfy the law $[x, \underbrace{y, \ldots, y}_{n_0}] = 1$.

25. Suppose that $\mathfrak{M}$ is some variety of groups, $G \in \mathfrak{M}$ and N is a normal subgroup of G such that G/N is a free group in $\mathfrak{M}$. Prove that there is a subgroup $H \leq G$ such that $G = NH$ and $H \cap N = 1$. [*Hint:* Choose some preimages g_i of the free generators $\bar{g}_i$ of G/N and consider the homomorphisms given by $g_i \to \bar{g}_i$ and $\bar{g}_i \to g_i$.]

Chapter 2
Automorphisms and their fixed points

First, semidirect products are used to represent both a group and its automorphisms within a larger group. Then we discuss how automorphisms of abelian groups may be regarded as linear transformations; the Jordan Normal Form Theorem is applied to produce important lemmas about fixed points. Finally, we consider induced automorphisms of factor-groups and their fixed points.

§ 2.1. Semidirect products

Suppose that a group G is a product $G = NH$ of a normal subgroup N and a subgroup H with trivial intersection $H \cap N = 1$. Then G is the *internal semidirect product* of N and H. Every element $g \in G$ has a unique representation in the form $g = nh$, $n \in N$, $h \in H$. Indeed, if $n_1 h_1 = n_2 h_2$, then $N \ni n_2^{-1} n_1 = h_2 h_1^{-1} \in H$, where both sides are trivial since $H \cap N = 1$, and hence $n_1 = n_2$ and $h_1 = h_2$. Since N is normal, every element $h \in H$ acting on N by conjugation induces an automorphism $\varphi(h) = \tau_h|_N$ of N. Since $(x^{h_1})^{h_2} = x^{h_1 h_2}$ for any $x \in N$, $h_1, h_2 \in H$, the mapping $\varphi : H \to \operatorname{Aut} N$ is a homomorphism.

Conversely, suppose that a group H acts as a group of automorphisms on a group N (not necessarily faithfully), which means that there is a homomorphism $\varphi : H \to \operatorname{Aut} N$. Then we can form the *external semidirect product* of H and N, denoted by $N \rtimes H$ (or $N \rtimes H$), which is the set of pairs $N \rtimes H = \{(n, h) \mid n \in N, \ h \in H\}$ with multiplication defined as follows:

$$(n_1, h_1) \cdot (n_2, h_2) = (n_1 n_2^{\varphi(h_1^{-1})}, h_1 h_2). \tag{2.1}$$

(The idea is to deal with pairs (n, h) as if they were products of elements $n \in N$ and $h \in H$ in a larger group, with the action of H by conjugation on a normal subgroup N: then we would have $n_1 h_1 \cdot n_2 h_2 = n_1 h_1 n_2 h_1^{-1} h_1 h_2 = n_1 n_2^{h_1^{-1}} h_1 h_2$, with $n_1 n_2^{h_1^{-1}} = n_1 n_2^{\varphi(h_1^{-1})} \in N$ and $h_1 h_2 \in H$.)

Lemma 2.2. *The set $N \rtimes H$ is a group with respect to the operation (2.1).*

Proof. We check that the group axioms are satisfied. Indeed, $(1, 1)$ is obviously the neutral element. To find the right inverse for (n_1, h_1), we need $(1, 1)$ on the right in (2.1), that is, $h_1 h_2 = 1 \Leftrightarrow h_2 = h_1^{-1}$ and $n_1 n_2^{\varphi(h_1^{-1})} = 1 \Leftrightarrow n_2 = (n_1^{-1})^{\varphi(h_1)}$. Similarly, the left inverse can be computed for (n_2, h_2): the right-hand side in (2.1) is $(1, 1) \Leftrightarrow h_1 = h_2^{-1}$ and $n_1 = (n_2^{-1})^{\varphi(h_1^{-1})}$. It remains to verify the associative law; we have

$$
\begin{aligned}
((n_1, h_1) \cdot (n_2, h_2)) \cdot (n_3, h_3) &= (n_1 n_2^{\varphi(h_1^{-1})}, h_1 h_2) \cdot (n_3, h_3) \\
&= (n_1 n_2^{\varphi(h_1^{-1})} n_3^{\varphi(h_2^{-1} h_1^{-1})}, h_1 h_2 h_3).
\end{aligned}
$$

On the other hand,

$$
\begin{aligned}
(n_1, h_1) \cdot ((n_2, h_2) \cdot (n_3, h_3)) &= (n_1, h_1) \cdot (n_2 n_3^{\varphi(h_2^{-1})}, h_2 h_3) \\
&= (n_1 (n_2 n_3^{\varphi(h_2^{-1})})^{\varphi(h_1^{-1})}, h_1 h_2 h_3) \\
&= (n_1 n_2^{\varphi(h_1^{-1})} n_3^{\varphi(h_2^{-1} h_1^{-1})}, h_1 h_2 h_3),
\end{aligned}
$$

where we used the facts that φ is a homomorphism and $\varphi(h_1^{-1}) \in \operatorname{Aut} N$. The results are the same, as required. $\qquad\square$

Note that the structure of $N \rtimes H$ depends not only on N and H but also on the homomorphism φ (so the notation $N \rtimes H$ is ambiguous). In other words, if H acts in another way as automorphisms of N, then the resulting semidirect product is different. For example, let $\langle a \rangle$ be a cyclic group of order 4, and let $\langle b \rangle$ be a cyclic group of order 2. If $\varphi(b)$ is the automorphism of N of taking inverse elements, $a^{\varphi(b)} = a^{-1}$, then the semidirect product $\langle a \rangle \rtimes \langle b \rangle$ is a non-abelian group, the so-called dihedral group D_8 of order 8. If, however, we put $\varphi(b) = 1$, that is, if $\langle b \rangle$ acts trivially on N, then $\langle a \rangle \rtimes \langle b \rangle$ is an abelian group, the direct product of $\langle a \rangle$ and $\langle b \rangle$. The latter is true in general: if $\varphi(H) = 1$ in the construction above, then $N \rtimes H$ is isomorphic to $N \times H$. Another special case is where H acts faithfully as automorphisms on N, that is, $\operatorname{Ker} \varphi = 1$; here, again, the product depends on the embedding φ, since there may be different subgroups in $\operatorname{Aut} N$ isomorphic to H. When we use semidirect products $N \rtimes H$, it will usually be clear from the context how H acts on N.

It follows from (2.1) that the subset $\hat{N} = \{(n, 1) \mid n \in N\}$ is a subgroup isomorphic to N; moreover, $\hat{N}$ is normal in $N \rtimes H$, since

$$
(n_1, h_1)^{-1}(n, 1)(n_1, h_1) = ((*, h_1^{-1})(*, 1))(*, h_1) = (*, h_1^{-1})(*, h_1) = (*, 1),
$$

where the star denotes some elements in N. Next, the subset $\hat{H} = \{(1, h) \mid h \in H\}$ is a subgroup isomorphic to H. Since $(n, 1)(1, h) = (n, h)$, we have $N \rtimes H = \hat{N}\hat{H}$; it is also clear that $\hat{N} \cap \hat{H} = 1$. Therefore, $N \rtimes H$ is an internal semidirect product of $\hat{N}$ and $\hat{H}$. It is natural to identify H with $\hat{H}$ and N with $\hat{N}$ and consider both H and N as subgroups of $N \rtimes H$, which can then be denoted simply by NH. Under this identification, the action of $h \in H$ by conjugation on N coincides with the automorphism $\varphi(h)$: for all $n \in N$

$$
h^{-1}nh = (1, h^{-1})(n, 1)(1, h) = (n^{\varphi(h)}, h^{-1})(1, h) = (n^{\varphi(h)}, 1) = n^{\varphi(h)}.
$$

In other words, $\tau_h|_N = \varphi(h)$; this justifies the notation n^h for $n^{\varphi(h)}$.

We saw already that any external semidirect product can be regarded as an internal one; the converse is also true (and going back produces the same group).

Lemma 2.3. *Suppose that a group G is an internal semidirect product $G = HN$ where $N \trianglelefteq G$ and $N \cap H = 1$. Then G is isomorphic to the external semidirect product $N \rtimes H$ with respect to the homomorphism $\varphi : h \to \tau_h|_N$.*

Proof. We saw above that every element of G has a unique representation in the form nh, where $n \in N$, $h \in H$. Hence the mapping $\sigma : nh \to (n, h)$ of G onto $N \rtimes H$ is a correctly defined bijection. It remains to show that σ preserves the operation: for any $n_1, n_2 \in N$, $h_1, h_2 \in H$, we have

$$(n_1 h_1 n_2 h_2)^\sigma = (n_1 (h_1 n_2 h_1^{-1}) h_1 h_2)^\sigma = (n_1 n_2^{h_1^{-1}} h_1 h_2)^\sigma$$

$$= (n_1 n_2^{h_1^{-1}}, h_1 h_2) = (n_1, h_1)(n_2, h_2) = (n_1 h_1)^\sigma (n_2 h_2)^\sigma. \qquad \square$$

We shall freely use the equivalence of external and internal semidirect products. In particular, the automorphism group of a group G will be considered as a subgroup of the natural semidirect product $G \rtimes \operatorname{Aut} G$. In general, whenever a group A acts as automorphisms on a group G, we shall regard A as a subgroup of the semidirect product $GA = G \rtimes A$. In this way other notation introduced earlier can be used, such as

$$C_G(a) = \{g \in G \mid g^a = g\} \quad \text{and} \quad [G, a] = \langle [g, a] \mid g \in G \rangle = \langle g^{-1} g^a \mid g \in G \rangle.$$

For $a \in A$, elements of the subgroup $C_G(a)$ are often called *fixed elements* or *fixed points* of a. A subgroup $H \leq G$ is a-invariant if and only if $a \in N_{GA}(H)$, which is equivalent to $[H, a] \leq H$ by Lemma 1.16. By Lemma 1.3, if H is an A-invariant subgroup of GA, then $C_G(H)$ is A-invariant too.

Example 2.4. Suppose that $\varphi \in \operatorname{Aut} G$. Then $C_G(\varphi^n)$ is a φ-invariant subgroup, since $C_G(\varphi^n) = C_G(\langle \varphi^n \rangle)$ and $\langle \varphi^n \rangle$ is a φ-invariant subgroup.

A semidirect product $N \rtimes H$ is often called a *split extension of N by H*. Any group K with a normal subgroup N and the factor-group $K/N \cong H$ is called an *extension of N by H*. But not every extension splits (and there may be both split and non-split extensions for the same N and H). For example, the dihedral group D_8 is a split extension of a cyclic group of order 4 by a group of order 2; so is also their direct product. But a cyclic group of order 8 is a non-split extension of a cyclic group of order 4 by a group of order 2 (and so is the non-abelian quaternion group Q_8, defined below in § 2.3).

§ 2.2. Automorphisms as linear transformations

Suppose that G is a group of automorphisms of an abelian group V which we regard as a right $\mathbb{Z}G$-module. Then G (and the group ring $\mathbb{Z}G$) is also a subset of $\text{Hom}_\mathbb{Z} V$, the ring of all endomorphisms of V. More generally, if G acts as automorphisms on an abelian group V, there is a homomorphism of $\mathbb{Z}G$ into $\text{Hom}_\mathbb{Z} V$. With some abuse of notation, we can regard G as a subset of $\text{Hom}_\mathbb{Z} V$, where $\mathbb{Z}$-linear combinations of elements of G are defined by $a(g \pm h) = ag \pm ah$ for any $g, h \in G$, $a \in V$. The value of an integral polynomial $a_0 x^m + a_1 x^{m-1} + \cdots + a_m$ on an element $g \in G$ is defined to be $a_0 g^m + a_1 g^{m-1} + \cdots + a_m 1$, where 1 is simultaneously the identity mapping of V and the neutral element of G. For example, if $g^n = 1$ for $g \in G$, then g as a linear transformation of V is a root of the polynomial $x^n - 1$. This is, of course, similar to the situation, where G is a subgroup of $GL(V)$ for a vector space V so that the elements of G are non-singular linear transformations of V. If V is a KG-module for some commutative associative ring K with unity, then we have a homomorphism of KG into $\text{Hom}_K V$, and we can similarly take K-linear combinations of elements of G, or values of polynomials from $K[x]$ on the elements of G, meaning their images in $\text{Hom}_K V$.

Recall that an elementary abelian group E of prime exponent p can be viewed as a vector space over $\mathbb{F}_p$, the field of p elements (of residues modulo p, say). Then the automorphism group $\text{Aut}\, E$ can be viewed as $GL(E)$, the group of non-singular linear transformations of E. The Jordan Normal Form Theorem can be used to establish a connection between the rank of a finite abelian p-group and the number of fixed points of its p-automorphism. First we recall this fact from courses on linear algebra.

The Jordan Normal Form Theorem. *Suppose that ψ is a linear transformation of a finite-dimensional vector space V over a field K. If all eigenvalues of ψ belong to K, then V has a basis with respect to which the matrix of ψ is in the Jordan normal form, that is, block-diagonal $\text{diag}(J_1, \ldots, J_s)$ with Jordan blocks J_i of the form*

$$J_i = \begin{pmatrix} \lambda_i & 1 & & \\ & \lambda_i & \ddots & \\ & & \ddots & 1 \\ & & & \lambda_i \end{pmatrix},$$

where the λ_i are eigenvalues of ψ (not necessarily distinct).

Recall that the characteristic of a field, if positive, is always a prime number. We can now prove the following theorem.

Theorem 2.5. *Suppose that φ is an element of $GL(V)$ of order p^n, where V is a finite-dimensional vector space over a field K of characteristic $p > 0$.*

Then all eigenvalues of φ are equal to 1, and V has a basis with respect to which the matrix of φ is in the Jordan normal form, with blocks of the form

$$\begin{pmatrix} 1 & 1 & & & \\ & 1 & \ddots & & \\ & & \ddots & 1 & \\ & & & 1 & \end{pmatrix},$$

all having size at most $p^n \times p^n$ including at least one of size greater than $p^{n-1} \times p^{n-1}$.

Proof. Since $\varphi^{p^n} = 1$ by the hypothesis, φ is a root of the polynomial $x^{p^n} - 1$. Over a field of characteristic p, we have $(x-1)^{p^n} = x^{p^n} - 1$. Thus, the minimal polynomial for φ divides $(x-1)^{p^n}$, and hence all eigenvalues of φ are equal to 1 (and, of course, $1 \in K$. By the Jordan Normal Form Theorem, V has a basis with respect to which the matrix of φ is in the Jordan normal form, with blocks of the form

$$\begin{pmatrix} 1 & 1 & & & \\ & 1 & \ddots & & \\ & & \ddots & 1 & \\ & & & 1 & \end{pmatrix}.$$

Taking the mth power of a block-diagonal matrix is equivalent to taking the mth powers of the blocks. Easy induction on m yields

$$\begin{pmatrix} 1 & 1 & & & \\ & 1 & \ddots & & \\ & & \ddots & 1 & \\ & & & 1 & \end{pmatrix}^m = \begin{pmatrix} 1 & \binom{m}{1} & \binom{m}{2} & \cdots & \\ & 1 & \binom{m}{1} & \ddots & \vdots \\ & & \ddots & \ddots & \binom{m}{2} \\ & & & 1 & \binom{m}{1} \\ & & & & 1 \end{pmatrix}$$

(here $\binom{m}{j} = 0$ for all $j > m$). Let $m = p^s$ for some $s \in \mathbb{N}$. It can be easily shown that $\binom{p^s}{i}$ is a multiple of p whenever $i < p^s$, while $\binom{p^s}{p^s} = 1$. It is clear now that if the size of any of the blocks is greater than $p^n \times p^n$, then $\varphi^{p^n} \neq 1$, contrary to the hypothesis. If the sizes of all blocks were at most $p^{n-1} \times p^{n-1}$, we would have $\varphi^{p^{n-1}} = 1$, a contradiction with the equality $|\varphi| = p^n$. $\square$

Now we derive a corollary on the dimensions of the vector spaces.

Corollary 2.6. *Under the conditions of Theorem 2.5, $\dim V \leq dp^n$, where $d = \dim C_V(\varphi)$ and $C_V(\varphi) = \{v \in V \mid v\varphi = v\}$.*

Proof. In the chosen basis, according to the block structure of the Jordan normal matrix of φ, we have $V = \bigoplus_i U_i$, where each U_i is a φ-invariant subspace such that the matrix of $\varphi|_{U_i}$ is a Jordan block of size at most $p^n \times p^n$. The dimension of $C_V(\varphi)$ is equal to the sum of the dimensions of the $C_{U_i}(\varphi)$, since $C_V(\varphi) = \bigoplus_i C_{U_i}(\varphi)$.

We claim that $\dim C_{U_i}(\varphi) = 1$ for all i. Indeed, the coordinates of the vectors in $C_{U_i}(\varphi|_{U_i})$ with respect to the chosen basis of U_i are exactly the solutions of the system of linear equations

$$
\vec{x}
\begin{pmatrix}
1 & 1 & & \\
 & 1 & \ddots & \\
 & & \ddots & 1 \\
 & & & 1
\end{pmatrix}
= \vec{x}
\quad \Leftrightarrow \quad
\vec{x}
\begin{pmatrix}
0 & 1 & & \\
 & 0 & \ddots & \\
 & & \ddots & 1 \\
 & & & 0
\end{pmatrix}
= \vec{0},
$$

where $\vec{x} = (x_1, \ldots, x_{d_i})$, $d_i = \dim U_i$. The rank of the system is $d_i - 1$, so that the dimension of the space of solutions is exactly 1. As a result, $d = \dim C_V(\varphi)$ is equal to the number of the U_i, so that $\dim V = \sum_{i=1}^{d} d_i$. Since $d_i \leq p^n$ for all i, we have $\dim V \leq dp^n$, as required. $\qquad\square$

Now we derive a corollary for groups.

Corollary 2.7. *Let p be a prime number and let φ be an automorphism of order p^n of a finite abelian p-group A. If $|C_A(\varphi)| = p^m$, then the rank of A is not greater than mp^n.*

Proof. Since $\Omega_1(A)$ is a characteristic subgroup of A, it is φ-invariant. Using additive notation, we consider $\Omega_1(A)$ as a vector space over the field $\mathbb{F}_p$ of p elements and φ as a linear transformation of $\Omega_1(A)$. Since $|C_{\Omega_1(A)}(\varphi)| \leq |C_A(\varphi)| = p^m$, we have $\dim C_{\Omega_1(A)}(\varphi) \leq m$. The order of the restriction of φ to $\Omega_1(A)$ is a divisor of p^n. Corollary 2.6 yields $\dim \Omega_1(A) \leq mp^n$. But the dimension of $\Omega_1(A)$ is exactly the rank of $\Omega_1(A)$ as an abelian p-group, which coincides with the rank of A itself (see §1.1). $\qquad\square$

We can say something also in an infinite-dimensional situation, or for an infinite abelian p-group.

Corollary 2.8. (a) *Let φ be an element of $GL(V)$ of order p^n, where V is a vector space over a field of characteristic $p > 0$. Then $C_V(\varphi) \neq 0$.*

(b) *Let φ be an automorphism of order p^n of an abelian p-group A. Then $C_A(\varphi) \neq 1$.*

Proof. (a) Pick an arbitrary non-zero element $v \in V$; then the span U of the $\langle \varphi \rangle$-orbit $\{a, a^\varphi, \ldots, a^{\varphi^{p^n-1}}\}$ is a non-trivial φ-invariant subspace of V of dimension at most p^n. By the finite-dimensional Corollary 2.6, we have $0 < \dim U \leq dp^n$, where $d = \dim C_U(\varphi)$; hence $d \neq 0$ and $0 \neq C_U(\varphi) \leq C_V(\varphi)$.

(b) Again, consider $\Omega_1(A)$ as a vector space over the field $\mathbb{F}_p$ of p elements and φ as a linear transformation of $\Omega_1(A)$. The result follows from (a).   □

Note that for finite p-groups a different argument easily yields a stronger result:

Lemma 2.9. *Let p be a prime, and suppose that A is a group of order p^m acting as automorphisms on a group P of order p^n, for some $m, n \in \mathbb{N}$. Then $C_P(A) \neq 1$.*

Proof. It suffices to consider the orbits of A on P: their cardinalities should divide $|A| = p^m$, and there is at least one one-element orbit consisting of the identity element of P. Since the sum of the cardinalities of the orbits is $|P| = p^n$, the divisibility argument shows that there must be other one-element orbits, which consist precisely of non-trivial elements of $C_P(A)$.   □

§ 2.3. Induced automorphisms of factor-groups

Suppose that $\varphi \in \operatorname{Aut} G$ and N is a normal φ-invariant subgroup of G. The mapping $\bar{\varphi} : Ng \to (Ng)^\varphi$ is an automorphism of the factor-group G/N. Indeed, the mapping is well-defined, since $(Ng)^\varphi = N^\varphi g^\varphi = Ng^\varphi$ is again a coset of N. It is a bijection, since $(Ng)^\varphi = (Nh)^\varphi$ implies that $(gh^{-1})^\varphi \in N$ $\Leftrightarrow gh^{-1} \in N^{\varphi^{-1}} = N$, and $Ng = ((Ng)^{\varphi^{-1}})^\varphi$ for all $g, h \in G$. Finally, the operation is preserved:

$$(Ng \cdot Nh)^{\bar{\varphi}} = (Ngh)^{\bar{\varphi}} = N(gh)^\varphi = Ng^\varphi h^\varphi = Ng^\varphi Nh^\varphi = (Ng)^{\bar{\varphi}} (Nh)^{\bar{\varphi}}.$$

The automorphism $\bar{\varphi}$ of G/N is called the *induced automorphism* of G/N. One can say that $\langle \varphi \rangle$ acts on G/N as a group of automorphisms (not necessarily faithfully); the order of $\bar{\varphi}$ divides the order of φ. It is usual to denote $\bar{\varphi}$ simply by φ, when it is clear which factor-group is taken and no confusion arises. On the other hand, $\bar{\varphi}$ may denote the image of φ in the factor-group $G\langle \varphi \rangle / N$; then $\langle \bar{\varphi} \rangle$ acts as automorphisms on G/N, but not necessarily faithfully.

How are the fixed points of the induced automorphism $\bar{\varphi}$ related to the fixed points of φ? Of course, the image of a fixed point of φ is a fixed point of $\bar{\varphi}$, that is, $C_G(\varphi)N/N \leq C_{G/N}(\bar{\varphi})$. The reverse may not hold, though.

Example 2.10. One can check that the mapping $\varphi : u \to v, \ v \to u$ extends to an automorphism of the quaternion group

$$Q_8 = \{1, u, u^2, u^3, v, uv, u^2v, u^3v \mid u^4 = v^4 = 1; \ u^2 = v^2; \ u^v = u^3\}.$$

Then $\langle u^2 \rangle$ is a normal φ-invariant subgroup (for example, as the derived subgroup of Q_8). Let a bar denote images in the factor-group $Q_8/\langle u^2 \rangle$; then $Q_8/\langle u^2 \rangle$ is the direct product $\langle \bar{u} \rangle \otimes \langle \bar{v} \rangle$ of two cyclic groups of order 2. The induced automorphism $\bar{\varphi}$ interchanges the factors, and hence $\bar{u}\bar{v}$ is a fixed point of $\bar{\varphi}$. But this fixed point of $\bar{\varphi}$ is not an image of any fixed point of φ, since $C_{Q_8}(\varphi) = \langle u^2 \rangle$, as can be easily checked.

The phenomenon of an "uncovered" fixed point cannot occur if the order of an automorphism $\varphi \in \operatorname{Aut} G$ is coprime to the order of the finite group G.

Lemma 2.11. *Suppose that φ is an automorphism of finite order n of a group G and N is a finite normal φ-invariant subgroup of G such that $(|N|, |\varphi|) = 1$. Then $C_{G/N}(\bar{\varphi}) = C_G(\varphi)N/N$.*

Proof. We need only prove the inclusion $C_{G/N}(\bar{\varphi}) \leq C_G(\varphi)N/N$. In other words, we need to find an element of $C_G(\varphi)$ in every φ-invariant coset gN of N. We proceed by induction on $|\varphi|$. Suppose first of all that $|\varphi| = p$ is a prime number. Consider the action of $\langle \varphi \rangle$ on the set gN. The sizes of the $\langle \varphi \rangle$-orbits on gN divide $|\varphi| = p$ by 1.20 and hence are either p or 1; the disjoint union of these orbits is gN by 1.18. If all orbits were of size p, then p would divide $|gN| = |N|$, a contradiction with the hypothesis $(|N|, |\varphi|) = 1$. Hence there must be at least one orbit of size 1, consisting of a required element from $C_G(\varphi) \cap gN$.

Now let $|\varphi| = mn$ be a composite number with $m > 1$ and $n > 1$. By the induction hypothesis we have

$$C_{G/N}(\varphi^n) = C_G(\varphi^n)N/N \cong C_G(\varphi^n)/(C_G(\varphi^n) \cap N).$$

The φ-invariant coset gN is φ^n-invariant and hence there exists $g_0 \in C_G(\varphi^n)$ such that $g_0 N = gN$. Now $g_0^\varphi \in g_0 N$, and $g_0^\varphi \in C_G(\varphi^n)$ since $C_G(\varphi^n)$ is φ-invariant (Example 2.4). Hence $g_0^{-1} g_0^\varphi \in N \cap C_G(\varphi^n)$, that is, the coset $g_0(C_G(\varphi^n) \cap N)$ is also φ-invariant. But φ acts as an automorphism of order n on $C_G(\varphi^n)$; hence, by the induction hypothesis, the coset $g_0(C_G(\varphi^n) \cap N)$ contains an element g_1 from $C_G(\varphi)$. This element is what is required, since $g_0(C_G(\varphi^n) \cap N) \subseteq g_0 N = gN$. $\qquad\square$

It is also possible to restrict the number of fixed points in the general situation.

Lemma 2.12. *Suppose that φ is an automorphism of a finite group G and N is a normal φ-invariant subgroup. Then $|C_{G/N}(\bar{\varphi})| \leq |C_G(\varphi)|$.*

Proof. We consider the action of G by conjugation on the semidirect product $G\langle \varphi \rangle$. The cardinality of φ^G, the G-orbit containing φ, is equal to $|G : C_G(\varphi)|$ by Lemma 1.9, since $C_G(\varphi)$ is the stabilizer of the point φ. Let a bar denote the image in $G\langle \varphi \rangle/N$. For the same reasons, we have

$$|\bar{\varphi}^{\bar{G}}| = |\bar{G} : C_{\bar{G}}(\bar{\varphi})| = \frac{|G/N|}{|C_{G/N}(\bar{\varphi})|} = \frac{|G|}{|N| \cdot |C_{G/N}(\bar{\varphi})|}.$$

But $\bar{\varphi}^{\bar{g}} = \varphi^g N$ for any $g \in G$, which means that every element of the $\bar{G}$-orbit $\bar{\varphi}^{\bar{G}}$ is the image of an element from φ^G under the natural homomorphism. At most $|N|$ elements can be mapped onto a given element in $G\langle\varphi\rangle/N$. Hence $|\bar{\varphi}^{\bar{G}}| \cdot |N| \geq |\varphi^G|$. Substituting the expressions for these cardinalities, we obtain

$$\frac{|G|}{|C_{G/N}(\varphi)|} = \frac{|G/N|}{|C_{G/N}(\varphi)|} \cdot |N| \geq \frac{|G|}{|C_G(\varphi)|}.$$

It follows that $|C_{G/N}(\varphi)| \leq |C_G(N)|$. □

Remark 2.13. The result of Lemma 2.11 can be extended to the case where A is an arbitrary group of automorphisms of a finite group G of order coprime to the order of an A-invariant normal subgroup N: if $(|A|, |N|) = 1$, then $C_G(A)N/N = C_{G/N}(A)$. The proof of this extension requires the Schur–Zassenhaus Theorem and the solubility of either A or N (which is, however, always true since all groups of odd order are soluble by the celebrated Feit–Thompson Theorem).

Exercises 2

1. Prove that every semidirect product of the cyclic group of order 15 and the cyclic group of order 7 is the direct product.

2. Prove that the group of non-singular upper triangular $n \times n$ matrices over a field k,

$$T_n(k) = \left\{ \begin{pmatrix} a_1 & & * \\ & \ddots & \\ 0 & & a_n \end{pmatrix} \,\middle|\, * \in k, \ a_1 \cdots a_n \neq 0 \right\}$$

is a semidirect product of the group of upper unitriangular matrices

$$UT_n(k) = \left\{ \begin{pmatrix} 1 & & * \\ & \ddots & \\ 0 & & 1 \end{pmatrix} \,\middle|\, * \in k \right\}$$

and the group of non-singular diagonal matrices.

3. Prove that $|\operatorname{Aut} G| \neq 3$ for any group G. [*Hint:* First reduce to the case where G is abelian using Exercise 1.14; then reduce to the case where G has exponent 2 using the mapping $x \to x^{-1}$; finally, view G as a vector space over $\mathbb{F}_2$.]

4. Prove that $Z(P) \neq 1$ in any group of prime-power order $|P| = p^n$, where p is a prime and $n \in \mathbb{N}$. [*Hint:* Use Lemma 2.9.]

5. Prove an analogue of Maschke's Theorem: if a finite group G acts as automorphisms on a finite abelian group U of coprime order, $(|G|, |U|) = 1$, then every G-invariant direct summand of U has a G-invariant direct complement: that is, if $U = V \oplus T$ with G-invariant V, then there is a G-invariant subgroup $W \leq U$ such that $U = V \oplus W$. [*Hint:* Take the kernel of the endomorphism of V defined as $x\pi^* = \sum_{g \in G}(x^g\pi)^{g^{-1}}$ where π is the projection of U onto V with respect to T.]

6. Let G be a finite group, $\varphi \in \operatorname{Aut} G$, and let $(|\varphi|, |G|) = 1$. Suppose that G has a subnormal series $G = G_1 \geq \ldots \geq G_s \geq G_{s+1} = 1$ of φ-invariant subgroups such that φ induces the trivial automorphism on each factor-group G_i/G_{i+1}, $i = 1, 2, \ldots, s$. Prove that $\varphi = 1$.

7. In fact, the subnormality condition in the previous exercise is superfluous. Let G be a finite group, $\varphi \in \operatorname{Aut} G$, and let $(|\varphi|, |G|) = 1$. Suppose that G has a series $G = G_1 \geq G_2 \geq \ldots \geq G_s \geq G_{s+1} = 1$ such that every coset of G_{i+1} in G_i is φ-invariant, for all $i = 1, 2, \ldots, s$. (in particular, all G_i are φ-invariant). Prove that $\varphi = 1$.

8. Extend the result of Corollary 2.6 to infinite-dimensional vector spaces (that is, show that the dimension is necessarily finite).

9. Determine the order of the automorphism group of an elementary abelian group of order p^n.

10. Prove that the elementary abelian group E of order p^n has an automorphism of order $p^n - 1$. [*Hint:* Regard E as the additive group of the finite field $\mathbb{F}_{p^n}$ of order p^n and use the fact that the multiplicative group of $\mathbb{F}_{p^n}$ is cyclic.]

11. Prove that the Jordan normal form of a finite-dimensional linear transformation of order n is diagonal over any field of characteristic coprime to n (or 0).

12. Prove that if φ and ψ are linear transformations of a vector space such that $\varphi\psi = \psi\varphi$, then the eigenspaces for φ are ψ-invariant.

13. Suppose that A is a finite abelian group of linear transformations of a finite-dimensional vector space V over an algebraically closed field of characteristic p such that $(|A|, p) = 1$ (or $p = 0$). Prove that if V is irreducible, that is, has no proper A-invariant subspaces, then A is cyclic. [*Hint:* Use 11 and 12 and the fact that the finite subgroups of the multiplicative group of any field are cyclic.]

14. Use 13 to prove that if A is an abelian group of automorphisms of an elementary abelian p-group E such that $(|A|, p) = 1$, then $E = \prod_{A_1 < A} C_E(A_1)$, where the product is over all proper subgroups $A_1 < A$ such that A/A_1 is cyclic. Deduce that if A is non-cyclic, then also $E = \prod_{a \in A\setminus\{1\}} C_E(a)$. [*Hint:* Extend the ground field of E viewed as a vector space.]

Chapter 3
Nilpotent and soluble groups

After establishing some properties of commutator subgroups that hold in any group, we give the definitions of nilpotent and soluble groups and prove some of their basic properties. Then we prove a criterion for a variety to be soluble, and some criteria for soluble groups and varieties to be nilpotent.

§ 3.1. The lower central series

For the definitions of commutators and commutator subgroups, see § 1.1. We begin with one more commutator formula, in addition to those in 1.11.

Hall–Witt Identity 3.1. *For any elements a, b, c in any group*

$$[a, b^{-1}, c]^b \cdot [b, c^{-1}, a]^c \cdot [c, a^{-1}, b]^a = 1.$$

Proof. Expand by the definitions and cancel. □

This enables us to prove the following important lemma.

Three Subgroup Lemma 3.2. *Suppose that A, B, C are subgroups and N is a normal subgroup in a group G. If $[[A, B], C] \leq N$ and $[[B, C], A] \leq N$, then $[[C, A], B] \leq N$ too.*

Proof. The images of commutator subgroups in G/N are commutator subgroups of the images by (1.14), and the image of a subgroup in G/N is trivial if and only if the subgroup is contained in N. Hence we may assume that $N = 1 = [[A, B], C] = [[B, C], A]$. For every $a \in A$, $b \in B$, $c \in C$, the Hall–Witt Identity 3.1 holds:

$$[a, b^{-1}, c]^b \cdot [b, c^{-1}, a]^c \cdot [c, a^{-1}, b]^a = 1.$$

The first factor belongs to $[[A, B], C]^b = 1^b = 1$ and hence equals 1; the second also equals 1, since it belongs to $[[B, C], A]^c = 1$. Hence $[c, a^{-1}, b]^a = 1 \Rightarrow [c, a^{-1}, b] = 1^{a^{-1}} = 1$. Replacing a by a^{-1}, we get $[c, a, b] = 1$ for all $a \in A$, $b \in B$, $c \in C$, that is, $[c, a] \in C_G(B)$ for all $a \in A$, $c \in C$. Since $[C, A]$ is generated by the $[c, a]$ and $C_G(B)$ is a subgroup, we obtain $[C, A] \leq C_G(B)$, whence $[C, A, B] = 1$, as required. □

Corollary 3.3. *Suppose that A, B, C are normal subgroups in a group. Then $[[C, A], B] \leq [[A, B], C] \cdot [[B, C], A]$.*

Proof. Commutator subgroups of normal subgroups are normal by (1.14); hence it suffices to apply Lemma 3.2 with $N = [[A, B], C] \cdot [[B, C], A]$. □

Definition 3.4. The terms of the *lower central series* of a group G are, recursively, the commutator subgroups $\gamma_1(G) = G$, $\gamma_{k+1}(G) = [\gamma_k(G), G]$.

By induction and by Lemma 1.51, each $\gamma_k(G)$ is a verbal, and hence a fully invariant, and hence a characteristic subgroup; in particular, $\gamma_i(G) \geq \gamma_{i+1}(G)$ for all i. For the rest of the section we put $\gamma_k = \gamma_k(G)$, for short. The following is a corollary of the Three Subgroup Lemma.

Corollary 3.5. *In any group G, we have $[\gamma_m, \gamma_n] \leq \gamma_{m+n}$ for all $m, n \in \mathbb{N}$.*

Proof. We prove the inclusion by induction on n (for all m). If $n = 1$, then $[\gamma_m, \gamma_1] = [\gamma_m, G] = \gamma_{m+1}$ by the definition. For $n > 1$, we have

$$[G, \gamma_m, \gamma_{n-1}] = [\gamma_{m+1}, \gamma_{n-1}] \leq \gamma_{m+n}$$

by the definition and by the induction hypothesis, and

$$[\gamma_m, \gamma_{n-1}, G] \leq [\gamma_{m+n-1}, G] = \gamma_{m+n}$$

by the induction hypothesis and by the definition. By the Three Subgroup Lemma 3.2, we have also

$$[\gamma_m, \gamma_n] = [\gamma_n, \gamma_m] = [\gamma_{n-1}, G, \gamma_m] \leq \gamma_{m+n}.$$

□

We prove the following useful, if technical, lemma.

Lemma 3.6. *In any group G, for every $k \in \mathbb{N}$,*

(a) γ_k *contains all commutators of weights $\geq k$ in the elements of G;*

(b) γ_k *is generated by the simple commutators of weight k in the elements of G;*

(c) *if $G = \langle M \rangle$, then γ_k is generated by the simple commutators of weight $\geq k$ in the elements $m^{\pm 1}$, $m \in M$.*

Proof. (a) Since $\gamma_i \geq \gamma_{i+1}$ for all i, it suffices to show that every commutator of weight k belongs to γ_k. Induction on k; for $k = 1$ all elements of G are contained in $\gamma_1 = G$. A commutator c of weight $k > 1$ is equal to $[c_1, c_2]$, where c_1 and c_2 are commutators of weights $w_1, w_2 > 0$ with $w_1 + w_2 = k$. By the induction hypothesis, $c_1 \in \gamma_{w_1}$ and $c_2 \in \gamma_{w_2}$. By Corollary 3.5, $c = [c_1, c_2] \in [\gamma_{w_1}, \gamma_{w_2}] \leq \gamma_{w_1 + w_2} = \gamma_k$.

(b) We define the subgroups $N_k = \langle [g_1, \ldots, g_k] \mid g_i \in G \rangle$; by (a), $N_k \leq \gamma_k$, so we need only prove that $\gamma_k \leq N_k$. Induction on k; for $k = 1$ we have $\gamma_1 = G = N_1$. Let now $k > 1$. Since $[g_1, \ldots, g_k]^g = [g_1^g, \ldots, g_k^g] \in N_k$

for any $g \in G$ by (1.14), N_k is a normal subgroup. For any $g \in G$ and any generator $[g_1, \ldots, g_{k-1}]$ of N_{k-1}, we have $[[g_1, \ldots, g_{k-1}], g] \in N_k$, whence the image of $[g_1, \ldots, g_{k-1}]$ in G/N_k belongs to the centre. Hence the image of N_{k-1} lies in the centre of G/N_k, that is, $[N_{k-1}, G] \leq N_k$. By the induction hypothesis, we obtain $\gamma_k = [\gamma_{k-1}, G] \leq [N_{k-1}, G] \leq N_k$.

(c) Using (b), we substitute into $[g_1, \ldots, g_k]$ expressions for the g_i as products of the elements $m^{\pm 1}$, $m \in M$. Then we use repeatedly the formulae 1.11, $[ab, c] = [a, c][a, c, b][b, c]$ and $[a, bc] = [a, c][a, b][a, b, c]$, until we arrive at a required product of simple commutators of weights $\geq k$ in the $m^{\pm 1}$, $m \in M$.

$\square$

We knew already that γ_k is a verbal subgroup, but now Lemma 3.6(b) tells us that γ_k is verbal with respect to the single word $[x_1, \ldots, x_k]$.

§ 3.2. Nilpotent groups

There are several equivalent definitions of nilpotent groups; we need another central series for one of them.

Definition 3.7. The terms of the *upper central series* of a group G are defined recursively: $\zeta_1(G) = Z(G)$ is the centre of G, and $\zeta_{k+1}(G)$ is the full inverse image of $Z(G/\zeta_k(G))$ in G.

We write $\gamma_i = \gamma_i(G)$ and $\zeta_i = \zeta_i(G)$ throughout the section.

Definition 3.8. A series $G = G_1 \geq G_2 \geq \ldots \geq G_c \geq G_{c+1} = 1$ is *central*, if $[G_i, G] \leq G_{i+1}$ for all $i = 1, 2, \ldots, c$ (it follows that then each G_i is normal).

Now we give equivalent definitions of nilpotent groups.

Theorem 3.9. *For a group G, the following are equivalent:*

(a) $\gamma_{c+1} = 1$;

(b) G has a central series of length c;

(c) $[g_1, g_2, \ldots, g_{c+1}] = 1$ for all $g_i \in G$;

(d) $\zeta_c = G$.

Proof. By Lemma 3.6(b), we have (a) $\Leftrightarrow$ (c). The lower central series is central by the definition, so (a) $\Rightarrow$ (b). To prove (a) $\Leftarrow$ (b), we show that $\gamma_i \leq G_i$ for the G_i as in 3.8. (So the "lower" central series is the most rapidly descending one.) Induction on i; for $i = 1$ we have $\gamma_1 = G = G_1$. For $i > 1$, we have $\gamma_i = [\gamma_{i-1}, G] \leq [G_{i-1}, G] \leq G_i$ by the induction hypothesis. Then $\gamma_{c+1} \leq G_{c+1} = 1$.

The upper central series is central by the definition, so (d) $\Rightarrow$ (b). To prove (d) $\Leftarrow$ (b), we show that $\zeta_i \geq G_{c-i+1}$ for the G_i as in 3.8. (So the "upper" central series is the most rapidly ascending one.) Induction on i; for $i = 1$ we have $\zeta_1 = Z(G) \geq G_c$ since $[G_c, G] = 1$. For $i > 1$, we have $\zeta_{i-1} \geq G_{c-i+2}$ by the induction hypothesis; then the inclusion $[G_{c-i+1}, G] \leq G_{c-i+2} \leq \zeta_{i-1}$ implies that the image of G_{c-i+1} in G/ζ_{i-1} is contained in the centre, which means that $\zeta_i \geq G_{c-i+1}$. Then $\zeta_c \geq G_1 = G$. $\qquad\square$

Definition 3.10. If a group G satisfies the conditions in Theorem 3.9 then G is *nilpotent of class* $\leq c$; the least such number c is the *nilpotency class* of G. (A group is often said to have nilpotency class c if it has nilpotency class $\leq c$.)

Remark 3.11. The definition and Theorem 3.9(d) imply that a group G is nilpotent of class $c > 1$ if and only if $G/Z(G)$ is nilpotent of class $c - 1$.

The definition and Theorem 3.9(c) mean that the nilpotent groups of class $\leq c$ form a variety of groups, usually denoted by $\mathfrak{N}_c$. Therefore, according to the general theory (see §1.3), all subgroups, all homomorphic images, and all Cartesian products of nilpotent groups of class $\leq c$ (that is, from $\mathfrak{N}_c$) also belong to $\mathfrak{N}_c$. A free nilpotent group of class c is the factor-group of an "absolutely" free group F by the verbal subgroup $\gamma_{c+1}(F)$. (We shall obtain more information on the structure of F and $F/\gamma_{c+1}(F)$ in Chapter 9.)

Suppose that all groups in some variety of groups $\mathfrak{V}$ are nilpotent; then, in fact, there is an upper bound for the nilpotency classes of all groups in $\mathfrak{V}$, that is, $\mathfrak{V}$ is contained in $\mathfrak{N}_c$ for some $c \in \mathbb{N}$. Indeed, a countably generated (relatively) free group in $\mathfrak{V}$ is nilpotent of some class c by the hypothesis; hence all groups in $\mathfrak{V}$ are nilpotent of class $\leq c$ by Lemma 1.50.

We shall later consider varieties of algebraic systems that are groups with additional operations. By definition, such a system is nilpotent of class $\leq c$ if it satisfies the law $[x_1, \ldots, x_{c+1}] = 1$ (there may be difficulties with other equivalent definitions, because of other operations). The varietal arguments from the two preceding paragraphs apply to such systems too.

A nice (and, probably, unique) feature of the nilpotency identity is that it need be verified only on the generators of the group.

Theorem 3.12. *Suppose that* $G = \langle M \rangle$. *Then* G *is nilpotent of class* $\leq c$ *if* $[m_1, \ldots, m_{c+1}] = 1$ *for any* $m_i \in M$.

Proof. Induction on c; if $c = 1$, then by the hypothesis all generators of G commute, whence G is abelian, as required. Let $c > 1$. Since $[m_1, \ldots, m_c, m] = 1$ for any $m \in M$, the commutator $[m_1, \ldots, m_c]$ commutes with all generators of G and hence with all elements of G; in other words, $[m_1, \ldots, m_c] \in Z(G)$, for any $m_i \in M$. By the induction hypothesis, $G/Z(G)$ is nilpotent of class $\leq c - 1$ (the factor-group $G/Z(G)$ is generated by the image of M and the images of commutators are commutators of the images). Then G is nilpotent of class $\leq c$ by 3.11. $\qquad\square$

Corollary 3.13. *Suppose that G is a nilpotent group of class c; then, for any $g \in G$, the nilpotency class of the subgroup $\langle g, [G, G] \rangle$ is at most $c - 1$.*

Proof. The subgroup $\langle g, [G, G] \rangle$ is generated by g and the commutators $[h_1, h_2]$, $h_1, h_2 \in G$. Every commutator of weight c in these generators has weight at least $c + 1$ in the elements of G, unless all of its entries are g; in any case, it is trivial. The result follows from Theorem 3.12. $\qquad \square$

The following theorem collects some elementary properties of nilpotent groups, easily derived from the definition.

Theorem 3.14. *Suppose that G is a nilpotent group of class c, H is a subgroup of G and N is a normal subgroup of G. Then*

(a) *if $N \neq 1$, then $[N, G] < N$;*

(b) *if $N \neq 1$, then $N \cap Z(G) \neq 1$ (in particular, $Z(G) \neq 1$, if $G \neq 1$);*

(c) *H is a member of a subnormal series of G of length c;*

(d) *if $H[G, G] = G$, then $H = G$;*

(e) *if $H \neq G$, then $N_G(H) > H$.*

Proof. (a) If $[N, G] = N$, then repeated substitution yields

$$N = [N, G] = [[N, G], G] = \ldots = [\underbrace{\ldots [N, G], \ldots, G}_{c}] \leq \gamma_{c+1} = 1,$$

a contradiction to the hypothesis $N \neq 1$.

(b) Choose $s \in \mathbb{N}$ to be the least with $[\underbrace{\ldots [N, G], \ldots, G}_{s}] = 1$ (such an s exists since G is nilpotent). If $s = 1$, we have simply $N \leq Z(G)$. If $s > 1$, then $1 \neq [\underbrace{\ldots [N, G], \ldots, G}_{s-1}] \leq N \cap Z(G)$.

(c) For a subnormal series of length c containing H, we can take

$$G = H\gamma_1 \geq H\gamma_2 \geq \ldots \geq H\gamma_c \geq H\gamma_{c+1} = H.$$

To show that $H\gamma_{s+1} \trianglelefteq H\gamma_s$, it is sufficient to show that both H and γ_s are contained in $N_G(H\gamma_{s+1})$. This is clearly true for $H \leq H\gamma_{s+1}$. For γ_s we have $[\gamma_s, H\gamma_{s+1}] \leq [\gamma_s, G] = \gamma_{s+1} \leq H\gamma_{s+1}$, so that $\gamma_s \leq N_G(H\gamma_{s+1})$ by 1.16.

(d) Induction on c, the nilpotency class of G. For $c = 1$ we have $[G, G] = 1$ and the result follows. For $c > 1$, consider the factor-group $G/Z(G)$, which is nilpotent of class $c - 1$. The hypothesis holds for the images of H and $[G, G]$; hence, by the induction hypothesis, the image of H coincides with $G/Z(G)$. In other words, $HZ(G) = G$; in particular, H is normal in G, being normalized by both H and $Z(G)$. Since $[Z(G), G] = 1$, we have by 1.17

$$[G, G] = [HZ(G), HZ(G)] = [H, H][H, Z(G)][Z(G), Z(G)] \leq H.$$

As a result, $G = H[G, G] = H$, as required.

(e) This follows from (c). $\qquad\square$

By a theorem of J. Roseblade [1965], if, for some $s \in \mathbb{N}$, every subgroup of a group G is contained in some subnormal series of length s, then G is nilpotent of s-bounded class; this is a kind of converse of (c). The converses of the other parts of Theorem 3.14 may not be true, but may be used to define some generalizations of nilpotent groups.

Corollary 3.15. *Suppose that G is a nilpotent group. Then*

(a) $\gamma_i > \gamma_{i+1}$ *unless* $\gamma_i = 1$;

(b) $\zeta_i < \zeta_{i+1}$ *unless* $\zeta_i = G$.

Proof. The assertion (a) follows from Theorem 3.14(a); and (b) follows from Theorem 3.14(b) applied to the nilpotent factor-group G/ζ_i. $\qquad\square$

We shall later need some properties of torsion-free nilpotent groups and divisible subgroups in nilpotent groups. The following lemma shows that extracting roots (if possible) is a well-defined operation in a torsion-free nilpotent group.

Lemma 3.16. *If $x^n = y^n$, for some $n \in \mathbb{N}$, for elements x, y of a torsion-free nilpotent group H, then $x = y$.*

Proof. Induction on the nilpotency class c of the group. If $c = 1$, the group is abelian and we have $x^n = y^n \Rightarrow (xy^{-1})^n = 1 \Rightarrow xy^{-1} = 1 \Rightarrow x = y$. Now let $c > 1$. We have $(x^y)^n = (x^n)^y = (y^n)^y = y^n = x^n$. The elements $x^y = x[x, y]$ and x are both in the subgroup $\langle x \rangle [H, H]$, which is nilpotent of class $\leq c - 1$ by Corollary 3.13. Hence, by the induction hypothesis, we have $x^y = x$, which means that x and y commute. Then $(xy^{-1})^n = x^n y^{-n} = 1$, whence, since the group is torsion-free, $xy^{-1} = 1$, that is, $x = y$. $\qquad\square$

A group G is *divisible*, if, for every $k \in \mathbb{N}$ and every $g \in G$, there is a kth root of g in G: an element $h \in G$ such that $h^k = g$. An example: the additive group of $\mathbb{Q}$. It is clear that every homomorphic image of a divisible group is also divisible. We prove here the following technical lemma.

Lemma 3.17. *Suppose that H is a divisible subgroup of a nilpotent group G. Then the normal closure $\langle H^G \rangle$ is a divisible group too.*

Proof. Induction on the least integer s such that $[H, \underbrace{G, \ldots, G}_{s}] = 1$. If $s = 1$, then $H \leq Z(G)$ and $\langle H^G \rangle = H$.

Before making the induction step, we prove that for a fixed $g_0 \in G$ and a subgroup $U \leq G$ the mapping $u[U, G] \to [u, g_0][U, G, G]$ is a homomorphism of $U[U, G]/[U, G]$ into $[U, G]/[U, G, G]$. It is well-defined: for $v \in [U, G]$

we have $[uv, g_0] = [u, g_0][u, g_0, v][v, g_0] \in [u, g_0][U, G, G]$. The operation is preserved: for $u_1, u_2 \in U$ we have $[u_1 u_2, g_0] = [u_1, g_0][u_1, g_0, u_2][u_2, g_0] \in [u_1, g_0][u_2, g_0][U, G, G]$. In addition, it is clear that $[U, G]/[U, G, G]$ is generated by the elements $[u, g][U, G, G]$, $u \in U$, $g \in G$.

The above-defined homomorphisms of $[H, \underbrace{G, \dots, G}_{i-1}] \big/ [H, \underbrace{G, \dots, G}_{i}]$ into $[H, \underbrace{G, \dots, G}_{i}] \big/ [H, \underbrace{G, \dots, G}_{i+1}]$ show that

$$N = [H, \underbrace{G, \dots, G}_{s-1}] = [H, \underbrace{G, \dots, G}_{s-1}] \big/ [H, \underbrace{G, \dots, G}_{s}]$$

is a subgroup of $Z(G)$ generated by the elements $[h, g_1, \dots, g_{s-1}]$, $h \in H$, $g_i \in G$, and $[h^m, g_1, \dots, g_{s-1}] = [h, g_1, \dots, g_{s-1}]^m$ for them. Since H is divisible and N is abelian, it follows that N is divisible too.

Since HN/N is a divisible subgroup in G/N, by the induction hypothesis its normal closure is divisible. The full inverse image of this normal closure is $\langle H^G \rangle N = \langle H^G \rangle$, since $N \leq [H, G] \leq \langle H^G \rangle$. Now for any $a \in \langle H^G \rangle$ and any $n \in \mathbb{N}$ there is $b \in G$ such that $aN = b^n N$, that is, $a = b^n z$ for some $z \in N$. But N is divisible as shown above, so there is $z_1 \in N$ such that $z = z_1^n$. As a result, $a = b^n z_1^n = (bz_1)^n$, since $z_1 \in N \leq Z(G)$. □

We shall derive more properties of nilpotent groups using the associated Lie rings in Chapter 6. Most of them can be obtained by group-theoretic calculations as well, but we shall need the associated Lie rings anyway.

§ 3.3. Soluble groups and varieties

Apart from standard material, we prove here a criterion for a variety to be soluble.

Definition 3.18. The terms of the *derived series* of a group G are defined recursively to be $G^{(1)} = [G, G]$ and $G^{(k+1)} = [G^{(k)}, G^{(k)}]$.

By induction and by Lemma 1.51, each $G^{(k)}$ is a verbal and hence a fully invariant, and hence a characteristic subgroup. It is nice, however, to have a single word defining a verbal subgroup. Recall that $\delta_1(x_1, x_2) = [x_1, x_2]$ and $\delta_{d+1}(x_1, \dots, x_{2^{d+1}}) = [\delta_d(x_1, \dots, x_{2^d}), \delta_d(x_{2^d+1}, \dots, x_{2^{d+1}})]$.

Lemma 3.19. *In any group* G, *we have* $G^{(k)} = \langle \delta_k(x_1, \dots, x_{2^k}) \mid x_i \in G \rangle$ *for all* $k \in \mathbb{N}$.

Proof. We denote by $N_j = \langle \delta_j(x_1, \dots, x_{2^j}) \mid x_i \in G \rangle$ the verbal subgroups in question. Easy induction shows that $\delta_k(x_1, \dots, x_{2^k}) \in G^{(k)}$ for every k, so that $N_k \leq G^{(k)}$. To prove the reverse inclusion, we also use induction on k; if $k = 1$, then $G^{(1)} = [G, G] = \langle [x_1, x_2] \mid x_i \in G \rangle = N_1$ by the definitions. For $k > 1$, we consider the factor-group G/N_k. By the induction hypothesis, $(G/N_k)^{(k-1)}$ which equals $G^{(k-1)} N_k/N_k$ (by Lemma 1.47), coincides with $N_{k-1} N_k/N_k$. The commutators of any two generators $\delta_{k-1}(x_1, \ldots, x_{2^{k-1}})$ of N_{k-1} belong to N_k; in other words, their images commute in G/N_k. Hence the image of N_{k-1} is an abelian subgroup in G/N_k. Then

$$G^{(k)} N_k/N_k = (G/N_k)^{(k)} = [G^{(k-1)} N_k/N_k, \, G^{(k-1)} N_k/N_k]$$

$$\leq [N_{k-1} N_k/N_k, \, N_{k-1} N_k/N_k] = 1,$$

that is, $G^{(k)} \leq N_k$, as required. $\qquad\square$

Thus, $G^{(k)}$ is a verbal subgroup with respect to $\delta_k(x_1, \ldots, x_{2^k})$. We proceed with equivalent definitions of soluble groups.

Theorem 3.20. *The following are equivalent for a group G:*
  (a) $G^{(d)} = 1$;
  (b) $\delta_d(x_1, \ldots, x_{2^d}) = 1$ *for all $x_i \in G$;*
  (c) G *has a normal series of length d with abelian factors;*
  (d) G *has a subnormal series of length d with abelian factors.*

Proof. By Lemma 3.19, (a) $\Leftrightarrow$ (b). Since the $G^{(k)}$ are all normal and all factors $G^{(k)}/G^{(k+1)}$ are abelian, (a) $\Rightarrow$ (c). Clearly, (c) $\Rightarrow$ (d). It remains to show that (d) $\Rightarrow$ (a). Suppose that

$$G = G_0 \geq G_1 \geq \ldots \geq G_{d-1} \geq G_d = 1$$

is a subnormal series of length d with abelian factors. Then induction shows that $G^{(k)} \leq G_k$ for all k. For $k = 1$ this is because G/G_1 is abelian (see 1.15); if $G^{(j)} \leq G_j$, then

$$G^{(j+1)} = [G^{(j)}, G^{(j)}] \leq [G_j, G_j] \leq G_{j+1}$$

by the definition and because G_j/G_{j+1} is abelian. $\qquad\square$

Definition 3.21. If a group G satisfies the conditions in Theorem 3.20 then G is *soluble of derived length $\leq d$*; the least d to suit Theorem 3.20 is the *derived length* of G. (It is often said that a group is soluble of derived length d if the derived length is $\leq d$.) Groups of derived length 2 are often called *metabelian*.

It follows from Theorem 3.20(d) that any extension of a soluble group of derived length a by a soluble group of derived length b is soluble of derived length $a+b$, that is, if $G^{(a)} = 1$ and $G/N^{(b)} = 1$, then $G^{(a+b)} = 1$. The definition and Theorem 3.20(b) mean that the soluble groups of derived length $\leq d$

form a variety of groups, usually denoted by $\mathfrak{A}^d$. Therefore, all subgroups, all homomorphic images, and all Cartesian products of soluble groups of derived length $\leq d$ (that is, from $\mathfrak{A}^d$) also belong to $\mathfrak{A}^d$. A free soluble group of derived length d is the factor-group of an "absolutely" free group F by the verbal subgroup $F^{(d)}$.

Suppose that all groups in some variety of groups $\mathfrak{V}$ are soluble; then, in fact, there is an upper bound for the derived lengths of all groups in $\mathfrak{V}$, that is, $\mathfrak{V}$ is contained in $\mathfrak{A}^d$ for some $d \in \mathbb{N}$. Indeed, a countably generated (relatively) free group in $\mathfrak{V}$ is soluble of some derived length d by the hypothesis, and hence all groups in $\mathfrak{V}$ are soluble of derived length $\leq d$ by Lemma 1.50.

Suppose we have a variety of algebraic systems that are groups with additional operations. By definition, such a system is soluble of derived length $\leq d$ if it satisfies the law $\delta_d(x_1, \ldots, x_{2^d}) = 1$ (there may be difficulties with equivalent definitions because of the additional operations). The varietal arguments from the two preceding paragraphs apply to such systems, too.

In contrast to Theorem 3.12, it may *not* be sufficient to verify the solubility law $\delta_d(x_1, \ldots, x_{2^d}) = 1$ only on the generators of the group (Exercise 3.7).

Now we prove a useful criterion for a variety to be soluble, which appeared independently in [E. I. Khukhro and I. V. L'vov, 1978] and [Yu. A. Kolmakov, 1984], and, implicitly, in [M. R. Vaughan-Lee and J. Wiegold, 1981]. This criterion has, in fact, a "varietal" nature and holds for various classes of groups and Lie rings with additional operations. We shall use this criterion in Chapter 7 to prove solubility of some graded Lie rings. But in a book on groups it seems natural to prove it here.

Theorem 3.22. *If every non-trivial group in a variety of groups $\mathfrak{V}$ is distinct from its derived subgroup, then the variety is soluble (that is, $\mathfrak{V} \subseteq \mathfrak{A}^k$ for some k).*

Proof. Let F be a free group of the variety $\mathfrak{V}$ on countably many free generators $x_1, x_2, \ldots$. By Lemma 1.50, it is sufficient to prove that F is soluble. Suppose the opposite, that F is non-soluble. Let τ be the homomorphism of F into itself extending the mapping $x_i \to [x_{2i-1}, x_{2i}]$, $i = 1, 2, \ldots$. We consider the sequence of isomorphic copies F_j of F,

$$F_1 \xrightarrow{\tau} F_2 \xrightarrow{\tau} \ldots \xrightarrow{\tau} F_j \xrightarrow{\tau} F_{j+1} \xrightarrow{\tau} \ldots , \qquad (3.23)$$

with the "same" homomorphisms τ such that $x_i^\tau = [x_{2i-1}, x_{2i}]$, where $x_i \in F_j$ and $x_{2i-1}, x_{2i} \in F_{j+1}$ for all i, j.

The Cartesian product $\mathrm{Cr}_{j=1}^\infty F_j$ consists of all sequences $(g_1, g_2, \ldots)$, $g_j \in F_j$ (with coordinate-wise operations, see § 1.1). Since τ is a homomorphism, all *almost τ-threads*, elements of the form $(b_1, \ldots, b_s, a, a^\tau, a^{\tau^2}, \ldots, a^{\tau^k}, \ldots)$, with arbitrary initial segments of finite lengths $s \in \mathbb{N}$, constitute a subgroup $H \leq \mathrm{Cr}_{j=1}^\infty F_j$. Clearly, the set of all elements with only finitely many non-trivial coordinates is a normal subgroup N of $\mathrm{Cr}_{j=1}^\infty F_j$ (in fact, N is the direct product of the F_j). The section $G = HN/N$ is called the *direct limit of the spectrum (3.23)* (usually denoted by $\varinjlim F_j$). It is important that $G \neq 1$: for example, the element $(x_1, x_1^\tau, x_1^{\tau^2}, \dots) \in H$ does not belong to N, because $x_1^{\tau^n} \neq 1$ for all $n \in \mathbb{N}$. Indeed, induction on d shows that $x_1^{\tau^d} = \delta_d(x_1, \dots, x_{2^d})$ for all d. So if $x_1^{\tau^n} = 1$ for some n, then $\delta_n(x_1, \dots, x_{2^n}) = 1$, whence $\delta_n(g_1, \dots, g_{2^n}) = 1$ for all $g_i \in F$, after applying the endomorphism of F extending the mapping $x_i \to g_i$. This means that the solubility law of derived length n holds on F, contrary to our assumption that F is non-soluble.

Given an almost τ-thread $(b_1, \dots, b_s, a, a^\tau, a^{\tau^2}, \dots, a^{\tau^k}, \dots)$, let $a = w(x_j)$ be regarded as a group word $w(x_j) = w(x_1, x_2, \dots)$ in the x_j. Since $w(x_j)^{\tau^k} = w(x_j^{\tau^k})$ for all $k \in \mathbb{N}$, we have

$$(b_1, \dots, b_s, a, a^\tau, a^{\tau^2}, \dots) = (b_1, \dots, b_s, w(x_j), w(x_j^\tau), w(x_j^{\tau^2}), \dots)$$

$$= w((\underbrace{1, \dots, 1}_{s}, x_j, x_j^\tau, x_j^{\tau^2}, \dots)) \cdot (b_1, \dots, b_s, 1, 1, \dots).$$

Since $(b_1, \dots, b_s, 1, 1, \dots) \in N$, all elements in G are group words in the images of the following almost τ-threads: $(\underbrace{1, \dots, 1}_{t}, x_i, x_i^\tau, x_i^{\tau^2}, \dots)$, $t \in \mathbb{N}$; in other words, G is generated by their images. For every such thread, we have

$$(\underbrace{1, \dots, 1}_{t}, x_i, x_i^\tau, x_i^{\tau^2}, \dots) = (\underbrace{1, \dots, 1}_{t}, x_i, [x_{2i-1}, x_{2i}], [x_{2i-1}^\tau, x_{2i}^\tau], \dots)$$

$$= \Big[(\underbrace{1, \dots, 1}_{t+1}, x_{2i-1}, x_{2i-1}^\tau, \dots), (\underbrace{1, \dots, 1}_{t+1}, x_{2i}, x_{2i}^\tau, \dots) \Big] \cdot (\underbrace{1, \dots, 1}_{t}, x_i, 1, 1, \dots).$$

Since the second factor on the right belongs to N, each of the above generators of G equals a commutator from $[G, G]$; hence $G = [G, G]$. On the other hand, G belongs to $\mathfrak{V}$, being a section of the Cartesian product of the $F_i \in \mathfrak{V}$; hence $G \neq [G, G]$ by the hypothesis, a contradiction. □

Remark 3.24. Another form of Theorem 3.22: if a variety of groups is non-soluble, then it contains a non-trivial group which coincides with its derived subgroup. In this form, it is convenient to state a generalization of Theorem 3.22 for algebraic systems that are groups with additional operations: provided $f_k^{n_k}(1, \dots, 1) = 1$ for all operations $f_k^{n_k}$, any non-soluble variety $\mathfrak{V}$ of such systems ("non-soluble" here means not satisfying any of the identities $\delta_d(x_1, \dots, x_{2^d}) = 1$ with respect to the chosen group operation) contains a non-trivial system which coincides with its derived subgroup. The proof of Theorem 3.22 can be repeated verbatim, with homomorphisms in the sense of the algebraic systems and commutators with respect to the chosen group operation; the condition $f_k^{n_k}(1, \dots, 1) = 1$ ensures that N (the direct product of the F_i) is a normal subsystem of the Cartesian product of the F_i. For arriving at the equality $G = [G, G]$, it really does not matter whether additional operations were used for forming $[G, G]$ or not.

§ 3.4. Nilpotency criteria for soluble groups

It is easy to see that a nilpotent group is soluble.

Lemma 3.25. *In any group G, we have $G^{(d)} \leq \gamma_{2^d}(G)$. Therefore if a group G is nilpotent of class $\leq 2^k - 1$, then G is soluble of derived length $\leq k$.*

Proof. Induction on d: for $d = 1$, we have $G^{(1)} = [G, G] = \gamma_2(G)$; if $G^{(k)} \leq \gamma_{2^k}(G)$, then using 3.5 we obtain

$$G^{(k+1)} = [G^{(k)}, G^{(k)}] \leq [\gamma_{2^k}(G), \gamma_{2^k}(G)] \leq \gamma_{2^k + 2^k}(G) = \gamma_{2^{k+1}}(G).$$

$\square$

The converse may not be true: for example, $\mathbb{S}_3$ is soluble of derived length 2 (metabelian), but is not nilpotent. Under some conditions, solubility may imply nilpotency. The following theorem of P. Hall [1958] often facilitates the use of induction in proving that a soluble group is nilpotent.

Theorem 3.26. *Let N be a normal subgroup of a group G. If N is nilpotent of class k and $G/[N, N]$ is nilpotent of class c, then G itself is nilpotent of class at most $f(k, c) = (c - 1)k(k + 1)/2 + k$.*

Proof. Using the fact that $G/[N, N]$ is nilpotent of class c, we prove that $\gamma_{f(k,c)+1}(G) \leq \gamma_{k+1}(N)$ for all $k \in \mathbb{N}$, where $f(k, c) = (c - 1)k(k + 1)/2 + k$. We proceed by induction on k; for $k = 1$ the hypothesis gives $\gamma_{c+1}(G) \leq \gamma_2(N)$, as required.

Suppose that $\gamma_{f(k,c)+1}(G) \leq \gamma_{k+1}(N)$. For any $s \in \mathbb{N}$, we consider the commutator subgroup

$$
\begin{aligned}
\gamma_{f(k,c)+s+1}(G) &= [\gamma_{f(k,c)+1}(G), \underbrace{G, \ldots, G}_{s}] \\
&\leq [\gamma_{k+1}(N), \underbrace{G, \ldots, G}_{s}] \\
&= [\underbrace{N, \ldots, N}_{k+1}, \underbrace{G, \ldots, G}_{s}].
\end{aligned}
$$

Repeated application of Corollary 3.3 gives

$$[\underbrace{N, \ldots, N}_{k+1}, G] \leq [[N, G], N, \ldots, N] \cdot [N, [N, G], N, \ldots, N] \cdots [N, \ldots, N, [N, G]],$$

and furthermore

$$[\underbrace{N,\ldots,N}_{k+1},\underbrace{G,\ldots,G}_{s}]$$

$$\leq \prod_{i_1+\cdots+i_{k+1}=s} \left[[N,\underbrace{G,\ldots,G}_{i_1}],\ldots,[N,\underbrace{G,\ldots,G}_{i_{k+1}}]\right] \qquad (3.27)$$

(here, by definition, $[N,\underbrace{G,\ldots,G}_{0}] = N$). We put $s = (k+1)(c-1)+1$; then

at least one of the i_j in the sum $i_1 + \cdots + i_{k+1} = s$ is greater than $c-1$, and
hence each commutator in the product (3.27) contains a subcommutator

$$[N,\underbrace{G,\ldots,G}_{i_r}] \qquad \text{with} \quad i_r \geq c. \qquad (3.28)$$

Corollary 3.3 allows us to transpose any of the normal subgroups A, B, C
to the beginning of the commutator subgroups:

$$[[C,A],B] \leq [[A,B],C] \cdot [[B,C],A] = [[B,A],C] \cdot [[B,C],A].$$

Repeated application of this inclusion enables us to "pull out" the subcommutators of the form (3.28) to the beginning of each of the factors in (3.27), so
that we get that

$$\prod_{i_1+\cdots+i_{k+1}=s} \left[[N,\underbrace{G,\ldots,G}_{i_1}],\ldots,[N,\underbrace{G,\ldots,G}_{i_{k+1}}]\right] \leq [[N,\underbrace{G,\ldots,G}_{c}],\underbrace{N,\ldots,N}_{k}].$$

On the right, we have replaced all commutators $[N,\underbrace{G,\ldots,G}_{i_r}]$ with $i_r \geq c$ by

the larger subgroup $[N,\underbrace{G,\ldots,G}_{c}]$, and the other $[N,\underbrace{G,\ldots,G}_{i_j}]$ by N.

As a result, we have

$$[\gamma_{f(k,c)+1}(G),\underbrace{G,\ldots,G}_{s}] \leq [[N,\underbrace{G,\ldots,G}_{c}],\underbrace{N,\ldots,N}_{k}]$$

$$\leq [[\underbrace{G,\ldots,G}_{c+1}],\underbrace{N,\ldots,N}_{k}]$$

$$\leq [[N,N],\underbrace{N,\ldots,N}_{k}] = \gamma_{k+2}(N).$$

The last step again uses the hypothesis $\gamma_{c+1}(G) \leq [N,N]$. Since $f(k,c)+s =$
$(c-1)k(k+1)/2+k+(k+1)(c-1)+1 = (c-1)(k+1)(k+2)/2+k+1 = f(k+1,c)$,
we have arrived at $\gamma_{f(k+1,c)+1}(G) \leq \gamma_{k+2}(N)$, as required.   □

Corollary 3.29. *Suppose that in a variety of groups $\mathfrak{V}$ all soluble groups of derived length 2 are nilpotent of class $\leq c$. Then every soluble group in $\mathfrak{V}$ of derived length s is nilpotent of (s,c)-bounded class. (In other words, there exists a function $g(s,c)$ such that $\mathfrak{V} \cap \mathfrak{A}^2 \subseteq \mathfrak{N}_c$ implies $\mathfrak{V} \cap \mathfrak{A}^s \subseteq \mathfrak{N}_{g(s,c)}$.)*

Proof. Induction on the derived length s of the group. The case $s = 1$ is trivial, and the case $s = 2$ is covered by the hypothesis. Let F be a free countably-generated group of the variety $\mathfrak{V} \cap \mathfrak{A}^{s+1}$. It is sufficient to prove that F is nilpotent of class $\leq g(s+1,c)$ (Lemma 1.50). By the induction hypothesis, $[F,F]$ is nilpotent of class $\leq g(s,c)$, and $F/F^{(2)}$ is nilpotent of class $\leq c$ by the hypothesis. By Theorem 3.26, F is nilpotent of class $\leq (c-1)g(s,c)(g(s,c)+1)/2 + g(s,c)$. $\qquad\square$

Remark 3.30. An explicit expression for $g(s,c)$ which can be extracted from the proof of Corollary 3.29 is a polynomial in $c/2$ with leading term $2^{2^{s-2}+1}(c/2)^{2^{s-1}-1}$. Another proof is outlined in Exercise 3.8 which gives a much better bound $(c^s-1)/(c-1)$ for $g(s,c)$. A. G. R. Stewart [1966] found the best possible bound for the nilpotency class in the conclusion of Theorem 3.26: $ck + (c-1)(k-1)$; this may well further improve the bound in Corollary 3.29.

Analogues of Corollary 3.29 hold for some varieties of algebraic systems that are groups with additional operations. At least, if the terms of the abstract derived and lower central series are always normal subsystems in such a variety (as in varieties of groups with operators, say), then the proof of Corollary 3.29 works for this variety without any changes.

Exercises 3

1. Suppose that H is a subgroup of a nilpotent group G. Prove that the subgroups $H\zeta_k(G)$ form a subnormal series containing H.

2. Use the Three Subgroup Lemma to prove that $[\gamma_m(G), \zeta_n(G)] \leq \zeta_{n-m}(G)$ for $n \geq m$ in any group G (where $\zeta_0(G) = 1$).

3. Produce an example of a nilpotent group G of class c such that the terms of the upper central series, numbered as $G = G_1 = \zeta_c(G) > G_2 = \zeta_{c-1}(G) > \ldots > G_c = \zeta_1(G)$, do not satisfy the inclusions $[G_i, G_j] \leq G_{i+j}$.

4. Let A be a maximal (with respect to inclusion) abelian normal subgroup of a nilpotent group G. Prove that $A = C_G(A)$. [*Hint:* If false, choose an element $g \in C_G(A) \setminus A$ whose image in G/A belongs to $Z(G/A)$.]

5. Prove that the order of a finite nilpotent group is bounded in terms of the maximum of the orders of its abelian normal subgroups. [*Hint:* Use 4 and consider the action of the group by conjugation on a maximal abelian normal subgroup.]

6. Prove that a group G is soluble of derived length d if and only if $[G, G]$ is soluble of derived length $d - 1$.

7. Let G be a group generated by two elements x, y. Show that the law δ_2 of solubility of derived length 2 holds on the generators x, y (while G may not be soluble: for example, $\mathbb{S}_5$ is generated by a cycle of order 5 and a transposition).

8. Under the hypothesis of Corollary 3.29, prove that every soluble group in $\mathfrak{V}$ of derived length s is nilpotent of class at most $(c^s - 1)/(c - 1)$. [*Hint:* Use induction on s to prove that $[G^{(s-1)}, \underbrace{G, \ldots, G}_{c^{s-1}}] \leq G^{(s)}$ for $G \in \mathfrak{V}$.]

9. (L. Kaluzhnin) Suppose that $G = G_0 \geq G_1 \geq \ldots \geq G_{s+1} = 1$ is a normal series of a group G. If $A \leq \operatorname{Aut} G$ is such that $[G_i, A] \leq G_{i+1}$ for all i, then show that both A and $[G, A]$ are nilpotent of class s. [*Hint:* Define $A_j = \bigcap_i C_A(G_i/G_{i+j})$ (in particular, $A_1 = A$ by the hypothesis) and use the Three Subgroup Lemma to prove that the A_i form a central series for A, which is equivalent to $[G_i, [A_j, A]] \leq G_{i+j+1}$ for all i, j. To prove nilpotency of $[G, A]$, apply the proved assertion for A to the action of $[G, A]$ by conjugation on G (use the Three Subgroup Lemma again).]

10. (P. Hall) Suppose that $G = G_0 \geq G_1 \geq \ldots \geq G_s = 1$ is *any* series of a group G. If $A \leq \operatorname{Aut} G$ is such that $[G_i, A] \leq G_{i+1}$ for all i (in particular, each G_i is A-invariant), then A is nilpotent of class $s(s - 1)/2$.

11. Let G be a group with $\gamma_k(G)$ finite of order m. Prove that G has a subgroup of m-bounded index which is nilpotent of class $\leq k$. [*Hint:* Take $C_G(\gamma_k(G))$.]

12. Suppose that G has a normal series of length $s + n$ whose s factors are abelian and the other n factors are finite of order at most m. Prove that G has a subgroup of (m, n)-bounded index which is soluble of derived length at most $s + n$.

13. Suppose that G is a group with $[G, G]$ finite of order n. Prove that the index $|G : \zeta_2(G)|$ is finite and n-bounded.

14. Prove that, for any elements x, y in any nilpotent group of class 2, we have $(xy)^n = x^n y^n [y, x]^{n(n-1)/2}$ for any $n \in \mathbb{Z}$.

15. Suppose that G is a torsion-free nilpotent group. Prove that $G/Z(G)$ is also torsion-free. [*Hint:* Use 14 and induction to show that $\zeta_k(G)/Z(G)$ is torsion-free.]

16. Prove the following consequence of the Three Subgroup Lemma: for any normal subgroups $A, B_1, \ldots, B_s \trianglelefteq G$

$$[A, [B_1, \ldots, B_s]] \leq \prod_{\pi \in \mathbb{S}} [A, B_{1\pi}, \ldots, B_{s\pi}].$$

17. Suppose that both a group G and its normal subgroup N are subgroups of the direct product $H_1 \times \cdots \times H_n$ such that both the projection of G and the projection of N on H_i coincide with H_i, for each $i = 1, \ldots, n$. Prove that G/N is nilpotent of class $n - 1$.

18. (N. Ito) If a group $G = AB$ is a product of two abelian subgroups, then show that G is soluble of derived length 2.

19. Prove that if a nilpotent group G has cyclic derived factor-group $G/[G,G]$, then, in fact, $[G,G] = 1$.

20. (H. Fitting) Suppose that M and N are nilpotent normal subgroups of a group G. If m and n are the nilpotency classes of M and N, then show that MN is a nilpotent subgroup of class $\leq m + n$.

21. Suppose that G is a soluble group of derived length 2 generated by s elements and satisfying the law $[x, \underbrace{y, \ldots, y}_{n}] = 1$. Prove that G is nilpotent of (s,n)-bounded class.

22. Let G be a finite soluble group, $\varphi \in \mathrm{Aut}\, G$ and $(|\varphi|, |G|) = 1$. Suppose that $\varphi \neq 1$ but φ acts trivially on all φ-invariant proper subgroups of G. Prove that G is nilpotent of class 2. [*Hint:* Use Exercise 2.6 to show that $G = [G, \varphi]$. Then apply the Three Subgroup Lemma 3.2 to G, $[G,G]$ and $\langle \varphi \rangle$.]

23. Suppose that $x^m y^n = y^n x^m$ for some elements x, y in a torsion-free nilpotent group, for some $m, n \in \mathbb{N}$. Prove that $xy = yx$.

Chapter 4
Finite p-groups

Throughout this chapter, p denotes a prime number. We prove here some elementary properties of finite p-groups including the Burnside Basis Theorem. Then we prove a theorem of P. Hall on the orders of the lower central factors of a normal subgroup. Many other properties of finite p-groups will be proved later, some in Chapter 6 using the associated Lie rings, some in Chapter 10 using the Mal'cev–Lazard correspondence, some in Chapter 11 on powerful p-groups. The main results of the book in Chapters 8, 12, 13 and 14 are also about finite p-groups.

We shall freely use the fact that the homomorphic images of commutator subgroups are commutator subgroups of the images (1.14), the same being true for verbal subgroups, like $G^n = \langle g^n \mid g \in G \rangle$, by Lemma 1.47.

§ 4.1. Basic properties

By the definition from § 1.1, a group is a p-group if the orders of all of its elements are powers of p. By Lagrange's Theorem, any group of order p^n, $n \in \mathbb{N}$, is a finite p-group. The converse is also true by the Sylow Theorems. Thus, we can safely redefine finite p-groups as groups of order p^n, $n \in \mathbb{N}$. By Lagrange's Theorem, all factor-groups and all subgroups of a finite p-group are again finite p-groups. Note that every group of order p is cyclic, since every non-trivial element generates a subgroup of order that divides p and hence equals p. One can also show that every group of order p^2 is abelian (Corollary 4.2).

Theorem 4.1. *Every finite p-group is nilpotent.*

Proof. Suppose that P is a group of order p^n. The group P acts on itself by conjugation as a group of automorphisms (not necessarily faithfully). By Lemma 2.9, the fixed-point subgroup is non-trivial and, obviously, coincides with the centre $Z(P) \neq 1$. We can now use induction on $|P|$ to show that P is nilpotent. The order of $P/Z(P)$ is less than $|P|$. By the induction hypothesis, $P/Z(P)$ is nilpotent, and hence P is nilpotent too (Remark 3.11). □

Corollary 4.2. *If P is a group of order $p^n > p$, then P is nilpotent of class at most $n - 1$.*

Proof. The orders of all factors of the lower central series are at least p. It remains to show that the order of the first factor G/γ_2 cannot be less than p^2.

Indeed, if $|G/\gamma_2(G)| = p$, then $G/\gamma_2(G)$ is cyclic and hence $G = \langle g, \gamma_2(G)\rangle$ for some $g \in G \setminus \gamma_2(G)$. But the nilpotency class of $\langle g, \gamma_2(G)\rangle$ is less than that of G by Corollary 3.13, a contradiction. $\qquad\square$

Definition 4.3. If a group P of order p^n has nilpotency class exactly $n-1$, then P is a *p-group of maximal class*.

It can be proved that every finite (or periodic) nilpotent group is a direct product of its Sylow subgroups. So most of the theory of finite nilpotent groups amounts to that of finite p-groups. The general properties of nilpotent groups established in Chapter 3 hold for finite p-groups. Some special effects are due to the divisibility arguments. For example, it follows from Lagrange's Theorem that every subgroup of index p in a finite p-group is *maximal*, that is, is not properly contained in any other proper subgroup. The converse is also true, along with a kind of a dual result.

Lemma 4.4. *In any finite p-group* P,

    (a) *every maximal subgroup is normal and has index* p;

    (b) *every normal subgroup of order* p *is contained in the centre.*

Proof. (a) If $H \neq P$, then $N_P(H) > H$ by Theorem 3.14, since P is nilpotent. Hence, if H is maximal, then $N_P(H) = P$, that is, $H \trianglelefteq P$. We choose an element $\bar{g} \in P/H$ of order p; this can always be done, since for $|x| = p^k > p$ we have $|x^{p^{k-1}}| = p$. Then $H \neq \langle g \rangle H$, where g is any preimage of $\bar{g}$, whence $H\langle g\rangle = P$, since H is maximal. We see that $|P : H| = |P/H| = |\langle\bar{g}\rangle| = p$, as required.

(b) Let C be a normal subgroup of order p; since P is nilpotent, $[C, P] < C$ by Theorem 3.14. By Lagrange's Theorem, the order of $[C, P]$ is then a proper divisor of $|C| = p$; hence $[C, P] = 1$, that is, $C \leq Z(P)$. $\qquad\square$

Theorem 4.5. *In a finite p-group* P,

    (a) *every normal subgroup* N *can be included in a central series with factors of order* p;

    (b) *every subgroup* H *can be included in a subnormal series with factors of order* p.

Proof. (a) Induction on the order of P. Since a required series for $N = P$ is also a required series for $N = 1$, we may assume that $N \neq 1$. By Theorem 3.14, we have then $N \cap Z(P) \neq 1$; let z be an element of $N \cap Z(P)$ of order p. Since $|P/\langle z\rangle| < |P|$, by the induction hypothesis there is a central series in $P/\langle z\rangle$ containing $N/\langle z\rangle$, with factors of order p. The full inverse images of the members of that series in P, together with $\langle z\rangle$, form the required series for P. Indeed, all of its factors have order p by the Homomorphism Theorems, and the series is central by Lemma 4.4(b).

(b) By Theorem 3.14, P has a subnormal series containing H. By (a), each factor of this series has a central series with factors of order p. The full inverse images of all members of all of these series form a subnormal series of P containing H, with factors of order p. □

We introduce now an important characteristic subgroup of a finite p-group.

Definition 4.6. The *Frattini subgroup* $\Phi(P)$ of a finite p-group P is defined to be $\Phi(P) = P^p[P, P]$. (Recall that $P^p = \langle g^p \mid g \in P \rangle$.)

There are two other characterizations of $\Phi(P)$.

Theorem 4.7. *In a finite p-group P,*

  (a) *the Frattini subgroup $\Phi(P)$ coincides with the intersection of all maximal subgroups of P;*
  (b) *if $P = \langle M, \Phi(P) \rangle$, then $P = \langle M \rangle$.*

Proof. (a) By Corollary 4.4(a), every maximal subgroup H of P is normal and has index p. Hence the factor-group P/H is cyclic of order p and, in particular, is abelian of exponent p. Therefore, $[P/H, P/H] = [P, P]H/H = 1$ and $(P/H)^p = P^p H/H = 1$, which means that $[P, P] \leq H$ and $P^p \leq H$. Thus, $\Phi(P) = P^p[P, P]$ is contained in every maximal subgroup of P and hence in their intersection.

The factor-group $P/\Phi(P)$ is abelian since $\Phi(P) \geq [P, P]$, and $P/\Phi(P)$ has exponent p since $\Phi(P) \geq P^p$. Therefore, $P/\Phi(P)$ can be viewed as a vector space over $\mathbb{F}_p$ (see §1.1). It is easy to see that the intersection of all subspaces of codimension 1 is the zero subspace: each non-trivial vector $\vec{a}$ can be included in a basis $\{\vec{a} = \vec{a}_1, \vec{a}_2, \dots, \vec{a}_d\}$ and the subspace $_+\langle \vec{a}_2, \dots, \vec{a}_d \rangle$ of codimension 1 does not contain $\vec{a}$. The subspaces of codimension 1 in the vector space $P/\Phi(P)$ are precisely the subgroups of index p, that is, maximal subgroups of $P/\Phi(P)$. Thus, the maximal subgroups of $P/\Phi(P)$ have trivial intersection. The full inverse images of the maximal subgroups of $P/\Phi(P)$ are maximal subgroups of P (by the Homomorphism Theorems). Since the image of the intersection is contained in the intersection of the images, the intersection of the maximal subgroups of P is contained in $\Phi(P)$.

(b) If $\langle M \rangle$ is a proper subgroup of P, then $\langle M \rangle \leq H < P$ for some maximal subgroup H of P. But $\Phi(P) \leq H$, as proved in (a). Then $P = \langle M, \Phi(P) \rangle \leq H$, a contradiction. □

A *minimal system of generators* of a group is a set of generators such that no proper subset of it generates the whole group. There exist groups that have no minimal systems of generators (Exercise 1.11). Of course, every finite group has a minimal system of generators. The cardinalities of such systems may vary for a given group, even for a finite one (the cyclic group of order 15 has a minimal system consisting of two elements of order 3 and 5). But for a finite p-group the cardinality of a minimal system of generators is an invariant.

Burnside Basis Theorem 4.8. *A set of elements in a finite p-group P is a minimal system of generators if and only if the images of these elements form a basis of $P/\Phi(P)$ viewed as a vector space over $\mathbb{F}_p$.*

Proof. Suppose that $\{a_1, \dots, a_n\}$ is a minimal system of generators for P. Then the images $\bar{a}_i$ of the a_i in $P/\Phi(P)$, of course, generate $P/\Phi(P)$; in other words, the $\bar{a}_i$ span $P/\Phi(P)$ as a vector space. If the set $\{\bar{a}_1, \dots, \bar{a}_n\}$ is linearly dependent, then some proper subset $\{\bar{a}_{i_1}, \dots, \bar{a}_{i_s}\}$, for $s < n$, also spans $P/\Phi(P)$. Translating to the group language, we obtain that $P/\Phi(P) = \langle \bar{a}_{i_1}, \dots, \bar{a}_{i_s} \rangle$, and therefore $P = \langle a_{i_1}, \dots, a_{i_s}, \Phi(P) \rangle$. Then, by Theorem 4.7(b), $P = \langle a_{i_1}, \dots, a_{i_s} \rangle$, contrary to the minimality of the generating set $\{a_1, \dots, a_n\}$. Thus, $\{\bar{a}_1, \dots, \bar{a}_n\}$ is a basis of $P/\Phi(P)$, as required.

Conversely, suppose that the images $\bar{a}_1, \dots, \bar{a}_n$ of some elements $a_1, \dots, a_n$ form a basis of $P/\Phi(P)$. Then $P = \langle a_1, \dots, a_n, \Phi(P) \rangle$, which implies that $P = \langle a_1, \dots, a_n \rangle$ by Theorem 4.7(b). Any proper subset of $\{a_1, \dots, a_n\}$ does not generate the whole of P because its image in $P/\Phi(P)$ does not span $P/\Phi(P)$, since $\{\bar{a}_1, \dots, \bar{a}_n\}$ is a basis. $\square$

§ 4.2. A theorem of P. Hall

We shall need the following consequence of the Three Subgroup Lemma.

Lemma 4.9. *For any two normal subgroups A, B in an arbitrary group,* $[A, \gamma_s(B)] \leq [A, \underbrace{B, \dots, B}_{s}]$ *for any $k \in \mathbb{N}$.*

Proof. Induction on s; if $s = 1$, then the assertion is obvious. Let $s > 1$. By the induction hypothesis applied to $[B, A]$ instead of A, we have

$$[[B, A], \gamma_{s-1}(B)] \leq [[B, A], \underbrace{B, \dots, B}_{s-1}] = [[A, B], \underbrace{B, \dots, B}_{s-1}] = [A, \underbrace{B, \dots, B}_{s}].$$

Next, by the induction hypothesis

$$[A, \gamma_{s-1}(B), B] \leq [[A, \underbrace{B, \dots, B}_{s-1}], B] = [A, \underbrace{B, \dots, B}_{s}].$$

By the Three Subgroup Lemma 3.2,

$$[A, \gamma_s(B)] = [\gamma_s(B), A] = [[\gamma_{s-1}(B), B], A] \leq [A, \underbrace{B, \dots, B}_{s}].$$

$\square$

The following theorem from the seminal paper of P. Hall [1934] is remarkable by a surprisingly strong conclusion following from a rather innocent-looking hypothesis.

Theorem 4.10. *Suppose that N is a normal subgroup of a finite p-group P such that $N \leq \gamma_n(P)$. Then all factors of the lower central series of N, with the possible exception of the last one, have orders at least p^n (in other words, $|\gamma_k(N)/\gamma_{k+1}(N)| \geq p^n$ whenever $\gamma_{k+1}(N) \neq 1$).*

Proof. We can assume that $\gamma_{k+1}(N) \neq 1$, since there is nothing to prove otherwise. As a characteristic subgroup of a normal subgroup, $\gamma_{k+1}(N)$ is a normal subgroup of P by 1.8. By Theorem 4.5, P has a central series including $\gamma_{k+1}(N)$ with factors of order p. Hence we can choose a normal subgroup M of P contained in $\gamma_{k+1}(N)$ as a subgroup of index p such that

$$[\gamma_{k+1}(N), P] \leq M. \tag{4.11}$$

At the same time, $[\gamma_k(N), N] = \gamma_{k+1}(N) \nleq M$. Since $N \leq \gamma_n(P)$, the latter implies that $[\gamma_k(N), \gamma_n(P)] \nleq M$. By Lemma 1.9, the more so

$$[\gamma_k(N), \underbrace{P, \ldots, P}_{n}] \nleq M. \tag{4.12}$$

We consider the following series of length n formed by taking successively the mutual commutator subgroup with $\overline{P}$, where a bar denotes the image in $P/\gamma_{k+1}(N)$:

$$\overline{\gamma_k(N)} \geq [\overline{\gamma_k(N)}, \overline{P}] \geq [\overline{\gamma_k(N)}, \overline{P}, \overline{P}] \geq \ldots \geq [\overline{\gamma_k(N)}, \underbrace{\overline{P}, \ldots, \overline{P}}_{n-1}] \geq 1. \tag{4.13}$$

All terms in (4.13) are normal subgroups of $\overline{P}$ (Lemma 1.12); we claim that all of them are *non-trivial*. It suffices to show this for the last one:

$$[\overline{\gamma_k(N)}, \underbrace{\overline{P}, \ldots, \overline{P}}_{n-1}] = \overline{[\gamma_k(N), \underbrace{P, \ldots, P}_{n-1}]} \neq 1.$$

Suppose the opposite; then $[\gamma_k(N), \underbrace{P, \ldots, P}_{n-1}] \leq \gamma_{k+1}(N)$, and hence by (4.11)

$$[\gamma_k(N), \underbrace{P, \ldots, P}_{n}] = [[\gamma_k(N), \underbrace{P, \ldots, P}_{n-1}], P] \leq [\gamma_{k+1}(N), P] \leq M,$$

which contradicts (4.12).

Since all terms in (4.13) are non-trivial, all inclusions in (4.13) are, in fact, *strict*, by Theorem 3.14 applied to $[\overline{\gamma_k(N)}, \underbrace{\overline{P}, \ldots, \overline{P}}_{i}]$ and $\overline{P}$, for $i = 0, 1, \ldots, n-1$. Since there are n factors in (4.13), and each is a non-trivial p-group of order at least p, the order of $\gamma_k(N)/\gamma_{k+1}(N)$ is at least p^n (by Lagrange's Theorem). $\square$

Exercises 4

1. Prove that the set of all upper unitriangular $n \times n$ matrices $UT_n(\mathbb{F}_p)$ (see Exercise 2.2) is a finite p-group with respect to matrix multiplication. Determine the order, the terms of the lower central series, the terms of the derived series, the nilpotency class and the derived length of $UT_n(\mathbb{F}_p)$. Find in $UT_n(\mathbb{F}_p)$ a characteristic series of length $\leq 1 + \log_2 n$ with elementary abelian factors.

2. Let φ be an automorphism of a finite p-group P such that $p \nmid |\varphi|$. Suppose that φ acts trivially on a φ-invariant maximal abelian normal subgroup A. Prove that $\varphi = 1$. [*Hint:* Apply the Three Subgroup Lemma 3.2 to obtain that $[P, \varphi] \leq A$; then use Lemma 2.11.]

3. Let φ be an automorphism of a finite p-group P such that $p \nmid |\varphi|$. Suppose that φ acts trivially on the factor-group $P/\Phi(P)$. Prove that $\varphi = 1$.

4. (P. Hall) Suppose that P is a group of order p^n such that $|P/\Phi(P)| = p^r$. Then show that $|\operatorname{Aut} P|$ divides $p^{r(n-r)}|GL_r(\mathbb{F}_p)|$. [*Hint:* Let N be the kernel of the action of $\operatorname{Aut} P$ on $P/\Phi(P)$; then $(\operatorname{Aut} P)/N$ is isomorphic to a subgroup of $\operatorname{Aut}(P/\Phi(P)) \cong GL_r(\mathbb{F}_p)$. By 3, N is a p-group. To estimate the order of N, fix a minimal generating set of P: the images of its elements under any $\varphi \in N$ do not change mod $\Phi(P)$.]

5. Let $\langle a \rangle$ be a cyclic group of order p^n. Prove that $\operatorname{Aut}\langle a \rangle$ is an abelian group and determine its structure (cyclic decomposition). Warning: the answer looks different for $p = 2$.

6. Suppose that $p \neq 2$ and P is a finite p-group with $|\Omega_1(P)| = p$. Prove that P is cyclic. [*Hint:* By induction, P has a cyclic subgroup of index p; use 5.] (Note that Q_8 from Example 2.10 is a counterexample for $p = 2$.)

7. Suppose that a finite p-group P has an abelian subgroup of index p^2. Prove that P has a normal abelian subgroup of index p^2. [*Hint:* If A is abelian, then $A \cap A^g \leq Z(\langle A, A^g \rangle)$.]

8. Suppose that P is a finite p-group such that $Z([P, P])$ is cyclic. Prove that $[P, P]$ is cyclic. [*Hint:* If false, choose $N \trianglelefteq P$ such that $Z([P, P]) < N \leq [P, P]$ with $|N : Z([P, P])| = p$. By Exercise 1.14, N is abelian. If $|\Omega_1(N)| = p^2$, then P acts on $\Omega_1(N)$ as a group of order p, whence $\Omega_1(N) \leq Z([P, P])$, a contradiction. If $|\Omega_1(N)| = p$, then N is cyclic and $[P, P]$ acts trivially on N by 5, whence $N \leq Z([P, P])$, a contradiction.]

9. Suppose that P is a finite p-group such that $|P^{(1)}/P^{(2)}| \leq p^2$. Prove that $P^{(2)} = 1$. [*Hint:* If $P^{(2)} \neq 1$, choose $N \trianglelefteq P$ such that $N \leq P^{(2)}$ with $|P^{(2)} : N| = p$; then apply 8 to P/N.]

Chapter 5
Lie rings

Definitions and basic properties of Lie rings are discussed in § 5.1. Then nilpotent and soluble Lie rings are defined in § 5.2, and the analogues of group-theoretic results from Chapter 3 are proved for them. Free Lie rings are constructed within enveloping free associative $\mathbb{Q}$-algebras in § 5.3. This construction will be used in Chapter 9 in conjunction with free (nilpotent) groups and the Baker–Hausdorff Formula; the only consequence that we need earlier is that free Lie rings are multihomogeneous with respect to free generators.

§ 5.1. Definitions and basic properties

Lie rings are non-associative and without 1. The multiplication (often called the *Lie product*) is usually denoted by brackets $[x, y]$ (which are always necessary, since the multiplication is not associative). In addition to the laws common to all rings (abelian additive group and distributive laws), Lie rings must satisfy the following axioms:

$$[x, x] = 0 \qquad (anticommutative\ law);$$

$$[[x, y], z] + [[y, z], x] + [[z, x], y] = 0 \qquad (Jacobi\ identity),$$

for all $x, y, z \in L$. It follows from the anticommutative law that $[a, b] = -[b, a]$ (since $[a + b, a + b] = 0$ and $[a, a] = [b, b] = 0$).

Let K be a commutative associative ring with unity. A Lie ring L which is also a K-module with naturally agreed operations, $[a, b]k = [ak, b] = [a, bk]$ for all $a, b \in L$, $k \in K$, is a *Lie K-algebra*, with *ground ring K*. Every Lie ring is, of course, a Lie $\mathbb{Z}$-algebra.

If we know the pairwise products of some generators $a_1, a_2, \ldots$ of the additive group of a Lie ring L, then we can compute the products of any elements in L as linear combinations of the a_i, using the distributive and anticommutative laws. If L is a Lie K-algebra, it suffices to know the products of the generators of L as a K-module. The products $[a_i, a_j]$ are themselves linear combinations of the a_k:

$$[a_i, a_j] = \sum_k C_{ij}^k a_k, \qquad C_{ij}^k \in \mathbb{Z} \ (\in K).$$

The C_{ij}^k are called the *structural constants* of L (with respect to the a_i).

Examples. 5.1. Vectors of the three-dimensional real vector space $\mathbb{R}^3$ with respect to the usual addition of vectors and the vector multiplication

$[u, v] = u \times v$ form a Lie $\mathbb{R}$-algebra, the axioms being well-known properties of the vector product. We can choose an orthonormal basis e_1, e_2, e_3 such that the structural constants are $[e_1, e_2] = e_3, \quad [e_2, e_3] = e_1, \quad [e_3, e_1] = e_2$.

5.2. Let A be an associative ring. The structure of the Lie ring $A^{(-)}$ is defined on the same additive group A with Lie multiplication $[x, y] = xy - yx$. The axioms are easily verified: anticommutative law $[x, x] = xx - xx = 0$; the Jacobi identity

$$[[x, y], z] + [[y, z], x] + [[z, x], y]$$

$$= xyz - yxz - zxy + zyx + yzx - zyx - xyz + xzy + zxy - xzy - yzx + yxz = 0;$$

distributive laws follow from those for A. If A is a K-algebra, then $A^{(-)}$ is a Lie K-algebra. This is a universal example, since every Lie algebra can be represented as a subalgebra of $A^{(-)}$ for a suitable A (the Poincaré–Birkhoff–Witt Theorem). We shall prove this theorem in the special case of free Lie rings in § 5.3.

5.3. In particular, the set of all $n \times n$ matrices over a field forms a Lie algebra with respect to the usual component-wise addition and Lie multiplication $[A, B] = AB - BA$. The Ado–Iwasawa Theorem states that every finite-dimensional Lie algebra over a field can be represented as a subalgebra of a matrix Lie algebra.

5.4. The free Lie ring F on free generators $\{x_i \mid i \in I\}$ can be obtained according to § 1.3 as the factor-ring of an absolutely free ring by the verbal ideal defined by the laws of Lie rings. In other words, F is the collection of all formal linear combinations of all formal (non-associative) bracket monomials in the x_i, and two such linear combinations are identified if one of them can be transformed into the other by applying the axioms of Lie rings. Any mapping $x_i \to l_i$ of the free generators to arbitrary elements l_i in any Lie ring L extends to a homomorphism of F into L. However, this construction does not say much about the structure of F; for a better description of F see § 5.3.

As usual, a *Lie subring* of a Lie ring is a subset closed under all operations, a *homomorphism* of a Lie ring is a mapping preserving all operations, an *isomorphism* is a bijection that is a homomorphism, and an *automorphism* of a Lie ring is an isomorphism onto itself. It follows from the anticommutative law that every left or right ideal in a Lie ring is a two-sided ideal; recall the notation $I \trianglelefteq L$ for an ideal I of a Lie ring L. The Lie subring generated by a set X is denoted by $\langle X \rangle$ and consists of all linear combinations of all Lie monomials in the elements of X. We denote by $_+\langle X \rangle$ the additive subgroup generated by the set X, the *span of X*. The ideal generated by X is denoted by $_{\mathrm{id}}\langle X \rangle$. The additive group of $_{\mathrm{id}}\langle X \rangle$ is spanned by all Lie monomials that involve at least one element of X. For $n \in \mathbb{Z}$, we denote by nX the set $\{nx \mid x \in X\}$. Then, obviously,

$$_+\langle nX \rangle = n\,_+\langle X \rangle, \qquad _{\mathrm{id}}\langle nX \rangle = n\,_{\mathrm{id}}\langle X \rangle \tag{5.5}$$

(although we can only claim that $\langle nX \rangle \subseteq n\langle X \rangle$). The above terminology applies to Lie K-algebras with adjustments for the additional operations of multiplying by scalars from K (for example, we always have $\langle nX \rangle = \langle X \rangle$ in a Lie $\mathbb{Q}$-algebra).

Cartesian and direct sums of Lie rings are defined as for all algebraic systems; see § 1.3. Similarly to groups, there are internal and external definitions of direct sums of ideals which are equivalent. The notation $\bigoplus_i A_i$ may be used both for direct sums of Lie rings A_i (in which case the A_i are ideals), and for the additive subgroups of a Lie ring where the A_i may also be subrings or may not; the meaning must be clear from the context.

Commutators as formal bracket expressions and their (multi)weights were defined in § 1.1. We can compute the value of any (complex) commutator on elements of a Lie ring, treating $[a, b]$ as the Lie product of a, b. So the additive group of $\langle X \rangle$ is spanned by all commutators in the elements of X. Recall the simple commutator notation $[a_1, \dots, a_k] = [\dots[[a_1, a_2], a_3], \dots, a_k]$.

Lemma 5.6. (a) *Every commutator in the elements $a_1, \dots, a_k$ is a $\mathbb{Z}$-linear combination of simple commutators each having the same entry set and the same multiweight in the a_i.*

(b) *In addition, these simple commutators can be chosen all to begin with the same fixed element $a_{j_0} \in \{a_1, \dots, a_k\}$ involved in the original commutator.*

Proof. (a) Induction on the weight of the commutator; if the weight is 1 (or 2), the commutator itself is simple. Let c be a commutator of weight > 1; then $c = [c_1, c_2]$, where c_1 and c_2 are commutators of smaller weight. By the induction hypothesis and by the distributive laws, we may assume that both c_1 and c_2 are simple commutators. Then we apply induction on the weight of c_2; if the weight of c_2 is 1, then $c = [c_1, c_2]$ is simple. Otherwise $c_2 = [c_{21}, c_{22}]$, and we have

$$c = [c_1, [c_{21}, c_{22}]] = [[c_1, c_{21}], c_{22}] - [[c_1, c_{22}], c_{21}]$$

by the Jacobi identity. On the right, the initial segments $[c_1, c_{21}]$ and $[c_1, c_{22}]$ are linear combinations of simple commutators by induction on the weight. Since the weights of c_{21} and c_{22} are less than that of c_2, induction on the weight of the second factor finishes the proof. At every step the multiweight of the commutators involved remained the same.

(b) Using (a), we only need to consider a simple commutator of the form $[a_{i_1}, \dots, a_{i_k}, a_{j_0}]$ which equals $[a_{i_1}, \dots, a_{j_0}, a_{i_k}] - [a_{i_1}, \dots, [a_{j_0}, a_{i_k}]]$ by the Jacobi identity. The first summand is the required linear combination by induction on the weight applied to the initial segment $[a_{i_1}, \dots, a_{j_0}]$. If $[a_{j_0}, a_{i_k}]$ is regarded as a new variable, the second summand has smaller weight and hence is a linear combination of simple commutators beginning with $[a_{j_0}, a_{i_k}]$, which are also simple commutators beginning with a_{j_0}. $\qquad \square$

Let $L = \langle X \rangle$ be a Lie ring; the *homogeneous component* of L of weight k

(with respect to X) is the span L_k of all commutators of weight k in the elements of X. The additive subgroup (ideal, subring) H of L is *homogeneous*, if $H = \bigoplus_{k=1}^{\infty} H \cap L_k$. Let $X = \{x_1, x_2, \dots\}$; the *multihomogeneous component* $L_{\vec{k}}$ of multiweight $\vec{k} = (k_1, k_2, \dots)$ (of weight k_1 in x_1, k_2 in x_2, $\dots$) is the span of all commutators of weight k_1 in x_1, k_2 in x_2, $\dots$. The additive subgroup (ideal, subring) H of L is *multihomogeneous*, if $H = \bigoplus_{\vec{k}} H \cap L_{\vec{k}}$, where the sum is taken over all multiweights $\vec{k}$. It is clear that multihomogeneous implies homogeneous. If L is a Lie K-algebra, we take K-spans instead of $\mathbb{Z}$-spans.

Examples. **5.7.** In Example 5.1, for $L = \mathbb{R}^3 = \langle e_1, e_2, e_3 \rangle$, we have $L = L_1 = L_2 = \dots$. We also have $L = \langle e_1, e_2 \rangle$; then $L_1 = {}_+\langle e_1, e_2 \rangle$; $L_2 = {}_+\langle e_3 \rangle$; $L_3 = {}_+\langle e_1, e_2 \rangle$; $L_4 = {}_+\langle e_3 \rangle$; $\dots$. For either set of generators, L is not homogeneous.

5.8. A free Lie ring F can be shown to be multihomogeneous with respect to the free generators. It can be shown that every homogeneous component F_k of weight k is a free abelian groups; the so-called *basic commutators* can be chosen as its free generators. We shall obtain these results in § 5.3.

Definition 5.9. For subsets A and B in a Lie ring L, we define $[A, B] = {}_+\langle [a, b] \mid a \in A, \, b \in B \rangle$.

Although much of Lie ring notation is similar to that of groups, note that $[A, B]$ may not be a Lie subring, even if both A and B are subrings (while for groups, $[M, N]$ is always a normal subgroup in $\langle M, N \rangle$). On the other hand, many things are simpler in Lie rings because of linearity. For example, it follows from 5.9 by the distributive laws that

$$[A_1, A_2, \dots, A_s] = [\dots[A_1, A_2], \dots, A_s] = {}_+\langle [a_1, \dots, a_s] \mid a_j \in A_j \rangle. \quad (5.10)$$

Lemma 5.11. *If A and B are ideals of a Lie ring L, then $[A, B]$ is also an ideal of L.*

Proof. By the definition, $[A, B]$ is an additive subgroup of L. To show that $[A, B]$ is an ideal, by the distributive laws, we only need to show that $[[a, b], l] \in [A, B]$ for any $a \in A$, $b \in B$ and $l \in L$. By the Jacobi identity, we have $[[a, b], l] = [[a, l], b] + [a, [b, l]] \in [A, B]$ since $[a, l] \in A$ and $[b, l] \in B$. $\square$

In particular, $[L, L]$ is an ideal of L (often denoted by L^2). If $L = {}_+\langle X \rangle$, then $[L, L]$ is spanned by all commutators $[x, y]$, $x, y \in X$; if $L = \langle X \rangle$, then $[L, L]$ is spanned by all commutators of weights ≥ 2 in the elements of X.

The subset $Z(L) = \{z \in L \mid [z, x] = 0 \text{ for all } x \in L\}$ is an ideal called the *centre* of a Lie ring L. As with groups, we call a Lie ring *abelian (commutative)* if $Z(L) = L$, which is, of course, equivalent to $[L, L] = 0$.

The subset $C_L(M) = \{c \in L \mid [c, m] = 0 \text{ for all } m \in M\}$ is a Lie subring called the *centralizer* of a subset M of a Lie ring L. If $I \trianglelefteq L$, then $C_L(I) \trianglelefteq L$.

For $\bar{y} \in L/C_L(I)$ the mapping $\vartheta(\bar{y}) : x \to [x, y]$ is well-defined, where y is any preimage of $\bar{y} = y + C_L(I)$. It follows from the distributive laws that ϑ is an isomorphism of the additive factor-group $L/C_L(I)$ into $\operatorname{Hom}_{\mathbb{Z}} I$. Moreover, ϑ is an isomorphism of the Lie factor-ring $L/C_L(I)$ into $(\operatorname{Hom}_{\mathbb{Z}} I)^{(-)}$ (see 5.2), as follows from the Jacobi identity. For $I = L$ the $\vartheta(\bar{y})$ are known as *inner derivations* of L usually denoted by $\operatorname{ad}(y)$. One can also define the *normalizer* ("idealizer") of a Lie subring and prove an analogue of Lemma 1.9 for $N_L(H)/C_L(H)$. The inner derivations are analogous to inner automorphisms of groups; they are used to define external and internal semidirect sums of Lie rings; we leave this as an exercise to the reader.

Definition 5.12. Let A be an additively written abelian group. A Lie ring is said to be *A-graded*, if the additive group of L is the direct sum $L = \bigoplus_{g \in A} L_g$ of the additive subgroups L_g, $g \in A$, such that $[L_g, L_h] \leq L_{g+h}$ for all $g, h \in A$.

Graded Lie rings appear naturally in relation to automorphisms.

Example 5.13. Let φ be an automorphism of order n of a Lie algebra L over $\mathbb{C}$. Then the Jordan normal matrix of φ is diagonal, since for any Jordan block of size greater than 1, we have

$$\begin{pmatrix} \alpha & 1 & \\ & \ddots & \\ & & \end{pmatrix}^n = \begin{pmatrix} \alpha^n & n\alpha^{n-1} & \\ & \ddots & \\ & & \end{pmatrix},$$

and this cannot be the identity matrix. Thus, L decomposes into the direct sum of the eigenspaces,

$$L = \bigoplus_{i=0}^{n-1} L_i, \tag{5.14}$$

where $L_i = \{l \in L \mid l^\varphi = \omega^i l\}$, for a fixed primitive nth root of unity ω. For any $a \in L_i$, $b \in L_j$ we have $[a, b]^\varphi = [a^\varphi, b^\varphi] = [\omega^i a, \omega^j b] = \omega^{i+j}[a, b]$, so that $[L_i, L_j] \leq L_{i+j}$, where $i + j$ is taken modulo n since $\omega^n = 1$. Therefore (5.14) is a $(\mathbb{Z}/n\mathbb{Z})$-grading of L.

Now we discuss extending the ground ring. If $K \leq R$ are commutative associative rings with 1, and L is a Lie K-algebra, then the R-module $L \otimes_K R$ is a Lie R-algebra with respect to the Lie multiplication

$$[l_1 \otimes r_1, l_2 \otimes r_2] = [l_1, l_2] \otimes r_1 r_2, \qquad l_i \in L, \ r_i \in R.$$

Of course, $L \otimes_K R$ is also a K-algebra. Suppose that R as a K-module decomposes in the direct sum $R = K \oplus U$. Then $L \otimes R = L \otimes_K K \oplus L \otimes_K U$ by Lemma 1.35. In this situation L can be identified with the K-subalgebra $L \otimes_K K = L \otimes 1 = \{l \otimes 1 \mid l \in L\}$ of $L \otimes_K R$. Under this identification, for any subsets $A, B \subseteq L$, we have

$$[A \otimes_K R, \, B \otimes_K R] = [A, B] \otimes_K R; \qquad _{\mathrm{id}}\langle A \otimes_K R \rangle = {}_{\mathrm{id}}\langle A \rangle \otimes_K R. \qquad (5.15)$$

If G is a group of automorphisms of L, then G can be viewed as a group of automorphisms of the Lie R-algebra $L \otimes_K R$ acting as $(l \otimes r)^g = l^g \otimes r$, for $l \in L$, $r \in R$, $g \in G$.

Example 5.16. Let G be a group of automorphisms of a Lie ring ($\mathbb{Z}$-algebra) L, and let ω be a primitive nth root of unity. Then G acts naturally, as described above, on the Lie $\mathbb{Z}[\omega]$-algebra $\tilde{L} = L \otimes \mathbb{Z}[\omega]$. Because of the decomposition $\mathbb{Z}[\omega] = \mathbb{Z} \oplus \omega\mathbb{Z} \oplus \cdots \oplus \omega^{\varphi(n)-1}\mathbb{Z}$, where $\varphi(n)$ is the Euler's function we have $\tilde{L} = L \oplus (L \otimes \omega) \oplus (L \otimes \omega^2) \oplus \cdots \oplus (L \otimes \omega^{\varphi(n)-1})$. Every $\mathbb{Z}$-submodule $L \otimes \omega^i$ is G-invariant, so that $C_{\tilde{L}}(G) = \bigoplus_{i=0}^{\varphi(n)-1} C_{L \otimes \omega^i}(G)$, while $C_{L \otimes \omega^i}(G) = C_L(G) \otimes \omega^i$. In other words, we have $C_{\tilde{L}}(G) = C_L(G) \otimes \mathbb{Z}[\omega]$. (Here, of course, $C_N(G) = \{n \in N \mid n^g = n \text{ for all } g \in G\}$ denotes the fixed-point subalgebra.)

§ 5.2. Nilpotent and soluble Lie rings

Many of the definitions and properties of soluble and nilpotent Lie rings are quite similar to those of soluble and nilpotent groups, with addition in Lie rings taking the role of products in groups and the Lie products taking the role of commutators in groups. Moreover, many of the properties become easier to prove; for example, in place of the Hall–Witt Identity

$$[a, b^{-1}, c]^b \cdot [b, c^{-1}, a]^c \cdot [c, a^{-1}, b]^a = 1,$$

which we proved for groups, we simply have the Jacobi identity $[a, b, c] + [b, c, a] + [c, a, b] = 0$, which holds in any Lie ring by definition. The Jacobi identity implies an analogue of the Three Subgroup Lemma 3.2 for Lie rings. In place of the commutator formulae 1.11, we have distributive and anticommutative laws in Lie rings. There must be some caution; for example, the additive subgroup $[M, N]$ in a Lie ring may not be an ideal in the Lie ring $\langle M, N \rangle$. However, if both M and N are ideals, $[M, N]$ is an ideal analogous to the commutator subgroup (Lemma 5.11).

Definition 5.17. The terms of the *lower central series* of a Lie ring L are defined recursively to be $\gamma_1(L) = L$, $\gamma_{k+1}(L) = [\gamma_k(L), L]$. (Often $\gamma_k(L)$ is denoted by L^k for Lie rings.)

For the rest of the section we put $\gamma_k = \gamma_k(L)$, for short. We see at once that γ_k is a verbal ideal of L with respect to the word $[x_1, \ldots, x_k]$. By (5.10), γ_k is spanned by the simple commutators of weight k in the elements of L; and every commutator of weight $\geq k$ is contained in γ_k by Lemma 5.6. If $L = \langle M \rangle$, then γ_k is spanned by the commutators of weights $\geq k$ in the elements of M, and by Lemma 5.6 it suffices to take only simple commutators. The following is a corollary of the Jacobi identity.

Corollary 5.18. *In a Lie ring L, we have $[\gamma_m, \gamma_n] \leq \gamma_{m+n}$ for all $m, n \in \mathbb{N}$.*

Proof. Repeat the proof of Corollary 3.5, replacing "the Three Subgroup Lemma" by "the Jacobi identity". $\square$

Definition 5.19. A Lie ring L is said to be *nilpotent of class $\leq c$*, if $\gamma_{c+1} = 0$, or, equivalently, if the law $[x_1, \ldots, x_{c+1}] = 0$ holds on L. The least such number c is the *nilpotency class* of L. (A Lie ring is often said to have nilpotency class c if it has nilpotency class $\leq c$.)

One can define the (upper) central series for Lie rings as in 3.7 and 3.8 and prove an analogue of Theorem 3.9 for Lie rings. Since γ_k is spanned by the simple commutators of weights $\geq k$ in the generators, we immediately obtain the following fact.

Corollary 5.20. *Suppose that $L = \langle M \rangle$ is a Lie ring generated by the subset M. Then L is nilpotent of class $\leq c$ if and only if $[m_1, m_2, \ldots, m_{c+1}] = 0$ for any $m_i \in M$.* $\square$

The nilpotent Lie rings of class $\leq c$ form a variety, denoted by $\mathfrak{N}_c$. Therefore, all Lie subrings, all homomorphic images, and all Cartesian products of nilpotent Lie rings of class $\leq c$ (that is, from $\mathfrak{N}_c$) also belong to $\mathfrak{N}_c$. A free nilpotent Lie ring of class c is the factor-ring of an "absolutely" free Lie ring F by the verbal ideal $\gamma_{c+1}(F)$. We shall prove in §5.3 that F is (multi)homogeneous with respect to the free generators. Then $\gamma_{c+1}(F) = \sum_{i \geq c+1} F_i$, where F_i is the homogeneous component of weight i. Therefore, the free nilpotent Lie ring of $F/\gamma_{c+1}(F)$ is also (multi)homogeneous with respect to the free generators, and its homogeneous components of weight $i \leq c$ are isomorphic to those of F.

If all Lie rings in some variety $\mathfrak{V}$ are nilpotent, then there is an upper bound for the nilpotency classes of all Lie rings in $\mathfrak{V}$, that is, $\mathfrak{V}$ is contained in $\mathfrak{N}_c$ for some $c \in \mathbb{N}$. Indeed, a countably generated free Lie ring of the variety $\mathfrak{V}$ is nilpotent of some class c by the hypothesis, and hence all Lie rings in $\mathfrak{V}$ are nilpotent of class $\leq c$ (Lemma 1.50).

We shall later consider varieties of algebraic systems that are Lie rings with additional operations. By definition, such a system is nilpotent of class $\leq c$ if it satisfies the law $[x_1, \ldots, x_{c+1}] = 0$ (there may be difficulties with equivalent definitions because of the additional operations). The varietal arguments from the two preceding paragraphs apply to such systems, too.

The properties of nilpotent groups recorded in Theorem 3.14 have their analogues for Lie rings; we leave it as an exercise to the reader to formulate and prove them.

Definition 5.21. The terms of the *derived series* of a Lie ring L are defined recursively to be $L^{(1)} = [L, L]$ and $L^{(k+1)} = [L^{(k)}, L^{(k)}]$.

Clearly, $L^{(k)}$ is a verbal ideal with respect to the word $\delta_k(x_1, \ldots, x_{2^k})$, where, recall, $\delta_1(x_1, x_2) = [x_1, x_2]$ and

$$\delta_{n+1}(x_1, \ldots, x_{2^{n+1}}) = [\delta_n(x_1, \ldots, x_{2^n}), \delta_n(x_{2^n+1}, \ldots, x_{2^{n+1}})].$$

Definition 5.22. A Lie ring L is said to be *soluble of derived length $\leq d$* if $L^{(d)} = 0$ or, equivalently, if the law $\delta_d(x_1, \ldots, x_{2^d}) = 0$ holds on L. The minimal such number d is the *derived length* of L. (It is often said that a Lie ring is soluble of derived length d if the derived length is $\leq d$.)

An analogue of Theorem 3.20 holds: a Lie ring L is soluble of derived length d if and only if it has a series of (sub)ideals of length d with abelian factors. It may not be sufficient to verify the solubility law $\delta_d(x_1, \ldots, x_{2^d}) = 0$ only on the generators (Exercise 5.8).

The soluble Lie rings of derived length $\leq d$ form a variety, denoted by $\mathfrak{A}^d$. Hence all subrings, all homomorphic images, and all Cartesian products of soluble Lie rings of derived length $\leq d$ (that is, from $\mathfrak{A}^d$) also belong to $\mathfrak{A}^d$.

If all Lie rings in a variety $\mathfrak{V}$ are soluble, then there is an upper bound for the derived lengths of all Lie rings in $\mathfrak{V}$, that is, $\mathfrak{V}$ is contained in $\mathfrak{A}^d$ for some $d \in \mathbb{N}$. Indeed, a countably generated (relatively) free Lie ring in $\mathfrak{V}$ is soluble of some derived length d by the hypothesis, and hence all Lie rings in $\mathfrak{V}$ are soluble of derived length $\leq d$ (Lemma 1.50).

Suppose we have a variety of algebraic systems that are Lie rings with additional operations. By definition, such a system S is soluble of derived length $\leq d$ if it satisfies the law $\delta_d(x_1, \ldots, x_{2^d}) = 0$. The varietal arguments from the two preceding paragraphs apply to such systems, too.

We record for further reference several obvious formulae that hold for any Lie ring L:

$$(nL)^{(d)} = n^{2^d} L^{(d)} \quad \text{and} \quad \gamma_k(nL) = n^k \gamma_k(L), \quad \text{for all } d, k \in \mathbb{N}. \tag{5.23}$$

Suppose that $L = \bigoplus_i M_i$ with $M_i \trianglelefteq L$; then

$$L^{(d)} = \bigoplus_i M_i^{(d)} \quad \text{and} \quad \gamma_k(L) = \bigoplus_i \gamma_k(M_i) \quad \text{for all } d, k \in \mathbb{N}. \tag{5.24}$$

A criterion for a variety to be soluble, analogous to Theorem 3.22, holds for Lie rings too (and for some classes of Lie rings with additional operations). We shall use this criterion in Chapter 7 for proving solubility of some graded Lie rings.

Theorem 5.25. *If $L \neq [L, L]$ for every non-trivial Lie ring $L \neq 0$ in a variety of Lie rings $\mathfrak{V}$, then the variety $\mathfrak{V}$ is soluble: $\mathfrak{V} \subseteq \mathfrak{A}^k$ for some k.*

Proof. Repeat the proof of Theorem 3.22, replacing 1 by 0, products by sums, "(sub)group" by "Lie (sub)ring", "normal subgroup" by "ideal". □

Remark 5.26. Another form of Theorem 5.25: if a variety of Lie rings is non-soluble, then it contains a non-trivial Lie ring which coincides with its derived subring. We state a generalization of Theorem 5.25 for algebraic systems that are Lie rings with additional operations: provided $f_k^{n_k}(0, \dots, 0) = 0$ for all operations $f_k^{n_k}$, any non-soluble variety $\mathfrak{V}$ of such systems ("non-soluble" here means not satisfying any of the identities $\delta_d(x_1, \dots, x_{2^d}) = 0$ with respect to the chosen Lie ring operations) contains a non-trivial system which coincides with its derived Lie subring. The proof of Theorem 5.25 can be repeated verbatim, with homomorphisms in the sense of the algebraic systems, and sums and commutators with respect to the chosen Lie ring operation; the condition $f_k^{n_k}(0, \dots, 0) = 0$ ensures that N (the direct sum of the F_i) is a normal subsystem of the Cartesian sum of the F_i. For arriving at the equality $L = [L, L]$, it really does not matter whether additional operations were used for forming $[L, L]$ or not.

Of course, a nilpotent Lie ring is soluble, by an analogue of Lemma 3.25. The converse may not be true: for example, the Lie $\mathbb{Q}$-algebra with basis $\{a, b\}$ and structural constants $[a, b] = a$ is soluble but not nilpotent. The following theorem is analogous to Theorem 3.26 of P. Hall.

Theorem 5.27. *Let N be an ideal of a Lie ring L. If N is nilpotent of class k and the factor-ring $L/[N, N]$ is nilpotent of class c, then L itself is nilpotent of class at most $f(k, c) = (c - 1)k(k + 1)/2 + k$.*

Proof. Repeat the proof of Theorem 3.26, with certain simplifications (for example, $[[A, B], C] \leq [[C, A], B] + [[C, B], A]$ by the Jacobi identity). □

Corollary 5.28. *Suppose that in a variety of Lie rings $\mathfrak{V}$ all soluble Lie rings of derived length 2 are nilpotent of class $\leq c$. Then a soluble Lie ring from $\mathfrak{V}$ of derived length s is nilpotent of (s, c)-bounded class. (In other words, there exists a function $g(s, c)$ such that if $\mathfrak{V} \cap \mathfrak{A}^2 \subseteq \mathfrak{N}_c$, then $\mathfrak{V} \cap \mathfrak{A}^s \subseteq \mathfrak{N}_{g(s,c)}$.)*

Remarks. 5.29. A better bound $(c^s - 1)/(c - 1)$ for $g(s, c)$ in Corollary 5.28 can be obtained, similarly to Exercise 3.8. The better bound of A. G. R. Stewart [1966] indicated in Remark 3.30 must hold for Lie rings too.

5.30. Theorem 5.27 and Corollary 5.28 remain valid for some varieties of Lie rings with additional operations; at least, if the terms of the (abstract) derived and lower central series are verbal subsystems, the same proof works.

5.31. Although in many aspects Lie rings are similar to groups, some caution must be exercised in drawing analogies between them. For example, one can prove that the derived subalgebra of any soluble finite-dimensional Lie algebra of characteristic 0 is nilpotent, which may not be true for finite soluble groups (a better analogy is with linear (matrix) groups). On the other hand, one can produce a simple non-abelian Lie algebra of dimension 3 over $\mathbb{F}_p$, that is, of order p^3 (Exercise 5.5), while every group of order p^3 is nilpotent of class 2 (Corollary 4.2).

§ 5.3. Free Lie rings

Let A be a free associative $\mathbb{Q}$-algebra on free (non-commuting) generators $x_1, x_2, \ldots$ (when necessary, we shall take a well-ordered set of generators of any given cardinality). In this section, we construct a free Lie ring L as a subring of $A^{(-)}$. The advantage is in the fact that the structure of A is quite transparent (Example 1.43); many important properties of L will be derived merely from the linear independence of associative monomials in the free generators of A.

Recall that A has a basis consisting of all monomials $x_{i_1} \cdots x_{i_k}$ of degrees $k \in \mathbb{N}$ (no parentheses are needed because of the associative laws). The multiplication of monomials is juxtaposition:

$$(x_{i_1} \cdots x_{i_k}) \cdot (x_{j_1} \cdots x_{j_l}) = x_{i_1} \cdots x_{i_k} x_{j_1} \cdots x_{j_l},$$

where no cancellations are allowed; all other products are then defined via distributive laws. Thus, A is multihomogeneous with respect to the x_i; in particular, $A = \bigoplus_{i=1}^{\infty} A_i$, where A_i is the homogeneous component of degree i.

The bracket multiplication $[x, y] = xy - yx$ defines the structure of a Lie $\mathbb{Q}$-algebra $A^{(-)}$ on the additive group of A (Example 5.2). Every Lie monomial (or product, or commutator) of weight k in the x_i is also a linear combination of associative monomials of the same degree k in the x_i (and the same multi-degree). We denote by L the Lie ring ($\mathbb{Z}$-algebra) generated by the x_i in $A^{(-)}$. Then $\mathbb{Q}L = \{rl \mid r \in \mathbb{Q}, \ l \in L\}$ is the Lie $\mathbb{Q}$-algebra generated by the x_i. Note that $\mathbb{Q}L \neq A^{(-)}$: for example, $x_1 x_2 \notin L$. The additive group of L is generated by the Lie products (commutators) in the x_i. Our aim is to prove that L is a free Lie ring on free generators x_i. The main tool in the proof is the so-called basic Lie products. We shall prove that every Lie ring is spanned by the basic products in the generators, and that the basic products in the x_i are linearly independent in L; then L will have to be a free Lie ring. To prove linear independence, we shall use "projections" onto associative monomials. We are using the basic products of A. I. Shirshov [1958] (rather than the more traditional basic commutators of P. Hall and M. Hall).

Definitions 5.32. The following lexicographical order is defined on the associative monomials (words) in the x_i: we have a well-ordering of the generators $x_1 < x_2 < \ldots$, and $x_{i_1} \cdots x_{i_m} < x_{j_1} \cdots x_{j_n}$ either if, for some k, we have

$i_1 = j_1, \ldots, i_{k-1} = j_{k-1}$ and $i_k < j_k$, or if the right-hand side is an initial segment of the left-hand side (sic! a proper initial segment is greater than the word). A word u is *regular* if u is greater than any of its cyclic permutations: if $u = vw$ is any non-trivial decomposition, then $u > wv$.

To distinguish Lie products (commutators) in the x_i, we shall temporarily use the notation $[u]$ for them, while u will denote the *underlying associative word* which is obtained from $[u]$ by omitting all brackets; for example, if $[u] = [x_3, [x_3, x_1]]$, then $u = x_3 x_3 x_1$. There are many ways of bracketing a given associative word u; thus, while u is well-defined by $[u]$, this notation is ambiguous in the direction from u to $[u]$.

Definition 5.33. The *basic Lie products* in the x_i are defined as some commutators in the x_i by induction on the weight. The elements $[x_i] = x_i$ are the basic products of weight 1. The commutator $[[b_1], [b_2]]$ is a basic product if both $[b_1]$ and $[b_2]$ are basic products and the following two conditions are satisfied:

(1) $b_1 > b_2$ (for the underlying associative words) and

(2) if the weight of $[b_1]$ is greater than 1 and $[b_1] = [[b_{11}], [b_{12}]]$,
    then $b_{12} \leq b_2$ (for the underlying associative words).

For example, x_1, $[x_2, x_1]$, $[x_2, [x_2, x_1]]$, $[[x_3, x_1], [x_2, x_1]]$ are basic products. It follows from the definition that all subcommutators of a basic product are also basic. The independence of the basic products in L will be proved by "marking" each basic product $[u]$ by the greatest associative word in the decomposition of $[u]$ as a linear combination of associative monomials in the x_i (this greatest word is, in fact, the underlying word u).

Lemma 5.34. (a) *If $[u]$ is a basic product, then the underlying word u is regular.*

(b) *For each regular word u there is a unique basic product $[u]$.*

Proof. (a) We use induction on the weight of $[u]$. If the weight is 1, then $[u]$ is one of the letters x_i, which is a regular word. Let the weight of $[u]$ be greater than 1, and let x_m be the least letter involved in $[u]$. Then the first occurrence of x_m in u is after some $x_j > x_m$ and each occurrence of the subword $x_j x_m$ in u comes from a subcommutator $[x_j, x_m]$ in $[u]$. Indeed, the closest bracket to any of the x_m in $[u]$ cannot be a left one, like $[x_m$, for the right-most of such occurrences would contradict condition (1) in Definition 5.33. If now $[a, x_m]$ is a subcommutator with $x_m]$, then either a has weight 1 or $a = [[b], [c]]$. In the second case, $c \leq x_m$ according to (2) of 5.33; then c must begin with x_m, which implies $c = x_m$.

We introduce the new letter $[x_j, x_m]$ into the original alphabet, adjusting the order for the letters as $\ldots < x_{j-1} < [x_j, x_m] < x_j < \ldots$; then the lexicographical order is defined on the new alphabet in the same way. It is easy to see that the order on those associative words in the new alphabet that have no subwords $x_j x_m$ coincides with the order in the old sense after removing the brackets in all occurrences of the new letter $[x_j, x_m]$. Hence $[u]$ is a basic product in the new alphabet too, if we regard all subcommutators $[x_j, x_m]$ in $[u]$ as occurrences of the new letter. By the induction hypothesis, u is a regular word in the new alphabet. Therefore, removing the brackets from the new letter, we obtain that u is greater than any cyclic permutation of u that does not break the subwords $x_j x_m$. The cyclic permutations of u that do break some $x_j x_m$ begin with x_m and hence are less than u, since u cannot begin with x_m, as we saw above.

(b) Induction on the degree of a regular word u. If the degree is 1, then $u = [u]$ is a basic product, being one of the x_i. Let the degree of u be greater than 1 and let x_m be the least letter involved in u. The word u, being regular, cannot begin with x_m, for otherwise x_m is the only letter involved and then $u = x_m$, contrary to the assumption that the degree is not 1. The first occurrence of x_m in u is then after some $x_j > x_m$. We replace all subwords $x_j x_m$ in u by the commutators $[x_j, x_m]$ and consider the resulting word in the new alphabet with the new letter $[x_j, x_m]$ and the order as in the proof of (a). Then u becomes a regular word in the new alphabet. By the induction hypothesis, there is a unique basic product $[u]$ in the new alphabet, which is also a basic product in the old alphabet. It is unique by the induction hypothesis and due to the fact that each occurrence of the subword $x_j x_m$ in u must come from a subcommutator $[x_j, x_m]$ in any possible basic product $[u]$, as shown in the proof of (a).   □

We shall need another simple property of regular words.

Lemma 5.35. *If a is regular and $b > a$ for some word b, then $ba > ab$.*

Proof. Since $b > a$, either b is a proper initial segment of a, or, after some equal initial segments, the next letter in b is greater than the next letter in a. In the latter case, any extension of b (to the right) is greater than any extension of a. Therefore, if $ab \geq ba$, then $a = bz$ for some word z, so that $bzb \geq bbz$. This implies $zb \geq bz = a$, contrary to the regularity of a.   □

Now we are ready to "mark" the basic products with the underlying associative words.

Lemma 5.36. *If a basic product $[u]$ is expressed as a linear combination of associative monomials in the x_i, then the underlying word u is the unique greatest word among these monomials.*

Proof. The assertion is trivial for weight 1. If $[u] = [[v], [w]]$, then $v > w$

and, by the induction hypothesis, we have

$$[v] = v + \sum_i \alpha_i v_i \quad \text{and} \quad [w] = w + \sum_j \beta_j w_j, \qquad \alpha_i, \beta_j \in \mathbb{Z},$$

where $v > v_i$ and $w > w_j$ for all i, j. Then

$$[u] = [[v], [w]] = vw + \sum_i \alpha_i v_i w + \sum_j \beta_j v w_j + \sum_{i,j} \alpha_i \beta_j v_i w_j$$

$$- wv - \sum_i \alpha_i w v_i - \sum_j \beta_j w_j v - \sum_{i,j} \alpha_i \beta_j w_j v_i.$$

Since w is regular and $v > w$, we have $vw > wv$ by Lemma 5.35, and the other words on the right are obviously less than either vw or wv:

$$v_i w_j < v_i w < vw \quad \text{and} \quad w_j v_i < w_j v < wv,$$

for all i, j (note that $\deg w = \deg w_j$ and $\deg v = \deg v_i$ for all i, j). $\qquad\square$

Now suppose that $M = \langle g_1, g_2, \ldots \rangle$ is an arbitrary Lie ring with well-ordered set of generators $g_1 < g_2 < \ldots$. We define the basic Lie products in the g_i using Definition 5.33 with the same lexicographical order 5.32 on formal associative words in the g_i. (In other words, basic products in the g_i are obtained from the basic products in the x_i by substituting the g_i in place of the corresponding x_i.)

Lemma 5.37. *The basic products in the g_i span M.*

Proof. The commutators in the g_i span M. Hence it is sufficient to express any commutator $[k]$ in the g_i as a linear combination of the basic products in the g_i. If the weight of $[k]$ is 1, the assertion is trivial. Let $[k] = [[k_1], [k_2]]$; by induction on the weight and the distributive laws, both $[k_1]$ and $[k_2]$ may be assumed to be basic products, and by the anticommutative law we may also assume $k_1 > k_2$ for the underlying formal words in the g_i. If the weight of $[k_1]$ is 1, then $[k]$ is a basic product. Let $[k_1] = [[k_{11}], [k_{12}]]$, where $[k_{11}], [k_{12}]$ are basic too and $k_{11} > k_{12}$. If $k_{12} \leq k_2$, then $[k]$ is basic. If $k_{12} > k_2$, then we apply the Jacobi identity to get

$$[[[k_{11}], [k_{12}]], [k_2]] = [[[k_{11}], [k_2]], [k_{12}]] - [[[k_{12}], [k_2]], [k_{11}]]. \tag{5.38}$$

By Lemma 5.34, the underlying associative words k_{11}, k_{12} and k_2 are regular. We have $k_{11} > k_{12} > k_2$; hence, by Lemma 5.35, $k_{11} k_{12} > k_{12} k_{11}$ and $k_{11} k_2 > k_2 k_{11}$, whence $k = k_{11} k_{12} k_2 > k_{12} k_{11} k_2 > k_{12} k_2 k_{11}$. Similarly, we have $k_{12} k_2 > k_2 k_{12}$, whence $k = k_{11} k_{12} k_2 > k_{11} k_2 k_{12}$. We see that the underlying associative words of both summands on the right of (5.38) are smaller than k; hence by induction on the order for the given degree, both of them are linear combinations of basic products. $\qquad\square$

Finally, we are ready to prove the main result.

Theorem 5.39. (a) *The additive group of L is freely generated by the basic products in the x_i.*
(b) *The Lie ring L is a free Lie ring on free generators x_i.*

Proof. (a) By Lemma 5.37, the basic products in the x_i span L. To prove that they are linearly independent, suppose the opposite: $\sum_i \alpha_i [u_i] = 0$ for some distinct basic products $[u_i]$, with $\alpha_i \neq 0$ for all i. Let u_{i_0} be the greatest among the underlying words u_i; by Lemma 5.34, $[u_{i_0}]$ is the unique corresponding basic product. Expanding all of the $[u_i]$ as linear combinations of associative monomials, we obtain $\alpha_{i_0} u_{i_0} + \sum_j \beta_j v_j = 0$, where $u_{i_0} > v_j$ for all j by Lemma 5.36. Since the associative monomials are linearly independent in A, this implies $\alpha_{i_0} = 0$, a contradiction.

(b) Now let F be a free Lie ring on free generators f_i corresponding to the x_i. Ordering the f_i correspondingly, we form the basic products in the f_i, which span F by Lemma 5.37. The mapping $f_i \to x_i$ extends to a homomorphism of F onto L. The images of the basic products in the f_i under this homomorphism are the corresponding basic products in the x_i. Since the latter are linearly independent by (a), this is, in fact, an isomorphism. $\square$

Corollary 5.40. *The free Lie ring L on free generators x_i is multihomogeneous with respect to the x_i.*

Proof. We can view L as constructed above within $A^{(-)}$. Every Lie product of weight k in the x_i is a linear combination of associative monomials of the same multiweight in the x_i. Therefore, $L_{\vec{n}} = L \cap A_{\vec{n}}$ for any multiweight $\vec{n}$. Since A is multihomogeneous, it follows that $L = \bigoplus_{\vec{n}} L_{\vec{n}}$. $\square$

We shall need the following technical lemma. First, we define the *Dynkin operator* δ on associative monomials in the x_i as bracketing from the left:

$$\delta(x_{i_1} x_{i_2} \cdots x_{i_m}) = [\ldots [x_{i_1}, x_{i_2}], \ldots, x_{i_m}]$$

(where $\delta(x_i) = x_i$); then δ is extended to A by linearity.

Lemma 5.41. $\delta(x_{k+1}[x_1, \ldots, x_k]) = [x_{k+1}, [x_1, \ldots, x_k]]$.

Proof. Induction on k; the case $k = 2$ follows from the definition: $\delta(x_2 x_1) = [x_2, x_1]$. For $k > 2$, we have

$$\delta(x_{k+1}[x_1, \ldots, x_k]) = \delta(x_{k+1}[x_1, \ldots, x_{k-1}]x_k) - \delta(x_{k+1}x_k[x_1, \ldots, x_{k-1}])$$

$$= [\delta(x_{k+1}[x_1, \ldots, x_{k-1}]), x_k] - \delta(x_{k+1}x_k[x_1, \ldots, x_{k-1}]).$$

The first summand on the right equals $[[x_{k+1}, [x_1, \ldots, x_{k-1}]], x_k]$ by the induction hypothesis. In the second, applying δ to $x_{k+1}x_k[x_1, \ldots, x_{k-1}]$, we replace

$x_{k+1}x_k$ by the commutator $[x_{k+1}, x_k]$ and then regard this commutator as a new variable. By the induction hypothesis we then have

$$\delta(x_{k+1}x_k[x_1, \ldots, x_{k-1}]) = [[x_{k+1}, x_k], [x_1, \ldots, x_{k-1}]].$$

As a result, we have

$$\begin{aligned}
\delta(x_{k+1}[x_1, \ldots, x_k]) &= [[x_{k+1}, [x_1, \ldots, x_{k-1}]], x_k] - [[x_{k+1}, x_k], [x_1, \ldots, x_{k-1}]] \\
&= [x_{k+1}, [x_1, \ldots, x_{k-1}, x_k]]
\end{aligned}$$

by the Jacobi identity. □

When a formula is proved for the free generators of a free Lie ring, the same formula holds for any elements in any Lie ring.

Corollary 5.42. *Let $\alpha_\pi \in \mathbb{Z}$ be the coefficients defined by the decomposition $[x_1, \ldots, x_k] = \sum_{\pi \in S_k} \alpha_\pi x_{1\pi} \cdots x_{k\pi}$ as a linear combination of associative monomials in A. Then for any elements $a_1, \ldots, a_{k+1} \in G$ of any Lie ring G we have*

$$[a_{k+1}, [a_1, \ldots, a_k]] = \sum_{\pi \in S_k} \alpha_\pi [a_{k+1}, a_{1\pi}, \ldots, a_{k\pi}].$$

Proof. We apply the homomorphism of L into G that extends the mapping $x_i \to a_i$ to the equation given by Lemma 5.41 for $[x_{k+1}, [x_1, \ldots, x_k]]$, where, obviously, $\delta(x_{k+1}[x_1, \ldots x_k]) = \sum_{\pi \in S_k} \alpha_\pi [x_{k+1}, x_{1\pi}, \ldots, x_{k\pi}]$. □

One can prove an assertion analogous to Corollary 5.42 for any commutator c in the free generators x_i: its decomposition as a linear combination of associative monomials in the x_i determines, in a similar way, the decomposition of the Lie product $[x_{i_0}, c]$ as a linear combination of simple commutators (and hence the same equality holds for any elements in any Lie ring).

Exercises 5

1. Check that the structural constants $[e_i, e_j] = (i - j)e_{i+j}$ define a Lie $\mathbb{Q}$-algebra on the vector space over $\mathbb{Q}$ with countable basis $\{e_1, e_2, \ldots\}$.

2. Let $A = \langle x_1, x_2 \rangle$ and $B = \langle x_3 \rangle$ be the subrings of the free Lie ring L on free generators x_1, x_2, x_3. Show that $[A, B]$ is not a subring of L.

3. Let a be an element of a Lie $\mathbb{Q}$-algebra L such that $[x, \overbrace{a, \ldots, a}^{n}] = 0$ for all $x \in L$. Prove that the mapping

$$x \to x + \frac{[x, a]}{1!} + \frac{[x, a, a]}{2!} + \cdots + \frac{[x, \overbrace{a, \ldots, a}^{n-1}]}{(n-1)!}$$

   is an automorphism of L.

4. Suppose that φ is an automorphism of order 3 of a Lie $\mathbb{C}$-algebra L such that $C_L(\varphi) = 0$. Prove that $\gamma_3(L) = 0$. [*Hint:* See Example 5.13.]

5. Construct a Lie ring of p^3 elements, p a prime, which has no proper ideals. [*Hint:* Consider the structural constants in Example 5.1.]

6. Prove that the structural constants $[x_1, x_2] = px_3$, $[x_2, x_3] = px_1$, $[x_3, x_1] = px_1$ define a Lie ring L on the direct sum $\langle x_1 \rangle \oplus \langle x_2 \rangle \oplus \langle x_3 \rangle$ of the cyclic groups of order p^n, where p is a prime. Prove that L is nilpotent and find the nilpotency class and the derived length of L.

7. Prove the analogues of Theorems 3.9 and 3.14 for Lie rings. The terms of the *upper central series* of a Lie ring L are defined as follows: $\zeta_1(L) = Z(L)$, the centre of L (see §5.1), and $\zeta_{k+1}(L)$ is the full inverse image of $Z(L/\zeta_k(L))$.

8. Let L be a free Lie ring on two free generators x, y. Show that the law $\delta_2 = 0$ of solubility of derived length 2 holds on the generators x, y (while L is not soluble).

9. If all soluble Lie rings of derived length 2 in a variety of Lie rings $\mathfrak{V}$ are nilpotent of class $\leq c$, then show that every soluble Lie ring in $\mathfrak{V}$ of derived length s is nilpotent of class at most $(c^s - 1)/(c - 1)$. [*Hint:* Use induction on s to prove that $[L^{(s-1)}, \underbrace{L, \ldots, L}_{c^{s-1}}] \leq L^{(s)}$ for $L \in \mathfrak{V}$.]

10. A homomorphism φ of the additive group of a Lie ring L is a *derivation* of L if $[a, b]\varphi = [a\varphi, b] + [a, b\varphi]$ for all $a, b \in L$ Prove that the set $\mathrm{Der}\, L$ of all derivations of L is a Lie subring of $(\mathrm{Hom}_{\mathbb{Z}} L)^{(-)}$.

11. Prove that for a subring $H \leq L$ of a Lie ring L the centralizer $C_L(H)$ is always an ideal of $N_L(H)$ and $N_L(H)/C_L(H)$ is isomorphic to a Lie subring of $\mathrm{Der}\, H$.

12. [P. J. Higgins, 1954] If a soluble Lie $\mathbb{Q}$-algebra L of derived length d satisfies the n-Engel identity $[x, \underbrace{y, \ldots, y}_{n}] = 0$, then show that L is nilpotent of (d, n)-bounded class. [*Hint:* Use Corollary 5.28 to reduce to the case $d = 2$; linearize the identity by substituting $y = k_1 y_1 + \cdots + k_n y_n$, where the y_i are free generators of L and $k_i \in \mathbb{Q}$.]

13. Suppose that L is a Lie ring such that the factor-ring $\gamma_k(L)/\gamma_{k+1}(L)$ is finite of order n. Prove that $\gamma_{k+1}(L)/\gamma_{k+2}(L)$ is finite of (k, n)-bounded order. [*Hint:* Fix a bounded number of elements a_i such that every coset of γ_{k+1} in the additive group of γ_k is equal to one of the $[a_{i_1}, \ldots, a_{i_k}] + \gamma_{k+1}$. For any $b \in L$ express $[a_{i_1}, \ldots, a_{i_k}, b]$ as a linear combination of the commutators of the form $[b, a_{j_1}, \ldots, a_{j_k}]$; every such commutator is congruent to some $[a_{s_1}, \ldots, a_{s_k}, a_{j_k}] \mod \gamma_{k+2}$.]

14. Suppose that L is a nilpotent Lie ring of class c such that the factor-ring $\gamma_k(L)/\gamma_{k+1}(L)$ is finite of order n. Prove that $\gamma_k(L)$ is finite of (n, c)-bounded order. Use 11 to show that $C_L(\gamma_k(L))$ is then a nilpotent ideal of class k whose additive group has (n, k, c)-bounded index in L.

15. State and prove the Lie ring analogues of Exercises 3.1, 3.6, 3.11, 3.12, 3.16, 3.18, 3.20, 3.21.

Chapter 6
Associated Lie rings

The Hall–Witt Identity,

$$[a, b^{-1}, c]^b \cdot [b, c^{-1}, a]^c \cdot [c, a^{-1}, b]^a = 1,$$

which holds in any group, strikingly resembles the Jacobi identity

$$[a, b, c] + [b, c, a] + [c, a, b] = 0,$$

which is a law in Lie rings. The commutator formula $[ab, c] = [a, c]^c[b, c]$ is also similar to a distributive law. It is natural to try to define a Lie ring with addition based on the group multiplication, and with Lie products based on taking group commutators. Lie rings may be easier to study, as more linear objects; for example, an automorphism of a Lie algebra can be regarded as a linear transformation, which has eigenvectors over the extended ground field. A Lie ring method of studying groups consists of translating conditions into the Lie ring language, obtaining (or using) results on Lie rings, and then translating the conclusions into the group language.

In this chapter we introduce one of the Lie ring methods based on the so-called associated Lie rings. This will be one of our main tools in the subsequent chapters. One of the advantages of this method is that every nilpotent group has an associated Lie ring, which is nilpotent of exactly the same nilpotency class. We shall also discuss the difficulties that arise from the fact that the associated Lie ring "forgets" some important information about the group, like its derived length.

§ 6.1. Definition

First we fix some notation. Let G be a group and let $\gamma_i = \gamma_i(G)$ be the terms of the lower central series of G. Let $\bar{a} \in \gamma_i/\gamma_{i+1}$ denote the image of $a \in \gamma_i$ in γ_i/γ_{i+1}.

Definition 6.1. The additive group of the *associated Lie ring* $L(G)$ of a group G is the direct sum

$$L(G) = \bigoplus_{i=1}^{\infty} (\gamma_i/\gamma_{i+1})$$

of the additively written factors of the lower central series of G. In particular, for $\bar{x}_1, \bar{x}_2 \in \gamma_i/\gamma_{i+1}$, by definition, $\bar{x}_1 + \bar{x}_2 = x_1 x_2 \gamma_{i+1} = \overline{x_1 x_2} \in \gamma_i/\gamma_{i+1}$. The

Lie product of the elements $\bar{x} \in \gamma_i/\gamma_{i+1}$, $\bar{y} \in \gamma_j/\gamma_{j+1}$ is defined to be

$$[\bar{x}, \bar{y}] = [x, y]\gamma_{i+j+1} \in \gamma_{i+j}/\gamma_{i+j+1}.$$

To make this clear: the Lie product of the elements $\bar{x} = x\gamma_{i+1} \in \gamma_i/\gamma_{j+1}$ and $\bar{y} = y\gamma_{j+1} \in \gamma_j/\gamma_{j+1}$ on the left is defined as the image of the group commutator $[x, y]$ in $\gamma_{i+j}/\gamma_{i+j+1}$. (Note that it can well happen that $[\bar{x}, \bar{y}] = 0$ in $L(G)$, although $[x, y] \neq 1$ in G, simply when $[x, y] \in \gamma_{j+j+1}$.) Then this bracket multiplication is extended to the direct sum $L(G) = \bigoplus_i(\gamma_i/\gamma_{i+1})$ by the distributive laws.

Brackets are used both for commutators in the group G and for the Lie products in $L(G)$; it will, however, always be possible to recognize the meaning, even if both appear in the same formula. We have yet to prove that everything is all right in Definition 6.1.

Theorem 6.2. *Definition 6.1 correctly defines the structure of a Lie ring* $L(G)$.

Proof. Recall that $[\gamma_s, \gamma_t] \leq \gamma_{s+t}$ for all $s, t \in \mathbb{N}$ by Corollary 3.5. Therefore, for $x \in \gamma_i$ and $y \in \gamma_j$, the commutator $[x, y]$ belongs to $[\gamma_i, \gamma_j] \leq \gamma_{i+j}$, so that we really can take the image of $[x, y]$ in $\gamma_{i+j}/\gamma_{i+j+1}$. We must also show that the result does not depend on the choice of the representatives x and y of the cosets. In other words, if $x'\gamma_{i+1} = x\gamma_{i+1}$ and $y'\gamma_{j+1} = y\gamma_{j+1}$, we must show that $[x', y']\gamma_{i+j+1} = [x, y]\gamma_{i+j+1}$. We have $x' = xu$ for some $u \in \gamma_{i+1}$, and $y' = yv$ for some $v \in \gamma_{j+1}$. By the commutator formulae 1.11, we have

$$[x', y']\gamma_{i+j+1} = [xu, yv]\gamma_{i+j+1} = [x, yv][x, yv, u][u, yv]\gamma_{i+j+1}$$

$$= [x, yv]\gamma_{i+j+1}$$

$$= [x, v][x, y][x, y, v]\gamma_{i+j+1}$$

$$= [x, y]\gamma_{i+j+1},$$

because of the inclusions $[\gamma_s, \gamma_t] \leq \gamma_{s+t}$. Thus, the bracket multiplication in $L(G)$ is well-defined.

The distributive laws are used to extend the multiplication to the direct sum from the summands. Therefore, to verify these laws on $L(G)$, we need only take sums of elements in the same summand. Let $\bar{x}_1, \bar{x}_2 \in \gamma_i/\gamma_{i+1}$ and $\bar{y} \in \gamma_j/\gamma_{j+1}$. By the definition,

$$[\bar{x}_1 + \bar{x}_2, \bar{y}] = [x_1 x_2, y]\gamma_{i+j+1} = [x_1, y][x_1, y, x_2][x_2, y]\gamma_{i+j+1}$$

$$= [x_1, y][x_2, y]\gamma_{i+j+1}$$

$$= [\bar{x}_1, \bar{y}] + [\bar{x}_2, \bar{y}],$$

since $[x_1, y, x_2] \in \gamma_{i+j+i} \leq \gamma_{i+j+1}$. The other distributive law is verified in a similar way. It follows that $-\bar{x} = x^{-1}\gamma_{i+1}$ and $[x^k, y]\gamma_{i+j+1} = [x, y^k]\gamma_{i+j+1} = k[\bar{x}, \bar{y}]$ for all $k \in \mathbb{Z}$.

Both the anticommutative law and the Jacobi identity are linear in their arguments. It is therefore sufficient to check that they hold for the elements in the direct summands. For $\bar{x} \in \gamma_i/\gamma_{i+1}$ we have $[x, x] = 1$ in the group, whence $[\bar{x}, \bar{x}] = 0$ in $L(G)$. To prove the Jacobi identity for $\bar{x} \in \gamma_i/\gamma_{i+1}$, $\bar{y} \in \gamma_j/\gamma_{j+1}$, $\bar{z} \in \gamma_k/\gamma_{k+1}$, we use the Hall–Witt Identity 3.1:

$$1 \cdot \gamma_{i+j+k+1} = [x, y^{-1}, z]^y [y, z^{-1}, x]^z [z, x^{-1}, y]^x \gamma_{i+j+k+1}$$

$$= [x, y^{-1}, z][y, z^{-1}, x][z, x^{-1}, y]\gamma_{i+j+k+1},$$

where we could omit the conjugating elements, since the commutators involved belong to γ_{i+j+k} whose image in $G/\gamma_{i+j+k+1}$ is in the centre. Further, we gradually switch to operations in $L(G)$:

$$0 = [x, y^{-1}, z]\gamma_{i+j+k+1} + [y, z^{-1}, x]\gamma_{i+j+k+1} + [z, x^{-1}, y]\gamma_{i+j+k+1}$$

$$= [[x, y^{-1}]\gamma_{i+j+1}, \bar{z}] + [[y, z^{-1}]\gamma_{j+k+1}, \bar{x}] + [[z, x^{-1}]\gamma_{k+i+1}, \bar{y}]$$

$$= -[\bar{x}, \bar{y}, \bar{z}] - [\bar{y}, \bar{z}, \bar{x}] - [\bar{z}, \bar{x}, \bar{y}].$$

(The reader will recognize where the Lie ring brackets replace the brackets of group commutators.) Hence $[\bar{x}, \bar{y}, \bar{z}] + [\bar{y}, \bar{z}, \bar{x}] + [\bar{z}, \bar{x}, \bar{y}] = 0$, as required.   □

Remarks. 6.3. It is clear that $L(G/\bigcap_{i=1}^{\infty} \gamma_i(G)) = L(G)$. So the associated Lie ring reflects only the properties of the (residually) nilpotent factor-groups of the group. For example, if $G = [G, G]$, then $L(G) = \{0\}$.

6.4. The associated Lie ring $L = L(G)$ of a group G is always $\mathbb{Z}$-graded, if we put $L_k = \{0\}$ for all $k \leq 0$ and $L_k = \gamma_k/\gamma_{k+1}$ for $k \geq 1$ (or "$\mathbb{N}$-graded"). This follows directly from the definition of the multiplication in $L(G)$.

6.5. The associated Lie rings of two non-isomorphic nilpotent groups may be isomorphic, as in the following example.

Example 6.6. Let $D_8 = \langle a, b \mid a^4 = b^2 = 1, \ a^b = a^3 \rangle$ be the dihedral group of order 8. Put $c = [a, b] = a^2$; then $L(D_8)$ is a vector space over $\mathbb{F}_2$ with the basis $\{\bar{a}, \bar{b}, \bar{c}\}$ and structural constants $[\bar{a}, \bar{b}] = \bar{c}$, $[\bar{a}, \bar{c}] = [\bar{b}, \bar{c}] = 0$. The associated Lie ring of the quaternion group $Q_8 = \langle u, v \mid u^4 = v^4 = 1, \ u^2 = v^2, \ u^v = u^3 \rangle$ is a vector space over $\mathbb{F}_2$ with basis $\{\bar{u}, \bar{v}, \bar{w} = \overline{[u, v]}\}$ and structural constants $[\bar{u}, \bar{v}] = \bar{w}$, $[\bar{u}, \bar{w}] = [\bar{v}, \bar{w}] = 0$. We see that $L(Q_8)$ is isomorphic to $L(D_8)$, although the groups are not isomorphic.

§ 6.2. Basic properties

First we prove the following useful, if technical, lemma. We continue to use the convention on the notation $\bar{a} \in \gamma_i/\gamma_{i+1}$ and $\gamma_i = \gamma_i(G)$.

Lemma 6.7. *Let $L(G)$ be the associated Lie ring of a group G.*

(a) *For any $\bar{a}_j \in G/\gamma_2$ and for any commutator $\varkappa$ of weight k, we have*

$$\varkappa(\bar{a}_1, \ldots, \bar{a}_k) = \varkappa(a_1, \ldots, a_k)\gamma_{k+1},$$

where the right-hand side is the image of the group commutator $\varkappa(a_1, \ldots, a_k)$ in γ_k/γ_{k+1} and the left-hand side is the Lie ring commutator in the $\bar{a}_i$. In particular, $[\bar{a}_1, \ldots, \bar{a}_k] = [a_1, \ldots, a_k]\gamma_{k+1}$.

(b) *The additive subgroup γ_k/γ_{k+1} is spanned by the Lie products $[\bar{a}_1, \ldots, \bar{a}_k]$ where $\bar{a}_i \in \gamma_1/\gamma_2$, that is, $\gamma_k/\gamma_{k+1} = [\underbrace{\gamma_1/\gamma_2, \ldots, \gamma_1/\gamma_2}_{k}]$.*

(c) *The Lie ring $L(G)$ is generated by γ_1/γ_2.*

(d) $\gamma_k(L(G)) = \bigoplus_{s \geq k} (\gamma_s/\gamma_{s+1})$.

Proof. (a) We use induction on the weight k. If $k = 1$, then $\varkappa(\bar{a}_1) = \bar{a}_1 = a_1\gamma_2 = \varkappa(a_1)\gamma_2$. Let $\varkappa = [\varkappa_1, \varkappa_2]$ with $\varkappa_i$ of weight k_i, $i = 1, 2$. Then by the induction hypothesis

$$\varkappa(\bar{a}_1, \ldots, \bar{a}_k) = [\varkappa_1(a_{i_1}, \ldots, a_{i_{k_1}})\gamma_{k_1+1}, \varkappa_2(a_{j_1}, \ldots, a_{j_{k_2}})\gamma_{k_2+1}],$$

where on the right $\varkappa_i\gamma_{k_i+1}$ is the image of the corresponding group commutator in $\gamma_{k_i}/\gamma_{k_i+1}$, $i = 1, 2$. By the definition of Lie products in $L(G)$, the right-hand side is the image of the group commutator $\varkappa(a_1, \ldots, a_k)$ in γ_k/γ_{k+1}, as required. In the special case of the simple commutator, this argument can be written in one line, with outer brackets for Lie products and inner for group commutators:

$$[\bar{a}_1, \ldots, \bar{a}_k] = [[a_1, a_2]\gamma_3, \bar{a}_3, \ldots, \bar{a}_k] \quad = \quad [[a_1, a_2, a_3]\gamma_4, \bar{a}_4, \ldots, \bar{a}_k] = \ldots$$

$$\ldots \quad = \quad [a_1, \ldots, a_k]\gamma_{k+1}.$$

(b) By Lemma 3.6, the group γ_k is generated by the simple commutators $[a_1, \ldots, a_k]$, $a_i \in G$. Hence the additive subgroup γ_k/γ_{k+1} of $L(G)$ is generated by their images $[a_1, \ldots, a_k]\gamma_{k+1}$ in γ_k/γ_{k+1}, and the result follows from (a).

(c) This follows directly from (b), since $L(G) = \bigoplus_{i=1}(\gamma_i/\gamma_{i+1})$.

(d) We have $[\gamma_u/\gamma_{u+1}, \gamma_v/\gamma_{v+1}] \leq \gamma_{u+v}/\gamma_{u+v+1}$ for all $u, v \in \mathbb{N}$ by the definition of multiplication in $L(G)$, whence $\gamma_k(L(G)) \leq \bigoplus_{s \geq k}(\gamma_s/\gamma_{s+1})$. The reverse inclusion follows from (b): $\gamma_s/\gamma_{s+1} = [\underbrace{\gamma_1/\gamma_2, \ldots, \gamma_1/\gamma_2}_{s}] \leq \gamma_s(L(G)) \leq$

$\gamma_k(L(G))$ for all $s \geq k$. $\square$

Corollary 6.8. *The associated Lie ring $L(G)$ is homogeneous with respect to the generating set γ_1/γ_2, with γ_k/γ_{k+1} being the homogeneous component of weight k.* $\square$

The following theorem establishes some connections between a group and its associated Lie ring.

Theorem 6.9. *Suppose that G is a nilpotent group and let $L(G)$ be the associated Lie ring of G.*

(a) *Then $L(G)$ is nilpotent, and the nilpotency class of $L(G)$ is exactly the same as the nilpotency class of G.*

(b) *If G is finite, then $|G| = |L(G)|$.*

(c) *For every automorphism $\varphi \in \operatorname{Aut} G$, the action of φ on the factor-groups γ_i/γ_{i+1} induces by linearity an automorphism of $L(G)$.*

Proof. (a) By Lemma 6.7(d) $\gamma_k(L(G)) = \bigoplus_{s \geq k}(\gamma_s/\gamma_{s+1})$ for all $k \in \mathbb{N}$. If $\gamma_{c+1}(G) = 1$, then, of course, $\gamma_{c+1}(L(G)) = 0$. Conversely, if $\gamma_{d+1}(L(G)) = 0$, then $\gamma_{d+1}/\gamma_{d+2} = 0$ or, in other words, $\gamma_{d+1} = \gamma_{d+2}$. In a nilpotent group, this implies $\gamma_{d+1} = 1$ (Corollary 3.15).

(b) By the definition of $L(G) = \bigoplus_i(\gamma_i/\gamma_{i+1})$ and by Lagrange's Theorem, both the order of $L(G)$ and the order of G are equal to the product $\prod_{i=1}^{c}|\gamma_i/\gamma_{i+1}|$, where c is the nilpotency class.

(c) The induced mapping $\bar{\varphi}$ is an automorphism of the additive group $L(G) = \bigoplus_i(\gamma_i/\gamma_{i+1})$, since φ induces automorphisms of the direct summands. Since $\bar{\varphi}$ is extended to the sum by linearity, as the Lie multiplication is, it suffices to check that $\bar{\varphi}$ preserves Lie products of elements from the summands. For $\bar{x} \in \gamma_i/\gamma_{j+1}$ and $\bar{y} \in \gamma_j/\gamma_{j+1}$, we have

$$[\bar{x}, \bar{y}]^{\bar{\varphi}} = ([x,y]\gamma_{i+j+1})^{\bar{\varphi}} = [x,y]^{\varphi}\gamma_{i+j+1} = [x^{\varphi}, y^{\varphi}]\gamma_{i+j+1} = [\bar{x}^{\bar{\varphi}}, \bar{y}^{\bar{\varphi}}],$$

as required. $\square$

Remarks. 6.10. Unlike the nilpotency class, the derived length of $L(G)$ may not coincide with that of G. This sometimes makes it difficult to recover information about the group from the result on its associated Lie ring. One can, however, prove that the derived length of $L(G)$ is not greater than that of G (Exercise 6.1).

6.11. The induced automorphism of $L(G)$ may well be a trivial one, for a non-trivial $\varphi \in \operatorname{Aut} G$. For example, every inner automorphism of G induces a trivial automorphism of $L(G)$. This sometimes makes it difficult to translate a hypothesis on a group and its automorphism into the language of Lie rings;

another trouble is that the number of fixed points may become greater. In general, the order of the induced automorphism $\bar{\varphi}$ is a divisor of $|\varphi|$, since $\varphi \to \bar{\varphi}$ is a homomorphism of $\operatorname{Aut} G$ into $\operatorname{Aut} L(G)$. It is usual to denote the induced automorphism of $L(G)$ simply by φ; then $\langle \varphi \rangle$ acts on $L(G)$ as automorphisms (but not necessarily faithfully). If, however, the order of φ is coprime to the order of a finite nilpotent group G, then $\langle \varphi \rangle$ acts faithfully on $L(G)$ and $|C_{L(G)}(\varphi)| = |C_G(\varphi)|$ (Exercise 6.3).

§ 6.3. Some applications

We can now reap the fruits of the hard work done in verifying the definitions. Using the associated Lie rings, we prove here a few useful lemmas on linear properties of nilpotent group. Most of them could be proved directly, but using associated Lie rings is an easier way, with some "economy of thought". One can say that the required commutator calculations were carried out once and for all in verifying the definition of $L(G)$.

Lemma 6.12. *Let $\gamma_i = \gamma_i(G)$ denote the terms of the lower central series of a group G.*

(a) *If $(\gamma_1/\gamma_2)^n = 1$, then $(\gamma_k/\gamma_{k+1})^n = 1$ for all $k \in \mathbb{N}$.*

(b) *If G is nilpotent of class c and $(\gamma_1/\gamma_2)^n = 1$, then $G^{n^c} = 1$.*

(c) *If G is nilpotent of class c and G is generated by elements of order dividing n, then $G^{n^c} = 1$. If, in addition, G is generated by r elements, then the order of G is (n, c, r)-bounded.*

Proof. (a) The hypothesis can be rewritten in the additive group of the associated Lie ring $L(G)$ as $n\gamma_1/\gamma_2 = 0$. Then, for the $\bar{a}_i \in \gamma_1/\gamma_2$, we have

$$n[\bar{a}_1, \ldots, \bar{a}_k] = [n\bar{a}_1, \bar{a}_2, \ldots, \bar{a}_k] = [0, \bar{a}_2, \ldots, \bar{a}_k] = 0.$$

By Lemma 6.7(b), the additive group γ_k/γ_{k+1} is spanned by the $[\bar{a}_1, \ldots, \bar{a}_k]$. Hence $n\gamma_k/\gamma_{k+1} = 0$, or, in multiplicative notation, $(\gamma_k/\gamma_{k+1})^n = 1$.

(b) We prove by induction that $G^{n^k} \le \gamma_{k+1}$, the case $k = 1$ being the hypothesis. For $k > 1$, by the induction hypothesis and by (a),

$$G^{n^k} \le (G^{n^{k-1}})^n \le (\gamma_k)^n \le \gamma_{k+1},$$

as required. In particular, $G^{n^c} \le \gamma_{c+1} = 1$.

(c) The images of the generating elements in γ_1/γ_2 clearly generate this abelian group. Hence $(\gamma_1/\gamma_2)^n = 1$, and $G^{n^c} = 1$ by (b). When G is r-generated, to estimate the order, it suffices to bound the $|\gamma_i/\gamma_{i+1}|$. The additive subgroup γ_i/γ_{i+1} of $L(G)$ is generated by the simple commutators in the r generators of G/γ_2 by Lemma 6.7(b); there are r^i such commutators. Since $n\gamma_i/\gamma_{i+1} = 0$ by (a), we have $|\gamma_i/\gamma_{i+1}| \le n^{r^i}$. $\qquad\square$

Proof. Induction on the nilpotency class c; if $c = 1$, then the group is abelian and $a_1^m \cdots a_s^m = (a_1 \cdots a_s)^m$. For $c > 1$, we consider the subgroup $H = \langle a_1^m, \ldots, a_s^m \rangle$. By the induction hypothesis applied to $H/\gamma_c(H)$, we have

$$a_1^{m^c} \cdots a_s^{m^c} = (a_1^m)^{m^{c-1}} \cdots (a_s^m)^{m^{c-1}} = b^m g, \tag{6.16}$$

where $g \in \gamma_c(H)$. Since G is nilpotent of class c, we can regard $\gamma_c(H) = \gamma_c(H)/\gamma_{c+1}(H)$ as an additive subgroup of $L(H)$ spanned by commutators of weight c in the generators a_i^m (Lemma 6.7(b)). For any such commutator, we have, by Lemma 6.13,

$$[a_{i_1}^m, \ldots, a_{i_c}^m] = [a_{i_1}, \ldots, a_{i_c}]^{m^c} = h^m,$$

where $h \in \gamma_c(H) \leq Z(G)$. So $\gamma_c(H)$ is generated by the mth powers of some elements from $Z(G)$ and therefore consists of the mth powers of some elements from $Z(G)$. Then $g = z^m$ for some $z \in Z(G)$ in (6.16), and hence $a_1^{m^c} \cdots a_s^{m^c} = b^m g = b^m z^m = (bz)^m$, as required. $\quad\square$

Remark 6.17. N. Blackburn [1965] proved that for every prime number p and every $c \in \mathbb{N}$ there is a (p, c)-bounded number $b(p, c)$ such that in any nilpotent p-group of class c, for any $k \in \mathbb{N}$, every product of $p^{k+b(p,c)}$th powers of elements is a p^kth power of some element.

Exercises 6

1. Prove that the derived length of $L(G)$ is not greater than that of G.

2. Prove that $\varphi \to \bar{\varphi}$ is a homomorphism of $\operatorname{Aut} G$ into $\operatorname{Aut} L(G)$.

3. Suppose that G is a finite nilpotent group, $\varphi \in \operatorname{Aut} G$, and $(|\varphi|, |G|) = 1$. Prove that $|C_{L(G)}(\varphi)| = |C_G(\varphi)|$. [*Hint:* Use Lemma 2.11.]

4. Suppose that φ is an automorphism of a group G such that φ acts trivially on $G/\gamma_2(G)$. Prove that φ acts trivially on every factor-group γ_i/γ_{i+1}. [*Hint:* Consider the induced automorphism of $L(G) = \langle \gamma_1/\gamma_2 \rangle$.]

5. Suppose that $G = N \rtimes H$ is a semidirect product of $N \trianglelefteq G$ and $H \leq G$. Construct a subring of $L(G)$ which is naturally isomorphic to $L(H)$.

6. Prove that a homomorphism of a group G onto a group H induces naturally a homomorphism of the associated Lie ring $L(G)$ onto $L(H)$.

7. Suppose that G is a nilpotent group of class c such that $\gamma_k(G)/\gamma_{k+1}(G)$ is finite of order n. Prove that $\gamma_k(G)$ is finite of (n, c)-bounded order. [*Hint:* Apply Exercise 5.14 to $L(G)$.]

8. [B. Hartley and T. Meixner, 1980] Suppose that a finite p-group P of odd order admits an automorphism φ of order 2 with $|C_P(\varphi)| = p^m$. Prove that P has a subgroup of (p, m)-bounded index which is nilpotent of class 2. [*Hint:* Show that $P_1 = [P, \varphi]$ is generated by elements inverted by φ. Consider $L = L(P_1)$ and show that φ acts trivially on $\gamma_{2k}(L)/\gamma_{2k+1}(L)$ for all $k \in \mathbb{N}$. Use Lemma 2.11 to estimate the nilpotency class. Use 7.]

9. Prove that the associated Lie ring of the unitriangular group $UT_n(\mathbb{F}_p)$ (see Exercise 2.2) is isomorphic to the Lie ring of null-triangular matrices

$$
nt_n(\mathbb{F}_p) = \left\{ \left. \begin{pmatrix} 0 & & * \\ & \ddots & \\ 0 & & 0 \end{pmatrix} \right| * \in \mathbb{F}_p \right\}
$$

(where the addition is component-wise and the Lie product of matrices A, B is $[A, B] = AB - BA$). [*Hint:* A coset of $\gamma_{k+1}(UT_n(\mathbb{F}_p))$ has the form

$$
\left\{ \left. \begin{pmatrix} 1 & 0 & \cdots & 0 & a_1 & * \\ & 1 & & & & \ddots \\ & & \ddots & & & a_{n-k} \\ & & & & & 0 \\ 0 & & & & & \vdots \\ & & & & 1 & 0 \\ & & & & & 1 \end{pmatrix} \right| * \in \mathbb{F}_p \right\}.
$$

Map it to

$$
\begin{pmatrix} & a_1 & & 0 \\ & & \ddots & \\ & & & a_{n-k} \\ 0 & & & \end{pmatrix} \in nt_n(\mathbb{F}_p).]
$$

10. Produce an example of a nilpotent group G with $L(G)^{(d)} = 0$, but $G^{(d)} \neq 1$.

11. Let G be a group and p a prime number. Show that the p-isolators of the terms of the lower central series defined as $G_i = \{x \in G \mid x^{p^n} \in \gamma_i(G) \text{ for some } n = n(x)\}$ are subgroups (see Theorem 10.19) and satisfy the inclusions $[G_i, G_j] \leq G_{i+j}$. Define the Lie ring $L = \bigoplus_i (G_i/G_{i+1})$ based on the G_i in the same way as $L(G)$ is based on the $\gamma_i(G)$. Prove that if G is nilpotent and has no elements of order p, then L is nilpotent too, and the nilpotency class of L coincides with that of G.

Chapter 7

Regular automorphisms of Lie rings

The theorems of G. Higman, V. A. Kreknin and A. I. Kostrikin on regular automorphisms of Lie rings can be viewed as combinatorial facts about $(\mathbb{Z}/n\mathbb{Z})$-graded Lie rings: they are actually proved as such, and it is in this form that they are used in studying p-automorphisms of nilpotent p-groups. We shall first prove Kreknin's Theorem for graded Lie rings using the varietal criterion from § 5.2, which simplifies the proof to a few lines. (A longer version which gives an explicit upper bound for the derived length is indicated in the exercises.) Then nilpotency is derived from solubility in the case of the automorphism of prime order, again for graded Lie rings. Free Lie rings allow us to derive the required combinatorial consequences for arbitrary Lie rings. The theorems on Lie rings and finite nilpotent groups with regular automorphisms are also obtained as corollaries of these combinatorial facts.

§ 7.1. Graded Lie rings

For the definition of graded Lie rings, see § 5.1. We begin with a version of a theorem of V. A. Kreknin [1963].

Theorem 7.1. *Let n be a positive integer and suppose that $L = L_0 \oplus L_1 \oplus \cdots \oplus L_{n-1}$ is a $(\mathbb{Z}/n\mathbb{Z})$-graded Lie ring with components L_s satisfying $[L_i, L_j] \subseteq L_{i+j}$, where $i + j$ is a residue mod n. If $L_0 = 0$, then L is soluble of n-bounded derived length: $L^{(k(n))} = 0$ for some function $k(n)$ depending only on n.*

We shall refer to $k(n)$ (meaning the minimal possible value) as *Kreknin's function*. The varietal criterion of Theorem 5.25 simplifies the proof to a few lines, but gives only the existence of the function $k(n)$ (see Exercise 7.1 for another direct, if more technical, proof which gives an explicit upper bound for $k(n)$).

Before proving Theorem 7.1, we note that the class of $(\mathbb{Z}/n\mathbb{Z})$-graded Lie rings is a variety of algebraic systems that are Lie rings with additional unary operations of taking the components of the elements. More precisely, consider the class $\mathfrak{V}(n)$ of Lie rings L with unary operations f_i, $i = 0, 1, \ldots, n-1$, satisfying the following laws (in addition to the laws of Lie rings):

$$f_i(x \pm y) = f_i(x) \pm f_i(y) \qquad \text{for all } i;$$

$$x = f_0(x) + f_1(x) + \cdots + f_{n-1}(x);$$

$$f_i(f_j(x)) = 0 \qquad \text{for } i \neq j;$$

$$f_i(f_i(x)) = f_i(x) \qquad \text{for all } i;$$

$$f_{i+j}([f_i(x), f_j(x)]) = [f_i(x), f_j(x)], \qquad \text{where } i+j \text{ is a residue mod } n.$$

The first four laws ensure that $L = f_0(L) \oplus f_1(L) \oplus \cdots \oplus f_{n-1}(L)$ with $f_i(L)$ being additive subgroups of L. Indeed, by the first law, each f_i is an endomorphism of the additive group of L, and hence $f_i(L)$ is an additive subgroup. The sum of the $f_i(L)$ equals L by the second law, and the sum is direct:

$$f_j(x) = \sum_{i \neq j} f_i(y_i) \quad \Rightarrow \quad f_j(x) = f_j(f_j(x)) = \sum_{i \neq j} f_j(f_i(y_i)) = 0$$

by the third and fourth laws. The fifth law means that this decomposition is a $(\mathbb{Z}/n\mathbb{Z})$-grading with components $f_i(L)$.

Conversely, if L is a $(\mathbb{Z}/n\mathbb{Z})$-graded Lie ring, then every element $l \in L$ admits the unique decomposition $l = l_0 + l_1 + \cdots + l_{n-1}$ with components $l_i \in L_i$. It is easy to see that if we define $f_i(l) = l_i$, then these unary operations satisfy the above laws, so that $L \in \mathfrak{V}(n)$. We fix the notation l_i for $f_i(l)$ within this section.

Considering $(\mathbb{Z}/n\mathbb{Z})$-graded Lie rings as objects in $\mathfrak{V}(n)$, we are allowed to deal only with *homogeneous* subrings and ideals, that is, closed under all operations, so that such a subring or ideal H must be equal to the direct sum of the $H \cap L_i$. (The words "homogeneous" and "component" here have different meanings from those in § 5.1.) It is straightforward to see that if $H = \sum_{i=0}^{n-1} H \cap L_i$ and $K = \sum_{i=0}^{n-1} K \cap L_i$, then

$$[H, K] = \sum_{i,j=0}^{n-1} [H \cap L_i, K \cap L_j] = \sum_{s=0}^{n-1} [H, K] \cap L_s. \qquad (7.2)$$

In particular, the terms of the (abstract) derived and central series are homogeneous.

Proof of Theorem 7.1. The Lie ring L satisfying the hypothesis belongs to the subvariety $\mathfrak{W}(n)$ of $\mathfrak{V}(n)$ defined by the additional law $f_0(x) = 0$. It suffices to prove that the variety $\mathfrak{W}(n)$ is soluble, which includes the existence of an upper bound for the derived lengths, the required function $k(n)$ (see § 1.3 and § 5.2). By Theorem 5.25 (and Remark 5.26), we need only prove that every non-trivial Lie ring $L \neq 0$ in $\mathfrak{W}(n)$ is distinct from its derived Lie subring: $L \neq [L, L]$. Suppose the opposite: $L = [L, L] \neq 0$ for $L \in \mathfrak{W}(n)$. We shall use induction on $k = 0, 1, \ldots, n-1$ to show that then

$$L \subseteq \langle L_{k+1}, \ldots, L_{n-1} \rangle, \qquad (7.3)$$

Lemma 6.13. *Let $\varkappa$ be a commutator of weight s. For any elements $a_1, \ldots, a_s$ in any group G and any $k_1, \ldots, k_s \in \mathbb{Z}$ we have*

$$\varkappa(a_1^{k_1}, \ldots, a_s^{k_s}) = \varkappa(a_1, \ldots, a_s)^{k_1 \cdots k_s} \prod_j \varkappa_j,$$

where the $\varkappa_j$ are some commutators of weight $\geq s + 1$ in the a_i and their inverses. In particular, $[a_1^{k_1}, \ldots, a_s^{k_s}] \equiv [a_1, \ldots, a_s]^{k_1 \cdots k_s} \pmod{\gamma_{s+1}(G)}$. If G is nilpotent of class s, then $\varkappa(a_1^{k_1}, \ldots, a_s^{k_s}) = \varkappa(a_1, \ldots, a_s)^{k_1 \cdots k_s}$; in particular, then $[a_1^{k_1}, \ldots, a_s^{k_s}] = [a_1, \ldots, a_s]^{k_1 \cdots k_s}$.

Proof. Put $H = \langle a_1, \ldots, a_s \rangle$, and let $\bar{a}_i$ denote the image of a_i in $H/\gamma_2(H)$. For the Lie ring commutators in $L(H)$, we have

$$\varkappa(k_1 \bar{a}_1, \ldots, k_s \bar{a}_s) = k_1 \cdots k_s \varkappa(\bar{a}_1, \ldots, \bar{a}_s).$$

By Lemma 6.7(a), this can be rewritten in terms of group commutators as

$$\varkappa(a_1^{k_1}, \ldots, a_s^{k_s}) = \varkappa(a_1, \ldots, a_s)^{k_1 \cdots k_s} h, \qquad h \in \gamma_{s+1}(H).$$

The element $h \in \gamma_{s+1}(H)$ equals the required product $\prod_j \varkappa_j$ by Lemma 3.6(c)). $\square$

We record here the following elementary fact.

Lemma 6.14. *For any elements x, y in any nilpotent group of class ≤ 2, we have $(xy)^n = x^n y^n [y, x]^{n(n-1)/2}$ for any $n \in \mathbb{N}$.*

Proof. Induction on n; for $n = 1$ we have $xy = xy[y, x]^0$. For $n > 1$, we use the induction hypothesis, Lemma 6.13, and the fact that all commutators lie in the centre:

$$
\begin{aligned}
(xy)^n = (xy)^{n-1} xy &= x^{n-1} y^{n-1} [y, x]^{(n-1)(n-2)/2} xy \\
&= x^{n-1} y^{n-1} xy [y, x]^{(n-1)(n-2)/2} \\
&= x^n y^n [y^{n-1}, x][y, x]^{(n-1)(n-2)/2} \\
&= x^n y^n [y, x]^{n-1} [y, x]^{(n-1)(n-2)/2} \\
&= x^n y^n [y, x]^{n(n-1)/2}
\end{aligned}
$$

$\square$

The following lemma is from [A. I. Mal'cev, 1958].

Lemma 6.15. *Let G be a nilpotent group of class c. For any $m \in \mathbb{N}$, every product of m^cth powers of any elements in G is an mth power in G: for any $s \in \mathbb{N}$ and any $a_1, \ldots, a_s \in G$ there is $b \in G$ such that $a_1^{m^c} \cdots a_s^{m^c} = b^m$.*

the last step, for $k = n - 1$, being $L = 0$, a contradiction to $L \neq 0$. (The right-hand side of (7.3) is a Lie subring generated by $L_{k+1}, \dots, L_{n-1}$, which is homogeneous no matter whether the additional operations f_i are used or not.) We need an elementary number-theoretical lemma.

Lemma 7.4. *Suppose that a, b, c are integers such that $1 \leq a \leq n - 1$, $1 \leq b \leq n - 1$ and $1 \leq c \leq n - 1$. If $a + b \equiv c \,(\mathrm{mod}\, n)$, then either both $a > c$ and $b > c$, or both $a < c$ and $b < c$.*

Proof. Since $a < n$ and $b < n$, we have $a + b < 2n$. Therefore, either $a + b = c$, or $a + b = c + n$. In the first case, $a < c$ and $b < c$. In the second case, both numbers are greater than c, because if one of them were less than c, then their sum would be less than $c + n$, since the other is less than n. $\square$

Now we prove (7.3). For $k = 0$, we have $L \subseteq \langle L_1, \dots, L_{n-1} \rangle$ since $L_0 = 0$.

For the induction step, assuming that $L \subseteq \langle L_k, \dots, L_{n-1} \rangle$, we need only show that $L_k \subseteq \langle L_{k+1}, \dots, L_{n-1} \rangle$. Since $L = [L, L]$, every element is a linear combination of simple commutators of weight ≥ 2 in the generators from $L_k, \dots, L_{n-1}$ (see §5.1). In particular, every element $l_k \in L_k$ is a linear combination of simple commutators of the form

$$[x_{i_1}, \dots, x_{i_t}], \tag{7.5}$$

where $t \geq 2$, $x_{i_j} \in L_{i_j}$, $k \leq i_j \leq n - 1$ for $j = 1, \dots, t$ and $i_1 + \dots + i_t \equiv k \,(\mathrm{mod}\, n)$. To lighten notation, we adopt here the following convention.

Convention 7.6. *The indices are used only to indicate the components L_i which the elements x_i belong to (so the same symbol may denote different elements). Then a commutator in the x_i belongs to the component L_m such that m is the $\mathrm{mod}\, n$ sum of the indices of the elements involved in the commutator.*

For every commutator (7.5), let $y_s \in L_s$ denote its initial segment of length $t - 1$. If $s = 0$, then $y_s = 0$ and the commutator (7.5) equals 0. If $s \neq 0$, then $i_t \neq k$ and hence $i_t > k$. Then, by Lemma 7.4, we also have $s > k$, in which case $[y_s, x_{i_t}] \in \langle L_{k+1}, \dots, L_{n-1} \rangle$, as required. This completes the proof of Theorem 7.1. $\square$

Now we turn to the special case of Theorem 7.1 where $n = p$ is a prime number. Then a stronger conclusion holds: the Lie ring is nilpotent of p-bounded class [G. Higman, 1957]. In view of Theorem 7.1, it is sufficient to prove a version of a theorem of V. A. Kreknin and A. I. Kostrikin [1963] that solubility implies nilpotency.

Theorem 7.7. *Let p be a prime number and suppose that $L = L_0 \oplus L_1 \oplus \dots \oplus L_{p-1}$ is a $(\mathbb{Z}/p\mathbb{Z})$-graded Lie ring with components L_s satisfying*

$[L_i, L_j] \subseteq L_{i+j}$ *where* $i + j$ *is a residue* mod p. *If* $L_0 = 0$ *and* L *is soluble of derived length* d, *then* L *is nilpotent of* (p, d)-*bounded class.*

Proof. As above, the $(\mathbb{Z}/p\mathbb{Z})$-graded Lie rings with $L_0 = 0$ form the variety $\mathfrak{W}(p)$. By (7.2), the terms of the (abstract) lower central and derived series of Lie rings in $\mathfrak{W}(n)$ are homogeneous (normal subsystems). Therefore, by Corollary 5.28 and Remark 5.30, it is sufficient to prove that if $L \in \mathfrak{W}(p)$ is soluble of derived length 2, then L is nilpotent of p-bounded class. We shall actually prove that L is then nilpotent of class at most p. Since L is generated by the components L_i, it is sufficient to prove that

$$[x_{i_1}, x_{i_2}, \dots, x_{i_{p+1}}] = 0 \qquad (7.8)$$

for any $x_{i_k} \in L_{i_k}$. Here, again, we adopt convention 7.6.

The idea is in the fact that the initial segment $[x_{i_1}, x_{i_2}]$ belongs to the abelian ideal $[L, L]$, so that the rest of the elements may be permuted arbitrarily without changing the commutator (7.8). Indeed, $[a, b, c] = [a, c, b] + [a, [b, c]] = [a, c, b]$, if $a \in [L, L]$. It is an elementary fact about residues mod p (see Lemma 7.9 below) that, given any $p - 1$ non-zero residues, there is a subset of them with a prescribed sum mod p. So the value $-i_1 - i_2 \bmod p$ can be obtained as a sum of some of the residues $i_3, \dots, i_{p+1}$. (We allow an empty subset in the case $i_1 + i_2 \equiv 0 \,(\mathrm{mod}\, p)$.) Transferring the corresponding elements x_{i_k} to place them right after x_{i_2} in (7.8), we obtain a commutator equal to (7.8) that has an initial segment with zero mod p sum of the indices; this initial segment belongs to L_0 (see 7.6) and hence equals 0. □

It remains to prove the number-theoretic fact used above.

Lemma 7.9. *Let* p *be a prime number and let* $i_1, \dots, i_k$ *be non-zero elements of* $(\mathbb{Z}/p\mathbb{Z})$ *(not necessarily distinct). We form the set*

$$M = \left\{ \sum_{s \in S} i_s \mid S \subseteq \{1, 2, \dots, k\} \right\}$$

(where the sum is 0 for $S = \varnothing$*). Then either* $M = \mathbb{Z}/p\mathbb{Z}$ *or* $|M| \geq k + 1$. *In particular, if* $k = p - 1$, *then* $M = \mathbb{Z}/p\mathbb{Z}$.

Proof. We proceed by induction on k. Let $M(s)$ denote the set of all sums for $k = s$, that is, built on $\{i_1, \dots, i_s\}$. For $k = 1$, we have $|M(1)| = |\{0, i_1\}| = 2$, since $i_1 \neq 0$. Next, if any of the sums $s + i_{k+1}$, $s \in M(k)$, does not belong to $M(k)$, then $|M(k+1)| \geq |M(k)| + 1 \geq k + 2$ by the induction hypothesis. If, however, $s + i_{k+1} \in M(k)$ for all $s \in M(k)$, then starting from 0 we find that all elements $0, i_{k+1}, 2i_{k+1}, \dots, (p-1)i_{k+1}$ belong to $M(k)$. These elements are all distinct since $i_{k+1} \neq 0$ by the hypothesis; hence $|M(k)| = |M(k+1)| = p$, that is, $M(k+1) = \mathbb{Z}/p\mathbb{Z}$. □

Theorems 7.1 and 7.7 together yield the following [G. Higman, 1957].

Corollary 7.10. *Let p be a prime number and suppose that $L = L_0 \oplus L_1 \oplus \cdots \oplus L_{p-1}$ is a $(\mathbb{Z}/p\mathbb{Z})$-graded Lie ring with components L_s satisfying $[L_i, L_j] \subseteq L_{i+j}$, where $i + j$ is a residue mod p. If $L_0 = 0$, then L is nilpotent of p-bounded class, at most $h(p)$.*

We shall refer to $h(p)$ (meaning the minimal possible value) as *Higman's function*. The exact (best possible) values of $h(p)$ are known only for a few small primes $p = 2, 3, 5, 7$. We shall need the values $h(2)$ and $h(3)$ in Chapter 12.

Lemma 7.11. (a) $h(2) = 1$;
(b) $h(3) = 2$.

Proof. (a) If $L = L_1$ is a $(\mathbb{Z}/2\mathbb{Z})$-graded Lie ring with $L_0 = 0$, then $[L, L] = [L_1, L_1] \subseteq L_0 = 0$, so that $h(2) = 1$.

(b) Suppose that $L = L_1 \oplus L_2$ is a $(\mathbb{Z}/3\mathbb{Z})$-graded Lie ring with $L_0 = 0$. To prove that $h(3) \leq 2$, we need only show that $[x_{i_1}, x_{i_2}, x_{i_3}] = 0$ for any $x_{i_j} \in L_1$, L_2 (we use Convention 7.6). We have $[x_1, x_1, x_1] \in L_0 = 0$, $[x_2, x_2, x_2] \in L_0 = 0$, $[x_1, x_2, x_j] = -[x_2, x_1, x_j] \in [L_0, x_j] = 0$. The two remaining cases are dealt with using the Jacobi identity: $[x_1, x_1, x_2] = [x_1, x_2, x_1] + [x_1, [x_1, x_2]] = 0$, and $[x_2, x_2, x_1] = [x_2, x_1, x_2] + [x_2, [x_2, x_1]] = 0$. $\square$

Remarks. 7.12. The original proof of V. A. Kreknin [1963] gives an upper bound $k(n) \leq 2^{n-1}$ for $k(n)$ in Theorem 7.1 (Exercise 7.1).

7.13. D. J. Winter [1968] proved that a finite-dimensional $(\mathbb{Z}/n\mathbb{Z})$-graded Lie algebra L is soluble under a weaker condition $[L, \underbrace{L_0, \ldots, L_0}_{m}] = 0$. He raised the question whether there is a function of m and n bounding the derived length of such a Lie algebra. An effective proof, which is also valid for infinite-dimensional algebras, was found in [E. I. Khukhro and P. Shumyatsky, 1995] (Exercise 7.2).

7.14. Exercise 5.9 gives $(p^d - 1)/(p - 1)$ as an explicit upper bound for the nilpotency class in Theorem 7.7 based on the bound p for $d = 2$. This bound can be slightly improved to $((p-1)^d - 1)/(p-2)$, by a direct induction argument [V. A. Kreknin and A. I. Kostrikin, 1963].

7.15. The existence of $h(p)$ was proved by G. Higman [1957] as a pure existence theorem. We followed here the completely different effective proof from [V. A. Kreknin, 1963] and [V. A. Kreknin and A. I. Kostrikin, 1963]: the bounds from Remarks 7.12, 7.14 yield an explicit bound $h(p) \leq ((p-1)^{2^{p-1}} - 1)/(p-2)$. These bounds, however, are believed to be far from best possible. G. Higman [1957] constructed examples showing that $h(p) \geq (p^2 - 1)/4$ and conjectured that, in fact, the equality holds for odd p. He confirmed this conjecture for $p = 5$ (the easier cases $p = 2$ and $p = 3$ must have been known earlier, see

Lemma 7.11). I. Hughes [1985] and B. Scimemi (unpublished) confirmed that $h(7) = 12$.

§ 7.2. Combinatorial consequences

Basing them on the results on graded Lie rings, we derive consequences that are valid for any Lie rings. Then we derive combinatorial consequences for Lie rings with automorphisms of finite order. Recall that $\delta_n(x_1, \ldots, x_{2^n}) = 0$ is the law of solubility of derived length n (see § 5.2) and that $k(s)$ is Kreknin's function as in Theorem 7.1.

Theorem 7.16. *Let n be a positive integer. Suppose that $x_{i_1}, \ldots, x_{i_{2k(n)}}$ are (not necessarily distinct) elements of an arbitrary Lie ring L with formally assigned indices. Then the commutator $\delta_{k(n)}(x_{i_1}, \ldots, x_{i_{2k(n)}})$ is equal to a linear combination of commutators all with the same entry set and each having a subcommutator with zero $\bmod n$ sum of the indices.*

Proof. First, let L be a free Lie ring with free generators $x_{i_1}, x_{i_2}, \ldots, x_{i_{2k(n)}}$. Let L_i be the span of all commutators in the x_{i_k} such that the sum of the indices of the entries is $i \bmod n$. Since L is multihomogeneous with respect to the free generators (Corollary 5.40), we have the decomposition $L = L_0 \oplus L_1 \oplus \cdots \oplus L_{n-1}$. Obviously, $[L_i, L_j] \subseteq L_{i+j}$, where $i + j$ is taken $\bmod n$, so this decomposition is a $(\mathbb{Z}/n\mathbb{Z})$-grading. The ideal I generated by L_0 is also multihomogeneous; hence L/I is also $(\mathbb{Z}/n\mathbb{Z})$-graded, with components being the images of the L_i. Since $(L/I)_0 = (L_0 + I)/I = 0$, the image in L/I of $\delta_{k(n)}(x_{i_1}, \ldots, x_{i_{2k(n)}})$ is 0 by Theorem 7.1. In other words, $\delta_{k(n)}(x_{i_1}, \ldots, x_{i_{2k(n)}}) \in I$, and therefore $\delta_{k(n)}(x_{i_1}, \ldots, x_{i_{2k(n)}})$ is a linear combination of commutators each involving a subcommutator from L_0. Since L is multihomogeneous with respect to the x_{i_k}, all of these commutators can be assumed to have the same entry set.

To prove the same equality for arbitrary elements in any Lie ring, we apply the homomorphism, extending the mapping of the free generators to the elements with corresponding indices, to the equality proved for a free Lie ring. □

Essentially the same argument allows us to derive a similar consequence of Corollary 7.10 (recall that $h(p)$ is Higman's function).

Theorem 7.17. *Let p be a prime, and suppose that $x_{i_1}, x_{i_2}, \ldots, x_{i_{h(p)+1}}$ are elements of an arbitrary Lie ring L with formally assigned indices. Then the commutator $[x_{i_1}, \ldots, x_{i_{h(p)+1}}]$ is equal to a linear combination of commutators all with the same entry set and each having a subcommutator with zero $\bmod p$ sum of the indices.* □

Now let L be a Lie ring with an automorphism φ of finite order n. We extend the ground ring by a primitive nth root of unity ω, forming $\widetilde{L} = L \otimes_{\mathbb{Z}} \mathbb{Z}[\omega]$. Then φ can be regarded as an automorphism of $\widetilde{L}$ (see § 5.1). The above results on graded Lie rings can be applied because $\widetilde{L}$ "almost" decomposes into the sum of analogues of eigenspaces with respect to φ (similarly to Example 5.13). To wit, we define the following additive subgroups of $\widetilde{L}$:

$$L_i = \{l \in \widetilde{L} \mid l^{\varphi} = \omega^i l\}, \qquad i = 0, 1, \ldots, n-1.$$

These additive subgroups resemble components of a $(\mathbb{Z}/n\mathbb{Z})$-grading, as the following lemma shows.

Lemma 7.18. (a) $[L_i, L_j] \subseteq L_{i+j}$ where $i + j$ is a residue mod n;
(b) $n\widetilde{L} \subseteq L_0 + L_1 + \cdots + L_{n-1}$.

Proof. (a) For any $l_i \in L_i$ and $l_j \in L_j$ we have $[l_i, l_j]^{\varphi} = [l_i^{\varphi}, l_j^{\varphi}] = [\omega^i l_i, \omega^j l_j] = \omega^{i+j}[l_i, l_j]$. Since $\omega^n = 1$, this means that $[l_i, l_j] \in L_{i+j}$, as required.

(b) For any $l \in \widetilde{L}$, put $l_i = \sum_{s=0}^{n-1} \omega^{-is} l^{\varphi^s}$, for $i = 0, 1, \ldots, n-1$. Then $l_i \in L_i$:

$$l_i^{\varphi} = \sum_{s=0}^{n-1} \omega^{-is} l^{\varphi^{s+1}} = \sum_{t=1}^{n} \omega^{-i(t-1)} l^{\varphi^t} = \omega^i \cdot \sum_{t=1}^{n} \omega^{-it} l^{\varphi^t} = \omega^i l_i,$$

since $\omega^{-in} l^{\varphi^n} = \omega^{-i0} l^{\varphi^0}$. We have

$$\sum_{i=0}^{n-1} l_i = \sum_{i=0}^{n-1} \sum_{s=0}^{n-1} \omega^{-is} l^{\varphi^s} = \sum_{s=0}^{n-1} \left(\sum_{i=0}^{n-1} \omega^{-is} \right) l^{\varphi^s} = nl,$$

whence (b) follows. Here we used the facts that $\sum_{i=0}^{n-1} \omega^0 = n$ and that $\sum_{i=0}^{n-1} \omega^{-is} = 0$ if $s \not\equiv 0 \,(\mathrm{mod}\, n)$ (to see the latter, note that the sum does not change when multiplied by $\omega^s \neq 0$). $\qquad\square$

Now we derive the combinatorial consequences for Lie rings with automorphisms of finite order that will be used for proving the main results on p-automorphisms of finite p-groups.

Corollary 7.19. *Suppose that L is a Lie ring with an automorphism φ of finite order n.*

(a) (**Kreknin's Theorem**) *Then $(nL)^{(k(n))} \subseteq \mathrm{id}\langle C_L(\varphi)\rangle$.*
(b) (**Higman's Theorem**) *If n is a prime, then $\gamma_{h(n)+1}(pL) \subseteq \mathrm{id}\langle C_L(\varphi)\rangle$.*

Proof. By Lemma 7.18(b), $(n\widetilde{L})^{(k(n))} \subseteq (L_0 + L_1 + \cdots + L_{n-1})^{(k(n))}$. The right-hand side is contained in the ideal $\mathrm{id}\langle L_0\rangle$ by Lemma 7.18(a) and Theorem 7.16 applied to commutators $\delta_{k(n)}$ in arbitrary elements $x_i \in L_i$ (under convention 7.6). Since $L_0 \subseteq C_{\widetilde{L}}(\varphi)$, we obtain that $(n\widetilde{L})^{(k(n))} \subseteq \mathrm{id}\langle C_{\widetilde{L}}(\varphi)\rangle$.

We have the following equalities under the identification of L as $L \otimes 1$ in $\widetilde{L}$ (see (5.15) and Example 5.16):

$$(n\widetilde{L})^{(k(n))} = (nL)^{(k(n))} \otimes_{\mathbb{Z}} \mathbb{Z}[\omega] \quad \Rightarrow \quad (nL)^{(k(n))} = (n\widetilde{L})^{(k(n))} \cap (L \otimes 1);$$

$$_{\mathrm{id}}\big\langle C_{\widetilde{L}}(\varphi)\big\rangle = {}_{\mathrm{id}}\big\langle C_L(\varphi)\big\rangle \otimes_{\mathbb{Z}} \mathbb{Z}[\omega] \quad \Rightarrow \quad {}_{\mathrm{id}}\big\langle C_L(\varphi)\big\rangle = {}_{\mathrm{id}}\big\langle C_{\widetilde{L}}(\varphi)\big\rangle \cap (L \otimes 1).$$

Hence

$$(nL)^{(k(n))} = (n\widetilde{L})^{(k(n))} \cap (L \otimes 1) \subseteq {}_{\mathrm{id}}\big\langle C_{\widetilde{L}}(\varphi)\big\rangle \cap (L \otimes 1) = {}_{\mathrm{id}}\big\langle C_L(\varphi)\big\rangle.$$

If n is a prime, then similarly, by Theorem 7.17 and Lemma 7.18, we have $\gamma_{h(n)+1}(n\widetilde{L}) \subseteq {}_{\mathrm{id}}\big\langle C_{\widetilde{L}}(\varphi)\big\rangle$. By (5.15) we also have

$$\gamma_{h(n)+1}(n\widetilde{L}) = \gamma_{h(n)+1}(nL) \otimes_{\mathbb{Z}} \mathbb{Z}[\omega],$$

so that

$$\gamma_{h(n)+1}(nL) = \gamma_{h(n)+1}(n\widetilde{L}) \cap (L \otimes 1) \subseteq {}_{\mathrm{id}}\big\langle C_{\widetilde{L}}(\varphi)\big\rangle \cap (L \otimes 1) = {}_{\mathrm{id}}\big\langle C_L(\varphi)\big\rangle.$$

$\square$

Remarks. 7.20. Of course, there is also a corresponding combinatorial consequence of Theorem 7.7: if a soluble Lie ring L of derived length d admits an automorphism φ of prime order p, then $\gamma_{f(d,p)}(pL) \subseteq {}_{\mathrm{id}}\langle C_L(\varphi)\rangle$ for some (d,p)-bounded number $f(d,p)$. In particular, if $d = 2$, then $\gamma_{p+1}(pL) \subseteq {}_{\mathrm{id}}\langle C_L(\varphi)\rangle$.

7.21. In Theorems 7.16 and 7.17 the commutators in the ambient linear combinations can be chosen to be simple, with the same entry set and each having an initial segment with zero mod n (or p) sum of the indices, by 5.6(b).

7.22. One can show that the sum $L_0 + L_1 + \cdots + L_{n-1}$ is "almost direct": if $l_0 + l_1 + \cdots + l_{n-1} = 0$, then $nl_i = 0$ for each i. We shall encounter a similar situation in Chapter 13.

7.23. Suppose that a cyclic group $\langle \varphi \rangle$ of finite order n acts as automorphisms on a Lie ring L (not necessarily faithfully). The same definition of the "eigenspaces" L_i for $\widetilde{L} = L \otimes_{\mathbb{Z}} \mathbb{Z}[\omega]$, where ω is a primitive nth root of unity, yields the same results: $(nL)^{(k(n))} \subseteq {}_{\mathrm{id}}\langle C_L(\varphi)\rangle$. In other words, $k(n_1) \leq k(n)$ for any n_1 dividing n (think of n_1 as the order of the automorphism of L induced by φ). For Higman's function an analogous statement holds trivially.

§ 7.3. Regular automorphisms

Recall that an automorphism is *regular* if it has no non-trivial fixed points ("non-trivial" means $\neq 0$ in a Lie ring, and $\neq 1$ in a group). In this section we derive the theorems on Lie rings with regular automorphisms of finite order and on nilpotent groups with regular automorphisms of prime order, although we shall need only the combinatorial facts from § 7.2 in the study of p-automorphisms of finite p-groups. Recall the fixed notation $k(n)$ and $h(p)$ for Kreknin's and Higman's functions from § 7.1. The following theorem was proved by G. Higman [1957]; a new effective proof was given in [V. A. Kreknin, 1963] and [V. A. Kreknin and A. I. Kostrikin, 1963].

Theorem 7.24. *If a Lie ring L admits a regular automorphism φ of prime order p, then L is nilpotent of class at most $h(p)$.*

Proof. By Corollary 7.19(b) (and by (5.23)), we have

$$p^{h(p)+1}\gamma_{h(p)+1}(L) = \gamma_{h(p)+1}(pL) \subseteq \mathrm{id}\langle C_L(\varphi)\rangle = 0.$$

So the additive group of $\gamma_{h(p)+1}(L)$ has finite exponent dividing $p^{h(p)+1}$. An automorphism of order p always has non-trivial fixed points on a non-trivial abelian p-group (Corollary 2.8). Hence $\gamma_{h(p)+1}(L) = 0$, as required. ☐

G. Higman [1957] also derived a consequence for locally nilpotent groups, of which we reproduce the easier finite case.

Theorem 7.25. *Suppose that G is a finite nilpotent group with a regular automorphism φ of prime order p. Then the nilpotency class of G does not exceed $h(p)$.*

Proof. By Lemma 2.12, φ induces regular automorphisms on all factor-groups $\gamma_i(G)/\gamma_{i+1}(G)$. Hence φ induces a regular automorphism of the associated Lie ring $L(G)$. By Theorem 7.24, $L(G)$ is nilpotent of class at most $h(p)$. By Theorem 6.9(a), the nilpotency class of G coincides with that of $L(G)$. ☐

The following is a theorem of V. A. Kreknin [1963].

Theorem 7.26. *If a Lie ring L admits a regular automorphism φ of finite order n, then L is soluble of derived length at most $2k(n)$.*

Proof. Note that if $nL = L$, then $L^{(k(n))} \subseteq \mathrm{id}\langle C_L(\varphi)\rangle = 0$ by Corollary 7.19(a).

In general, by Corollary 7.19(a),

$$n^{2^{k(n)}}L^{(k(n))} = (nL)^{(k(n))} \subseteq \mathrm{id}\langle C_L(\varphi)\rangle = 0.$$

So the additive group A of $L^{(k(n))}$ has finite exponent dividing $n^{2^{k(n)}}$. Let $p_1, \ldots, p_m$ be the primes dividing n. Then $A = A_{p_1} \oplus \cdots \oplus A_{p_m}$ where A_{p_i} is the Sylow p_i-subgroup of A. Each A_{p_i} is a φ-invariant ideal of L: if $p_i^s a = 0$, then $p_i^s[a, b] = [p_i^s a, b] = [0, b] = 0$ for any $b \in L$ and $p_i^s a^\varphi = (p_i^s a)^\varphi = 0$.

For every $p = p_i$, we have $\langle \varphi \rangle = \langle \varphi^{p^k} \rangle \times \langle \varphi^q \rangle$ where $n = p^k q$ with $p \nmid q$. We claim that the restriction of φ^{p^k} is a regular automorphism of A_p. Suppose that $C_{A_p}(\varphi^{p^k}) \neq 0$; then this is an additive p-subgroup which is φ^q-invariant because φ^q commutes with φ^{p^k} (see Example 2.4). Since $|\varphi^q| = p^k$, there are non-trivial fixed points of φ^q on the abelian p-group $C_{A_p}(\varphi^{p^k})$ (Corollary 2.8) and hence $C_{A_p}(\varphi) = C_{A_p}(\varphi^p) \cap C_{A_p}(\varphi^q) \neq 0$, a contradiction. Thus, the ideal A_p admits a regular automorphism φ^{p^k} of order dividing q. Since $qA_p = A_p$, this ideal is soluble of derived length $k(q) \leq k(n)$, by the remark made at the beginning of the proof. Thus, all ideals A_{p_i} are soluble of derived length $\leq k(n)$, and therefore their direct sum $L^{(k(n))} = A_{p_1} \oplus \cdots \oplus A_{p_m}$ is also soluble of derived length $\leq k(n)$ (see (5.24)). As a result,

$$L^{(2k(n))} = (L^{(k(n))})^{(k(n))} = 0,$$

as required.   □

Corollary 7.27. *If, under the hypothesis of Theorem 7.26, the additive group of L is torsion-free, then L is soluble of derived length at most $k(n)$.*

Proof. We saw above that the additive group of $L^{(k(n))}$ has finite exponent dividing $n^{2^{k(n)}}$ and hence $L^{(k(n))} = 0$.   □

Remarks. 7.28. By a deep theorem of J. G. Thompson [1959], every finite group admitting a regular automorphism of prime order is nilpotent.

7.29. It follows from the classification of the finite simple groups that every finite group with a regular automorphism of finite order n is soluble. It was also proved that such a group has a normal series with nilpotent factors of length bounded by the number of prime divisors of n, counting multiplicities (E. Shult, F. Gross, T. Berger; other generalizations are also due to J. G. Thompson, E. C. Dade, H. Kurzweil, A. Turull, B. Hartley, M. Isaacs, and others). Thus, the problem of studying such groups is, to a great extent, reduced to the case of nilpotent groups.

7.30. It is an open problem whether the analogue of Kreknin's Theorem 7.26 holds for a (locally) nilpotent or (locally) finite group G with a regular automorphism of finite order n: *is it true that G is soluble of n-bounded derived length?* As noted in 7.29, for (locally) finite groups the problem is reduced to the case where G is a finite nilpotent group with a regular automorphism. However, an application of Kreknin's Theorem 7.26 to the associated Lie ring $L(G)$ fails to produce a required result, since the derived length of $L(G)$ may be smaller than that of G. A positive answer is known only for regular automorphisms of prime order (Higman's Theorem 7.25) and for a regular automorphism of order 4 [L. G. Kovács, 1961]. For torsion-free nilpotent groups the analogue of Kreknin's Theorem will be proved in Chapter 10.

7.31. Theorem 7.25 for finite nilpotent groups was an immediate consequence of the Lie ring result. To extend Theorem 7.25 to arbitrary (locally) nilpotent groups (not necessarily periodic) with a regular automorphism of prime order [G. Higman, 1957] requires some additional effort, since the induced automorphism of the associated Lie ring may have non-trivial fixed points. However, another "strongly central" series, of p-isolators of the terms of the lower central series, reduces the proof to Theorem 7.24 on Lie rings (Exercise 7.7).

7.32. Unlike Lie rings, one has to impose some additional conditions on groups with regular automorphisms to obtain results like nilpotency or solubility. For example, the automorphism φ of the free group F on free generators x, y, defined by $x^\varphi = y$, $y^\varphi = x$, has order 2 and $C_F(\varphi) = 1$.

7.33. We had to double the function $k(n)$ in Theorem 7.26; is this really necessary?

7.34. The following generalization of Higman's Theorem was proved in [E. I. Khukhro, 1990, 1993b]: *if a Lie ring (algebra) admits an automorphism of prime order p with exactly m fixed points (with fixed-point space of dimension m), then there is a subring (subalgebra) of (p, m)-bounded index (codimension) which is nilpotent of p-bounded class.* No such generalization is known so far for Kreknin's Theorem, for automorphisms of composite order, apart from the special case of order four [E. I. Khukhro and N. Yu. Makarenko, 1996a, 1997]). The main results of this book are generalizations of the theorems on regular automorphisms of Lie rings to the "modular" case of p-automorphisms of finite p-groups.

7.35. There are simple Lie algebras with regular non-cyclic group of automorphisms of order 4. For example, let L be the Lie $\mathbb{F}_p$-algebra with basis $\{e_1, e_2, e_3\}$ and structural constants $[e_1, e_2] = e_3$, $[e_2, e_3] = e_1$, $[e_3, e_1] = e_2$. Then the linear transformations given in this basis by the matrices

$$\begin{pmatrix} -1 & 0 & 0 \\ 0 & -1 & 0 \\ 0 & 0 & 1 \end{pmatrix}, \qquad \begin{pmatrix} 1 & 0 & 0 \\ 0 & -1 & 0 \\ 0 & 0 & -1 \end{pmatrix}$$

generate an elementary abelian group $A \leq \operatorname{Aut} L$ of order 4 with $C_L(A) = 0$. By contrast, any finite group with a regular non-cyclic group of automorphisms of order 4 is soluble (G. Glauberman obtained a proof of this fact without using the Feit–Thompson Theorem) and has nilpotent derived subgroup [S. F. Bauman, 1966] of class bounded in terms of the derived length [P. Shumyatsky, 1988].

Exercises 7

1. [V. A. Kreknin, 1963] Use induction on s to prove that, under the hypothesis of Theorem 7.1, the following two inclusions hold:

   (a) $L^{(2^{s-1})} \subseteq \langle L_{s+1}, \ldots, L_{n-1} \rangle$;

   (b) $L^{(2^s - 1)} \subseteq \langle L_{s+1}, \ldots, L_{n-1} \rangle$.

Thus, $k(n) \leq 2^{n-1}$. [*Hint:* In the induction step, prove (a) first, using the induction hypothesis for both (a) and (b); then prove (b) using (a) and the induction hypothesis for (b).]

2. ([D. J. Winter, 1968], [E. I. Khukhro and P. Shumyatsky, 1995]) Prove that if L is a $(\mathbb{Z}/n\mathbb{Z})$-graded Lie ring such that $[L, \underbrace{L_0, \dots, L_0}_{m}] = 0$, then L is soluble of (m, n)-bounded derived length. [*Hint:* Using Theorem 5.25, it is only necessary to prove that $L \neq [L, L]$ for $L \neq 0$.] To obtain an explicit bound for the derived length, use induction on $s = 1, 2, \dots, n - 1$ to prove simultaneously the following two statements:

   (a) $L^{(m(m+1)^{s-1})} \cap L_s \subseteq \langle L_{s+1}, L_{s+2}, \dots, L_{n-1}, L_0 \rangle$;

   (b) $L^{((m+1)^s - 1)} \subseteq \langle L_{s+1}, L_{s+2}, \dots, L_{n-1}, L_0 \rangle$.

   As a result, $L^{((m+1)^{n-1} - 1)} \subseteq \langle L_0 \rangle = L_0$, while $\gamma_{m+1}(L_0) = 0$.

3. Suppose that L is a $(\mathbb{Z}/4\mathbb{Z})$-graded Lie ring with $L_0 = 0$. Prove that the Lie subring $\langle L_1, L_3 \rangle$ is a nilpotent ideal of class 3 with abelian factor-ring.

4. Let φ be a regular automorphism of order 4 of a Lie ring L. Prove that $\gamma_3(\gamma_2(L)) = 0$. [*Hint:* Use 3 to show that $\gamma_3(\gamma_2(4L)) \leq {}_{\mathrm{id}}\langle C_L(\varphi) \rangle$; for that extend the ground ring of L by $i = \sqrt{-1}$.]

5. Produce a regular automorphism of order 4 of the Lie algebra over $\mathbb{Q}$ with basis $\{a, b\}$ and structural constants $[a, b] = a$ (which is not nilpotent).

6. [L. G. Kovács, 1961] Let φ be a regular automorphism of order 4 of a finite nilpotent group G. Prove that $\gamma_3(\gamma_2(G)) = 1$. [*Hint:* Reduce the proof to the case where $G = [G, \varphi^2]$ and $2 \nmid |G|$; then consider the associated Lie ring of G and use 3.]

7. [G. Higman, 1957] Let G be an arbitrary nilpotent group (not necessarily periodic) with a regular automorphism φ of prime order p. Show that G has no elements of order p, and that φ acts regularly on the Lie ring constructed as in Exercise 6.11. Deduce that the nilpotency class of G is at most $h(p)$.

8. Let p and q be distinct primes. Suppose that G is a finite q-group of derived length 2 admitting a regular automorphism φ of order p^n such that $C_G(\varphi^{p^{n-1}}) \leq [G, G]$. Prove that G is nilpotent of (p, n)-bounded class.

9. Prove Remarks 7.14, 7.20, 7.22, 7.35.

Chapter 8

Almost regular automorphism of order p: almost nilpotency of p-bounded class

Now we are in a position to prove the first of the main theorems on finite p-groups with p-automorphisms having few fixed points. If such an automorphism is "almost regular", with p^m fixed points, then the group is "almost nilpotent": it has a subgroup of (p, m)-bounded index and of nilpotency class at most $h(p)$, where h is Higman's function. This bound for the nilpotency class of a subgroup of (p, m)-bounded index is best possible, if required to depend on the order of the automorphism only. The result of this chapter will be used in Chapters 13 and 14.

Theorem 8.1. *If a finite p-group P admits an automorphism φ of prime order p with exactly p^m fixed points, then P has a characteristic subgroup of (p, m)-bounded index which is nilpotent of class at most $h(p)$, where $h(p)$ is the value of Higman's function.*

The proof relies on Higman's Theorem from § 7.2 on regular automorphisms of Lie rings in its combinatorial form and on the use of the associated Lie rings. Note that, at a first glance, an application of Higman's Theorem to $L(P)$ and the induced automorphism φ cannot give us much information, since not only is φ not regular, but the number of fixed points of φ on $L(P)$ can be much greater than on P, by a factor equal to the nilpotency class, say. Another important tool in the proof is a theorem of P. Hall from § 4.2.

We shall freely use the facts that if M is a characteristic subgroup of N and N is a characteristic (normal) subgroup of G, then M is a characteristic (normal) subgroup of G, and that if M and N are characteristic (normal) subgroups, then $[M, N]$ and $M^n = \langle m^n \mid m \in M \rangle$ are also characteristic (normal) subgroups (see (1.14) and Lemma 1.8). The automorphisms of φ-invariant sections induced by φ will be denoted by the same letter. The following lemma makes use of the elementary material from Chapter 2; henceforth P and φ satisfy the hypothesis of Theorem 8.1.

Lemma 8.2. *The rank of any φ-invariant abelian section M/N of P is at most pm.*

Proof. We have $|C_{M/N}(\varphi)| \leq p^m$ by Lemma 2.12; hence the result follows by Corollary 2.7. □

Proof of Theorem 8.1. Consider the associated Lie ring $L(P)$. Then φ

induces the automorphism of $L(P)$, which we denote by the same letter. We have $|C_{\gamma_i(P)/\gamma_{i+1}(P)}(\varphi)| \leq p^m$ for all i by Lemma 2.12, whence in additive notation $p^m C_{\gamma_i(P)/\gamma_{i+1}(P)}(\varphi) = 0$ by Lagrange's Theorem. By the definition of the action of φ on $L(P)$, we have $C_{L(P)}(\varphi) = \bigoplus_i C_{\gamma_i(P)/\gamma_{i+1}(P)}(\varphi)$. Hence $p^m C_{L(P)}(\varphi) = 0$, which implies (by (5.5)) that

$$p^m \,_{\mathrm{id}}\!\big\langle C_{L(P)}(\varphi)\big\rangle = \,_{\mathrm{id}}\!\big\langle p^m C_{L(P)}(\varphi)\big\rangle = 0. \tag{8.3}$$

We fix the notation $h = h(p)$ for the value of Higman's function for the rest of the proof. Applying Corollary 7.19(b), we obtain $\gamma_{h+1}(pL(P)) \leq \,_{\mathrm{id}}\!\big\langle C_{L(P)}(\varphi)\big\rangle$. The left-hand side is equal to $p^{h+1}\gamma_{h+1}(L(P))$ by (5.23), and the right-hand side is annihilated by p^m by (8.3). Hence

$$p^{h+1+m}\gamma_{h+1}(L(P)) = 0. \tag{8.4}$$

Since $\gamma_{h+1}(L(P)) = \bigoplus_{i \geq h+1}(\gamma_i(P)/\gamma_{i+1}(P))$ (Lemma 6.7), equality (8.4) can be rewritten in terms of the group P as $(\gamma_i(P)/\gamma_{i+1}(P))^{p^{h+m+1}} = 1$ for all $i \geq h+1$. For every i, the rank of $\gamma_i(P)/\gamma_{i+1}(P)$ is at most pm by Lemma 8.2. Together, the bounds for the exponent and the rank of a finite abelian group yield a bound for the order: $|\gamma_i(P)/\gamma_{i+1}(P)| \leq p^{pm(h+m+1)}$, for all $i \geq h+1$. If Q is any φ-invariant subgroup of P, the same kind of a hypothesis holds: $|C_Q(\varphi)| \leq p^m$. Thus, we have proved the following lemma.

Lemma 8.5. *If Q is a φ-invariant subgroup of P, then*

    (a) $(\gamma_{h+1}(Q)/\gamma_{h+2}(Q))^{p^{h+m+1}} = 1$;

    (b) $|\gamma_{h+1}(Q)/\gamma_{h+2}(Q)| \leq p^{pm(h+m+1)}$.   □

The idea is to collide the inequality of Lemma 8.5(b) with an opposite one given by Theorem 4.10. If we put $H = \gamma_{pm(h+m+1)+1}(P)$, then, by Theorem 4.10, $|\gamma_i(H)/\gamma_{i+1}(H)| \geq p^{pm(h+m+1)+1}$ for all i, unless $\gamma_{i+1}(H) = 1$. On the other hand, $|\gamma_{h+1}(H)/\gamma_{h+2}(H)| \leq p^{pm(h+m+1)}$ by Lemma 8.5(b). We see that the only way to avoid a contradiction, is to admit that $\gamma_{h+2}(H) = 1$.

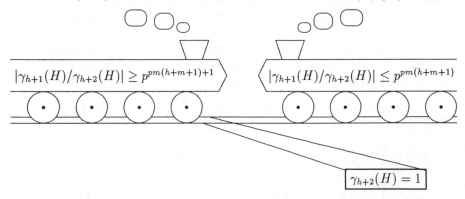

As a result, we have proved that $\gamma_{h+2}(\gamma_{pm(h+m+1)+1}(P)) = 1$. This already gives an upper bound for the derived length of P in terms of p and m; more precisely, P is an extension of a nilpotent group of class $\leq h+1$ by a nilpotent group of class $\leq pm(h+m+1)$.

We can, however, choose a subgroup in a better way. The semidirect product $P\langle\varphi\rangle$ also admits φ as the inner automorphism with exactly p^{m+1} fixed points. (Indeed, $C_{P\langle\varphi\rangle}(\varphi) = C_P(\varphi)\langle\varphi\rangle$.) Therefore, if we replace m by $m+1$ in the above formulae and put $H_1 = \gamma_{p(m+1)(h+m+2)+1}(P\langle\varphi\rangle)$, then the same conflict of inequalities yields the same kind of result: $\gamma_{h+2}(H_1) = 1$, that is, H_1 is nilpotent of class at most $h+1$.

The advantage here lies in the fact that φ acts, of course, trivially on the factors of the lower central series of the group $P\langle\varphi\rangle$ which φ belongs to. By Lemma 2.12, the orders of these factors are at most p^{m+1}, since $|C_{P\langle\varphi\rangle}(\varphi)| = p^{m+1}$. Moreover, all of these factors, excepting the first one, are, in fact, φ-invariant sections of P. Hence all of them, with a possible exception of the first one, have orders at most p^m; and for the first factor, of order at most p^{m+1}, we have $|P\langle\varphi\rangle : \gamma_2(P\langle\varphi\rangle)| = p|P : \gamma_2(P\langle\varphi\rangle)|$ since $\gamma_2(P\langle\varphi\rangle) \leq P$. Therefore, the index of H_1 in P is at most

$$(p^m)^{p(m+1)(h+m+2)} = p^{mp(m+1)(h+m+2)}.$$

Thus, H_1 is the subgroup of P of (p,m)-bounded index which is nilpotent of class at most $h(p)+1$.

We have yet to improve the bound for the nilpotency class down to $h(p)$. Since H_1 is nilpotent of class $h+1$, we have $\gamma_{h+1}(H_1)^{p^{h+m+1}} = 1$ by Lemma 8.5(a), and, by Lemma 6.13,

$$[a_1^{p^t}, \ldots, a_{h+1}^{p^t}] = [a_1, \ldots, a_{h+1}]^{p^{(h+1)t}}$$

for any $a_i \in H_1$, $t \in \mathbb{N}$. So, if we put $r = [(h+m+1)/(h+1)]+1$ (brackets denote the integral part), which is a (p,m)-bounded number, then

$$[a_1^{p^r}, \ldots, a_{h+1}^{p^r}] \in \gamma_{h+1}(H_1)^{p^{h+m+1}} = 1$$

for any $a_i \in H_1$. We claim that the subgroup $H_2 = H_1^{p^r}[H_1, H_1]$ is nilpotent of class at most $h(p)$. By Theorem 3.12, it is sufficient to verify the nilpotency law on the generators. Since H_2 is generated by the generators of $H_1^{p^r}$ and $[H_1, H_1]$, it is sufficient to check that $[b_1, \ldots, b_{h+1}] = 1$, where every b_i is of the form either a^{p^r} or $[c,d]$, $a,c,d \in H_1$. If there is at least one commutator among the b_j, then $[b_1, \ldots, b_{h+1}] \in \gamma_{h+2}(H_1) = 1$ (by Lemma 3.6), and if all of the b_j are p^rth powers, the result was established above.

The index of H_2 in H_1 is at most p^{pmr}, since H_1/H_2 is an abelian group of exponent dividing p^r and of rank at most pm (Lemma 8.2); the index of H_1

in P is also (p,m)-bounded. Thus, H_2 is a subgroup of (p,m)-bounded index in P which is nilpotent of class at most $h(p)$. To produce a characteristic subgroup, we simply take $P^{p^f} \leq H_2$, for some (p,m)-bounded f (say, $p^f = |P:H_2|$). Since the number of generators, the exponent and the nilpotency class of P/P^{p^f} are (p,m)-bounded, the order of P/P^{p^f} is (p,m)-bounded as well (Lemma 6.12(c)). Thus, P^{p^f} is the required characteristic subgroup of (p,m)-bounded index which is nilpotent of class $\leq h(p)$. $\square$

Every element of a group acts by conjugation as an inner automorphism; hence we immediately obtain the following corollary.

Corollary 8.6. *Suppose that p^m is the minimal order of a centralizer of an element of order p in a finite p-group P. Then P has a subgroup of (p,m)-bounded index which is nilpotent of class at most $h(p)$.* $\square$

Remarks. 8.7. The construction of H in the proof Theorem 8.1, which gives a "weak" bound, in terms of p and m, for the derived length of P, is due to J. Alperin [1962]. The construction of H_1 is due to E. I. Khukhro [1985], and the last step, from H_1 to H_2, was made by N. Yu. Makarenko [1992].

8.8. One can show by means of examples that the bound $h(p)$ for the nilpotency class in the conclusion of Theorem 8.1 is best possible (if we require that the bound depends on p only and assume that $h(p)$ is the best possible Higman's function as in § 7.1). It is not clear how realistic is the bounds for the index that could be extracted from the proof. (Of course, all bounds obtained involve Higman's function $h(p)$, good estimates of which are yet to be found, but one can seek the bounds in terms of $h(p)$.)

8.9. In the particular case of $m = 1$, that is, $|C_P(\varphi)| = p = |\varphi|$, C. R. Leedham-Green and S. McKay [1976] and R. Shepherd [1971] proved that P has a subgroup of p-bounded index which is nilpotent of class ≤ 2. We shall prove this result in Chapter 13. The study of p-groups of maximal class amounts to this situation: N. Blackburn [1958] proved that every p-group P of maximal class contains a subgroup P_1 of index p and an element a with $|C_{P_1}(a)| = p$; the inner automorphism induced by a has order p, since the pth power of a must be in the centre of P.

8.10. The results on p-groups of maximal class enabled me to conjecture that there exists a function $f(m)$ such that every finite p-group admitting an automorphism of order p with exactly p^m fixed points has a subgroup of (p,m)-bounded index which is nilpotent of class at most $f(m)$. The result cited in 8.9 means that we can indeed take $f(1) = 2$. This conjecture was recently confirmed by Yu. Medvedev [1994b], whose theorem is proved in Chapter 14.

8.11. It is worth mentioning that there is no universal constant c (independent of p and m) such that every finite p-group P admitting an automorphism of order p with p^m fixed points has a subgroup of (p,m)-bounded index which is nilpotent of class c, even for P soluble of derived length 2 (Exercise 8.3).

8.12. The result of this chapter is the "modular case" of the following general theorem: *if a nilpotent group admits an automorphism of prime order p with a finite number m of fixed points, then there is a subgroup of (p, m)-bounded index which is nilpotent of p-bounded class.* The "coprime component" of this theorem is much more difficult. In [E. I. Khukhro, 1990] this theorem was proved for periodic nilpotent groups, on the basis of a rather intricate connection with the analogous theorem on Lie rings (see Remark 7.34). Then Yu. Medvedev [1994c] extended the result to arbitrary nilpotent groups, essentially by reduction (although by no means trivial) to the case of periodic groups. (See also the book [E. I. Khukhro, 1993b].) For $p = 2$ the result was proved earlier in [B. Hartley and T. Meixner, 1980].

8.13. The enormous advantage of the "modular case", where a p-automorphism acts on a p-group, is the bound for the rank (dimension). This is especially true for Lie rings: suppose that φ is an automorphism of order p^n of a Lie ring L whose additive group is a p-group, and let p^m be the number of fixed points of φ. Corollary 7.19(a) immediately implies that $p^{m+n2^k} L^{(k)} = 0$, where $k = k(p^n)$. Then $p^{[m/2^k]+n+1} L$ is a soluble ideal of derived length $\leq k(p^n)$ which has (p, m, n)-bounded index in the additive group of L, since the rank of the latter is at most mp^n.

If in this situation $|\varphi| = p$, then $p^{m+h+1} \gamma_{h+1}(L) = 0$, where $h = h(p)$, by Corollary 7.19(b). Then $p^{[m/(h+1)]+2} L$ is a nilpotent ideal of class $\leq h(p)$ which has (p, m)-bounded index in the additive group of L. If L is in addition soluble of derived length d, then Remark 7.20 yields a subring $p^{f(d,p,m)} L$ of possibly smaller index which is nilpotent of possibly smaller, (p, d)-bounded class; in particular, if L is soluble of derived length 2, then $p^{[m/(p+1)]+2} L$ is a nilpotent ideal of class $\leq p$.

Exercises 8

1. Compute an explicit upper bound (in terms of $h(p)$) for the index of the subgroup in the conclusion of Theorem 8.1.

2. Use Remarks 7.14 and 7.20 to prove that if a group P satisfying the hypothesis of Theorem 8.1 is soluble of derived length d, then P has a subgroup of (p, m)-bounded index which is nilpotent of class $\leq (p^d - 1)/(p - 1)$. Compute the explicit upper bound for the index of the subgroup.

3. [E. I. Khukhro, 1985] Let p be a prime and let ω be a primitive pth root of unity. For a positive integer s put $K_s = \mathbb{Z}[\omega]/p^s \mathbb{Z}[\omega]$ and consider the free K_s-module U on free generators $u_1, \ldots, u_{p-1}$. Every element $u =$

$\sum_{i=1}^{p-1} k_i u_i \in U$ can be viewed as a vector $(k_1, \ldots, k_{p-1})$, $k_i \in K_s$. The group of matrices

$$
A = \left\{ \left(\begin{array}{cccc} 1 & a_1 & \cdots & a_{p-2} \\ & 1 & \ddots & \vdots \\ & & \ddots & a_1 \\ 0 & & & 1 \end{array} \right) \,\middle|\, a_i \in K_s \right\}
$$

is commutative and acts on the vectors from U by right multiplication as a group of automorphisms of the additive group of U. Let $P = U \rtimes A$ be the semidirect product of U and A. The matrix $\varphi = \mathrm{diag}(\omega, \omega^2, \ldots, \omega^{p-1})$ also acts on U by right multiplication and on A by conjugation. Hence φ can be regarded as an automorphism of $P = U \rtimes A$. Show that

$$
|C_P(\varphi)| = |C_U(\varphi)| \cdot |C_A(\varphi)| = p^{p-1} \cdot p^{p-2} = p^{2p-3}.
$$

Prove that the minimum of the indices of the subgroups of P that are nilpotent of class $p-1$ increases unboundedly with the growth of s. Thus, there is no constant c (independent of p and m) such that every finite p-group admitting an automorphism of order p with p^m fixed points has a subgroup of (p, m)-bounded index which is nilpotent of class c.

Chapter 9

The Baker–Hausdorff Formula
and nilpotent Q-powered groups

We construct both free nilpotent groups and free nilpotent Lie Q-algebras within associative Q-algebras. The Baker–Hausdorff Formula is proved to be a Lie polynomial which links the operations in the group and the Lie algebra. The construction is used to embed any torsion-free nilpotent group in its Q-powered "hull". In Chapter 10, all this will be applied to establish the Mal'cev Correspondence between nilpotent Q-powered groups and nilpotent Lie Q-algebras and the Lazard Correspondence for nilpotent p-groups and Lie rings of class $\leq p - 1$.

§ 9.1. Free nilpotent groups

In § 5.3 we used a free associative Q-algebra $\mathcal{A}$ to construct a free Lie ring $\mathcal{L}$ as a subring of $\mathcal{A}^{(-)}$. We use new "calligraphic" letters for these objects, since here we prefer to denote by $A = \mathcal{A}/\mathcal{A}^{c+1}$ and $L = \mathcal{L}/\gamma_{c+1}(\mathcal{L})$ the free nilpotent factor-algebras. In this section, we construct a free nilpotent group F within A with adjoined outer unity; A is the common ground for both L and F, which helps to establish connections between them.

We recall some definitions and basic properties. Let A be a free nilpotent associative Q-algebra of nilpotency class c with free (non-commuting) generators $x_1, x_2, \ldots$ (when necessary, we shall take a well-ordered set of generators of any given cardinality); "nilpotent of class c" means that every product of any $c + 1$ elements is 0. Thus, A has a basis consisting of all monomials $x_{i_1} \cdots x_{i_k}$ of degrees $k \leq c$ (no parentheses are needed because of the associative law). The multiplication of monomials is juxtaposition:

$$(x_{i_1} \cdots x_{i_k}) \cdot (x_{j_1} \cdots x_{j_l}) = x_{i_1} \cdots x_{i_k} x_{j_1} \cdots x_{j_l},$$

where no cancellations are allowed, but all monomials of degree $\geq c + 1$ are equal to 0. It is clear that A is homogeneous: $A = A_1 \oplus \cdots \oplus A_c$, where A_i is the homogeneous component of A of degree i.

In fact, A is the factor-algebra of a free associative Q-algebra $\mathcal{A}$ by the ideal $\mathcal{A}^{c+1} = \bigoplus_{i \geq c+1} \mathcal{A}_i$. The bracket multiplication $[x, y] = xy - yx$ defines the structure of the Lie Q-algebra $A^{(-)}$ on the additive group of A. We denote by L the Lie ring ($\mathbb{Z}$-algebra) generated by the x_i in $A^{(-)}$. Then $\mathbb{Q}L = \{rl \mid$

$r \in \mathbb{Q}$, $l \in L\}$ is the Lie $\mathbb{Q}$-algebra generated by the x_i. By Theorem 5.39, the Lie subring $\mathcal{L}$ of $\mathcal{A}^{(-)}$ generated by the free generators of $\mathcal{A}$ is a free Lie ring (and $\mathbb{Q}\mathcal{L}$ is a free Lie $\mathbb{Q}$-algebra). Since $\mathcal{L}$ is a homogeneous subspace of $\mathcal{A}$, it follows that $L = \mathcal{L}/\gamma_{c+1}(\mathcal{L}) = \mathcal{L}/(\mathcal{L} \cap \mathcal{A}^{c+1})$ is a free nilpotent Lie ring of class c with free generators x_i (and $\mathbb{Q}L$ is a free nilpotent $\mathbb{Q}$-algebra). Both L and $\mathbb{Q}L$ are homogeneous with components $L_k = L \cap A_k$ and $\mathbb{Q}L_k = \mathbb{Q}L \cap A_k$, and even multihomogeneous with respect to the free generators x_i.

Adjoining an outer unity 1 to A we form the associative $\mathbb{Q}$-algebra $A_0 \oplus A$, where A_0 is the one-dimensional subspace spanned by 1. We are going to construct a free nilpotent group within the set $1 + A = \{1 + a \mid a \in A\}$. In fact, $1 + A$ is a group with respect to the associative multiplication: 1 is the neutral element, and $1 - a + a^2 - a^3 + \cdots + (-1)^c a^c$ (which also belongs to $1 + A$) is the inverse of $1 + a$. We need, however, a subgroup G of $1 + A$, generated by the elements $1 + x_i + y_i$, where the y_i are some fixed elements from $A_2 \oplus \cdots \oplus A_c$, that is, each y_i is a linear combination of monomials of degrees ≥ 2. We shall specify the y_i later; the proof of the fact that G is a free nilpotent group with free generators $1 + x_i + y_i$ does not depend on the choice of the y_i. (In fact, for some purposes, it is simplest to take the $1 + x_i$ as the generators of G, but we shall choose another set for some other reasons.)

We denote by A^n the sum $A_n \oplus \cdots \oplus A_c$, so that the elements of A^n are linear combinations of monomials of degrees $\geq n$. We use brackets to denote both Lie products in L and group commutators in G; the meaning will always be clear from the context.

Lemma 9.1. (a) *Suppose that $\varkappa$ is a commutator of weight k. Then the group commutator $\varkappa(1 + x_i + y_i)$, the value of $\varkappa$ on the elements $1 + x_i + y_i$ in G, is equal to $1 + \varkappa(x_i) + \lambda$, where $\lambda \in A^{k+1}$ and $\varkappa(x_i)$ is the corresponding Lie ring commutator, the value of $\varkappa$ on the x_i in L.*

(b) *Suppose that $g \equiv \prod_j \varkappa_j^{\alpha_j} \pmod{\gamma_{k+1}(G)}$, $\alpha_j \in \mathbb{Z}$, where the $\varkappa_j = \varkappa_j(1 + x_i + y_i)$ are group commutators of weight k in the $1 + x_i + y_i$. Then $g = 1 + \sum_j \alpha_j \varkappa_j(x_i) + \lambda$, where $\lambda \in A^{k+1}$ and the $\varkappa_j(x_i)$ are the corresponding Lie ring commutators in the x_i.*

Proof. (a) Induction on the weight of $\varkappa$. If the weight is 1, the group commutator is one of the $1 + x_i + y_i$; then x_i is the same commutator in L while $y_i \in A^2$, as required. If the weight k of $\varkappa$ is greater than 1, then $\varkappa = [\varkappa_1, \varkappa_2]$, where $\varkappa_j$ has weight k_j with $k_1 + k_2 = k$. By the induction hypothesis, we have $\varkappa_j(1 + x_i + y_i) = 1 + \varkappa_j(x_i) + \lambda_j$, where $\lambda_j \in A^{k_j+1}$, $j = 1, 2$. Then

$$\varkappa_j(1 + x_i + y_i)^{-1} = 1 - (\varkappa_j(x_i) + \lambda_j) + (\varkappa_j(x_i) + \lambda_j)^2 - \cdots = 1 - \varkappa_j(x_i) + \mu_j,$$

where $\mu_j \in A^{k_j+1}$, $j = 1, 2$.

Now we can compute

$$\varkappa(1 + x_i + y_i)$$

$$= [\varkappa_1(1 + x_i + y_i), \varkappa_2(1 + x_i + y_i)]$$

$$= \varkappa_1(1 + x_i + y_i)^{-1} \varkappa_2(1 + x_i + y_i)^{-1} \varkappa_1(1 + x_i + y_i) \varkappa_2(1 + x_i + y_i)$$

$$= (1 - \varkappa_1(x_i) + \mu_1)(1 - \varkappa_2(x_i) + \mu_2)(1 + \varkappa_1(x_i) + \lambda_1)(1 + \varkappa_2(\bar{x}_i) + \lambda_2)$$

$$= 1 + \varkappa_1(x_i)\varkappa_2(x_i) - \varkappa_2(x_i)\varkappa_1(x_i) + \lambda.$$

On the right, the sum of all products of two Lie ring kappas is precisely the Lie ring commutator $[\varkappa_1(x_i), \varkappa_2(x_i)]$. All other products of two, three or four elements $\varkappa_s(x_i)$, λ_t, μ_u involving different indices 1, 2 are in $A^{k_1+k_2+1} = A^{k+1}$ and their linear combination is denoted by λ. The linear combination of the products of one or two elements $\varkappa_j(x_i)$, λ_j, μ_j with the same index j equals 0 since it is the same as in the product $(1 - \varkappa_j(x_i) + \mu_j)(1 + \varkappa_j(x_i) + \lambda_j) = 1$.

(b) We have

$$g = \prod_j \varkappa_j^{\alpha_j} \cdot \prod_s \tau_s^{\beta_s},$$

where τ_s are group commutators of weights $\geq k + 1$ in the $1 + x_i + y_i$, and $\beta_s \in \mathbb{Z}$. By (a), $\varkappa_j = 1 + \varkappa_j(x_i) + \lambda_j$ for some $\lambda_j \in A^{k+1}$, and $\tau_s = 1 + \nu_s$ for some $\nu_s \in A^{k+1}$. Then also $\varkappa_j^{-1} = 1 - \varkappa_j(x_i) + \lambda'_j$ for some $\lambda'_j \in A^{k+1}$, and $\tau_s^{-1} = 1 + \nu'_s$ for some $\nu'_s \in A^{k+1}$. Substituting these expressions in the above product and collecting all terms of degree $\leq k$ we get $1 + \sum_j \alpha_j \varkappa_j(x_i)$, as required. □

Now it is obvious that every commutator of weight $c + 1$ in the generators $1 + x_i + y_i$ is equal to 1, which means that the group G is nilpotent of class c. Moreover, G is a free nilpotent group of class c.

Theorem 9.2. *The group G is free nilpotent of class c with free generators $1 + x_i + y_i$.*

Proof. Let U be a free nilpotent group of class c with free generators u_i (with i in the same index set as for the elements $1 + x_i + y_i$). Then the mapping $u_i \rightarrow 1 + x_i + y_i$ extends to a homomorphism ϑ of U onto G. We claim that ϑ is an isomorphism. Suppose the opposite, and let u be a non-trivial element in $\operatorname{Ker}\vartheta$; choose $k \leq c$ such that $u \in \gamma_k(U) \setminus \gamma_{k+1}(U)$. Then

$$u \equiv \prod_j \varkappa_j^{\alpha_j} \not\equiv 1 \, (\operatorname{mod} \gamma_{k+1}(U)), \tag{9.3}$$

where the $\varkappa_j$ are some commutators of weight k in the u_i, and $\alpha_j \in \mathbb{Z}$. By Lemma 6.7(a), in the associated Lie ring $L(U)$ we have $\sum_j \alpha_j \varkappa_j(\bar{u}_i) \neq 0$, where $\varkappa_j(\bar{u}_i)$ are the same Lie ring commutators in the $\bar{u}_i$, the images of the u_i in $U/\gamma_2(U)$. Since L is a free nilpotent Lie ring of class $c \geq k$ with free generators x_i, this implies that $\sum_j \alpha_j \varkappa_j(x_i) \neq 0$ in L (because of the homomorphism of L onto $L(U)$ extending the mapping $x_i \to \bar{u}_i$).

Applying ϑ to (9.3), we obtain $\vartheta(u) \equiv \prod_j \vartheta(\varkappa_j)^{\alpha_j} \pmod{\gamma_{k+1}(G)}$, where $\vartheta(\varkappa_j)$ are the same group commutators $\varkappa_j$ in the $1 + x_i + y_i$. By Lemma 9.1(b), $\vartheta(u) = 1 + \sum_j \alpha_j \varkappa_j(x_i) + \lambda$, where $\lambda \in A^{k+1}$ and $\varkappa_j(x_i)$ are the same Lie ring commutators in the x_i. Since $\sum_j \alpha_j \varkappa_j(x_i) \neq 0$ and A is homogeneous, it follows that $\vartheta(u) \neq 1$, a contradiction. $\square$

Remarks. 9.4. It can be shown that L is naturally isomorphic to the associated Lie ring $L(G)$, the isomorphism being induced by the mapping $x_i \to (1 + x_i + y_i)\gamma_2(G)$.

9.5. Let $\mathcal{A}$ be a free associative algebra freely generated by the ξ_i. Let $\hat{\mathcal{A}}$ be the algebra of formal power series in the ξ_i, the Cartesian sum of the homogeneous components $\mathcal{A}_i$. Then the elements $1 + \xi_i$ freely generate an (absolutely) free group $\mathcal{F}$. An analogue of Lemma 9.1 holds for $\hat{\mathcal{A}}$ too, which is a way to prove that $\bigcap_{i=1}^{\infty} \gamma_i(\mathcal{F}) = 1$ and $L(\mathcal{F}) \cong \mathcal{L}$.

§ 9.2. The Baker–Hausdorff Formula

In this section, we choose a special set of generators of a free nilpotent group. For any $a \in A$, we define the formal exponent

$$e^a = 1 + \frac{a}{1!} + \frac{a^2}{2!} + \cdots + \frac{a^c}{c!},$$

which is, of course, an element of $1 + A$. In particular, the elements e^{x_i} have the form $1 + x_i + y_i$ with $y_i \in A^2$. By Theorem 9.2, the e^{x_i} freely generate a free nilpotent group F of class c. In fact, all elements in $1 + A$ can be represented in the form e^a, $a \in A$. For any $a \in A$, we define the formal logarithm

$$\log(1 + a) = a - \frac{a^2}{2} + \frac{a^3}{3} - \cdots + (-1)^{c-1}\frac{a^c}{c}.$$

Although A is not a commutative ring, the powers of the same element clearly commute, which makes it easy to prove by a direct calculation, just as for "real" exponents and logarithms, that $\log(e^a) = a$ and $e^{\log(1+a)} = 1 + a$ for any $a \in A$. One of the reasons for representing elements of the group F in the form e^a, $a \in A$, is the fact that it is extremely easy to take powers of such elements: $(e^a)^k = e^{ka}$ for any $k \in \mathbb{Z}$. (This is, again, easy to prove by a direct calculation, since the powers of a commute.)

Definition 9.6. The *Baker–Hausdorff Formula* is $H(x_1, x_2) = \log(e^{x_1} e^{x_2})$ regarded as a polynomial in two non-commuting variables; equivalently, this is a polynomial $H(x_1, x_2)$ such that

$$e^{H(x_1, x_2)} = e^{x_1} e^{x_2}. \tag{9.7}$$

(Actually, it is the theorem below stating that $H(x_1, x_2) \in \mathbb{Q}L$ that is usually referred to as the Baker–Hausdorff Formula.) The homogeneous component of $H(x_1, x_2)$ of degree n is denoted by $H_n(x_1, x_2) \in A_n$, so that $H(x_1, x_2) = \sum_{n=1}^{c} H_n(x_1, x_2)$.

Direct calculations can give some of the first terms of $H(x_1, x_2)$. For example, $H_1(x_1, x_2) = x_1 + x_2$ and $H_2(x_1, x_2) = \frac{1}{2} x_1 x_2 - \frac{1}{2} x_2 x_1 = \frac{1}{2} [x_1, x_2]$. By Lemma 9.1(a), $[e^{x_1}, e^{x_2}] = 1 + [x_1, x_2] + \cdots = e^{[x_1, x_2] + \cdots}$, where the dots denote summands of degrees ≥ 3.

We need to make the following remark on the connections between free nilpotent algebras of different nilpotency classes. It is clear that A_c is an ideal of $A_0 \oplus A$, and the images of A, L, $\mathbb{Q}L$, and F in $(A_0 \oplus A)/A_c$ are naturally isomorphic to the corresponding objects constructed for the nilpotency class $c - 1$. Hence the components $H_k(x_1, x_2)$ of degrees $k \leq c - 1$ coincide with the components of the Baker–Hausdorff Formula defined in the same way for the free nilpotent algebra of class $c - 1$. Thus, with some abuse of notation, *the Baker–Hausdorff Formula is the same for all nilpotency classes.* (It may be more natural to define $H(x_1, x_2)$ as an infinite formal power series in the context of Remark 9.5, but we prefer to deal with finite polynomials in nilpotent algebras instead.)

Another remark makes use of the fact that A is a free nilpotent algebra. Let B be an arbitrary nilpotent associative $\mathbb{Q}$-algebra of class $\leq c$. For any $b \in B$, the exponents e^b can be defined in the same way in $B_0 \oplus B$, where B_0 is spanned by the outer unity. Then (9.7) implies that the same Baker–Hausdorff Formula holds for them:

$$e^{H(b_1, b_2)} = e^{b_1} e^{b_2}, \tag{9.8}$$

for any $b_1, b_2 \in B$: simply apply to (9.7) the homomorphism of $A_0 \oplus A$ onto $B_0 \oplus B$ extending the mapping $x_i \to b_i$, $1 \to 1$.

Since we aim at proving that $H(x_1, x_2) \in \mathbb{Q}L$, we shall need a criterion for an element in A to belong to $\mathbb{Q}L$. Recall that in §5.3 the Dynkin operator δ was defined on associative monomials as bracketing from the left:

$$\delta(x_{i_1} x_{i_2} \cdots x_{i_m}) = [\ldots[x_{i_1}, x_{i_2}], \ldots, x_{i_m}]$$

(where $\delta(x_i) = x_i$); then δ is extended to A by linearity.

Lemma 9.9. *A homogeneous element $a \in A$ of degree k belongs to $\mathbb{Q}L$ if and only if $\delta(a) = ka$.*

Proof. It follows from the definition that $\delta(a) \in \mathbb{Q}L$ for any $a \in A$; hence if $\delta(a) = ka$, then $a = \frac{1}{k}\delta(a) \in \mathbb{Q}L$.

Since the elements of $\mathbb{Q}L$ are linear combinations of simple commutators in the x_i (Lemma 5.6), to prove the converse we need only to show that $\delta(a) = ka$ for any simple commutator $a = [x_{i_1}, \dots, x_{i_k}]$ of weight k in the x_i. If $k = 1$, the assertion is obvious. For $k > 1$ we use the induction hypothesis and the equality $\delta(x_{i_k}[x_{i_1}, \dots, x_{i_{k-1}}]) = [x_{i_k}, [x_{i_1}, \dots, x_{i_{k-1}}]]$, which holds by Corollary 5.42:

$$
\begin{aligned}
\delta([x_{i_1}, \dots, x_{i_k}]) &= \delta([x_{i_1}, \dots, x_{i_{k-1}}]x_{i_k} - x_{i_k}[x_{i_1}, \dots, x_{i_{k-1}}]) \\
&= [\delta([x_{i_1}, \dots, x_{i_{k-1}}]), x_{i_k}] - \delta(x_{i_k}[x_{i_1}, \dots, x_{i_{k-1}}]) \\
&= (k-1)[[x_{i_1}, \dots, x_{i_{k-1}}], x_{i_k}] - [x_{i_k}, [x_{i_1}, \dots, x_{i_{k-1}}]] \\
&= k[x_{i_1}, \dots, x_{i_{k-1}}, x_{i_k}].
\end{aligned}
$$

$\square$

For any $u \in A$, we denote by $\mathrm{ad}(u)$ the operator of Lie ring multiplication by u, that is, $a\,\mathrm{ad}(u) = [a, u]$ for $a \in A$. The operator $\mathrm{ad}(u)$ is an element of the associative $\mathbb{Q}$-algebra $\mathrm{Hom}\,A$ of all $\mathbb{Q}$-linear transformations of A. If $f = f(x_i)$ is an arbitrary element of A regarded as a linear combination of associative monomials in the x_i, we can form $f(\mathrm{ad}(x_i)) \in \mathrm{Hom}\,A$ by replacing the x_i by $\mathrm{ad}(x_i)$. Note that we have $af(\mathrm{ad}(x_i)) = \delta(af(x_i))$ by the definition of the Dynkin operator.

For any $u \in A$, we denote by $\mathrm{l}(u) \in \mathrm{Hom}\,A$ the operator of left multiplication by u in A, that is, $a\,\mathrm{l}(u) = ua$ for $a \in A$; similarly, $\mathrm{r}(u) \in \mathrm{Hom}\,A$ is the right multiplication by u. It is clear that $\mathrm{ad}(u) = \mathrm{r}(u) - \mathrm{l}(u)$ in $\mathrm{Hom}\,A$. Let B denote the subalgebra of $\mathrm{Hom}\,A$ generated by the $\mathrm{l}(u), \mathrm{r}(u), \mathrm{ad}(u)$ for all $u \in A$. Applying any of the $\mathrm{l}(u), \mathrm{r}(u), \mathrm{ad}(u)$ to an element from A^k produces an element in A^{k+1}. Since $A^{c+1} = 0$, it follows that the subalgebra B is nilpotent of class $c - 1$. Therefore, the same Baker-Hausdorff Formula (9.8), $e^{H(b_1, b_2)} = e^{b_1}e^{b_2}$, holds for any $b_1, b_2 \in B$, where e^b is defined for $b \in B$ with 1 being the identity mapping of A.

It is clear that $\mathrm{l}(u)\mathrm{r}(v) = \mathrm{r}(v)\mathrm{l}(u)$ for any $u, v \in A$. Hence $e^{\mathrm{ad}(u)} = e^{\mathrm{r}(u)-\mathrm{l}(u)} = e^{\mathrm{r}(u)}e^{-\mathrm{l}(u)}$ and $e^{\mathrm{l}(u)}e^{\mathrm{r}(v)} = e^{\mathrm{r}(v)}e^{\mathrm{l}(u)}$ for any $u, v \in A$. It is obvious that $\mathrm{r}(a)\mathrm{r}(b) = \mathrm{r}(ab)$ and $\mathrm{l}(a)\mathrm{l}(b) = \mathrm{l}(ba)$ for any $a, b \in A$. It follows that $H(\mathrm{r}(a), \mathrm{r}(b)) = \mathrm{r}(H(a, b))$, while $H(\mathrm{l}(b), \mathrm{l}(a)) = \mathrm{l}(H(a, b))$. These remarks enable us to prove the following crucial lemma. To lighten notation, we set $x = x_1$, $y = x_2$ for the rest of the section.

Lemma 9.10. $H(\text{ad}(x), \text{ad}(y)) = \text{ad}(H(x, y))$.

Proof. We have

$$
\begin{aligned}
e^{H(\text{ad}(x), \text{ad}(y))} &= e^{\text{ad}(x)} e^{\text{ad}(y)} = e^{r(x) - l(x)} e^{r(y) - l(y)} \\
&= e^{r(x)} e^{-l(x)} e^{r(y)} e^{-l(y)} = e^{r(x)} e^{r(y)} e^{-l(x)} e^{-l(y)} \\
&= e^{H(r(x), r(y))} \left(e^{l(y)} e^{l(x)} \right)^{-1} = e^{H(r(x), r(y))} e^{-H(l(y), l(x))} \\
&= e^{r(H(x, y))} e^{-l(H(x, y))} = e^{\text{ad}(H(x, y))}.
\end{aligned}
$$

Taking logarithms of both sides of the equation $e^{H(\text{ad}(x), \text{ad}(y))} = e^{\text{ad}(H(x, y))}$ obtained, we get $H(\text{ad}(x), \text{ad}(y)) = \text{ad}(H(x, y))$. □

Now we prove the following fact often referred to as the "Baker–Hausdorff Formula".

Theorem 9.11. $H(x, y) \in \mathbb{Q}L$.

Proof. Given the nilpotency class c of A, we shall prove that $H_n(x, y) \in \mathbb{Q}L$ for all degrees $n \le c-1$. Since c is arbitrary, this will imply that $H_n(x, y) \in \mathbb{Q}L$ for all n, in view of the remark above on the uniqueness of $H(x, y)$.

We consider the image of $z = x_3$ under both parts of the equation of Lemma 9.10 (the reader will distinguish the right operator notation and products in A):

$$
z H(\text{ad}(x), \text{ad}(y)) = z \, \text{ad}(H(x, y)) = [z, H(x, y)] = z H(x, y) - H(x, y) z.
$$

Then

$$
z H_n(\text{ad}(x), \text{ad}(y)) = z H_n(x, y) - H_n(x, y) z \tag{9.12}
$$

for each $n \le c - 1$, since A is homogeneous. Both parts of (9.12) are homogeneous elements of A of degree $n + 1$; we apply the Dynkin operator to them:

$$
\begin{aligned}
(n + 1) z H_n(\text{ad}(x), \text{ad}(y)) \\
= \ & \delta(z H_n(\text{ad}(x), \text{ad}(y))) \\
= \ & \delta(z H_n(x, y)) - [\delta(H_n(x, y)), z] \\
= \ & z \, H_n(\text{ad}(x), \text{ad}(y)) - \delta(H_n(x, y)) z + z \, \delta(H_n(x, y)).
\end{aligned}
$$

We applied Lemma 9.9 to the left-hand side of (9.12), which is in $\mathbb{Q}L$, and used the fact that $\delta(z H_n(x, y)) = z H_n(\text{ad}(x), \text{ad}(y))$. As a result, we have

$$
n z H_n(\text{ad}(x), \text{ad}(y)) = -\delta(H_n(x, y)) z + z \, \delta(H_n(x, y)),
$$

or, applying (9.12) to the left-hand side,

$$nzH_n(x,y) - nH_n(x,y)z = -\delta(H_n(x,y))\,z + z\,\delta(H_n(x,y)). \qquad (9.13)$$

Since the associative monomials are linearly independent, we must have the equality for those parts of (9.13) which involve only the monomials beginning with z. This gives us

$$nzH_n(x,y) = z\,\delta(H_n(x,y)),$$

whence, for the same reason, $nH_n(x,y) = \delta(H_n(x,y))$. So, by Lemma 9.9, $H_n(x,y) \in \mathbb{Q}L$ for all $n \leq c-1$. Varying the nilpotency class c of A, we obtain that $H_n(x,y) \in \mathbb{Q}L$ for all n. $\qquad\square$

Corollary 9.14. *The subset $e^{\mathbb{Q}L} = \{e^u \mid u \in \mathbb{Q}L\}$ is a nilpotent group of class c.*

Proof. This is, indeed, a subgroup of $1+A$: for any $u, v \in \mathbb{Q}L$ we have, obviously, $e^{-u} \in e^{\mathbb{Q}L}$, and $e^u e^v = e^{H(u,v)} \in e^{\mathbb{Q}L}$ by the Baker–Hausdorff Formula since $H(u,v) \in \mathbb{Q}L$ as the image of $H(x_1,x_2) \in \mathbb{Q}L$ under the endomorphism of $\mathbb{Q}L$ extending the mapping $x_1 \to u$, $x_2 \to v$. To prove that $[e^{u_1},\dots,e^{u_c+1}] = 1$ for any $u_i \in \mathbb{Q}L$, consider the endomorphism ϑ of $A_0 \oplus A$ extending the mapping $1 \to 1$, $x_i \to u_i$. We have $[e^{u_1},\dots,e^{u_c+1}] = [e^{x_1},\dots,e^{x_c+1}]^{\vartheta} = 1$. $\qquad\square$

Since $e^{x_i} \in e^{\mathbb{Q}L}$, we obtain that F is a subgroup of $e^{\mathbb{Q}L}$: if $e^u \in F$, then $u \in \mathbb{Q}L$. But not every element in $e^{\mathbb{Q}L}$ belongs to F, for example, $e^{x_1/2} \notin F$.

We record here a technical lemma on the structure of the first terms of $H(x,y)$.

Lemma 9.15. (a) $H(x,y) = x + y + \sum_j t_j \varkappa_j$, *where $t_j \in \mathbb{Q}$ and the $\varkappa_j$ are Lie ring commutators in x and y of weight ≥ 2;*

(b) $[e^x, e^y] = e^{K(x,y)}$, *where $K(x,y) = [x,y] + \sum_j u_j \varkappa_j$, where $u_j \in \mathbb{Q}$ and the $\varkappa_j$ are Lie ring commutators in x and y of weight ≥ 3.*

Proof. (a) It is clear that $e^x e^y = 1 + x + y + \lambda$ with $\lambda \in A^2$. Then $H(x,y) = \log(1 + x + y + \lambda) = (x + y + \lambda) - (x + y + \lambda)^2 + \cdots = x + y + \lambda'$ with $\lambda' \in A^2$. Since $H(x,y) \in \mathbb{Q}L$, we also have $\lambda' \in \mathbb{Q}L \cap A^2$ which implies the result.

(b) By Lemma 9.1(a), $[e^x, e^y] = 1 + [x,y] + \mu$ with $\mu \in A^3$. Then $K(x,y) = \log(1 + [x,y] + \mu) = [x,y] + \mu'$ with $\mu' \in A^3$. Since $K(x,y) \in \mathbb{Q}L$, we have $\mu' \in \mathbb{Q}L \cap A^3$, and the result follows. $\qquad\square$

Applying the definitions, one can compute directly the explicit form of

$H(x, y)$ as a linear combination of associative monomials:

$$H(x, y) = \log(e^x e^y) = \log\left(1 + \sum_{\substack{r_i, s_i \in \mathbb{N} \cup \{0\} \\ r_i + s_i > 0}} \frac{x^{r_i} y^{s_i}}{r_i! s_i!}\right)$$

$$= \sum_{m \in \mathbb{N}} \sum_{\substack{r_i, s_i \in \mathbb{N} \cup \{0\} \\ r_i + s_i > 0}} \frac{(-1)^{m-1}}{m} \frac{x^{r_1} y^{s_1} \cdots x^{r_m} y^{s_m}}{r_1! s_1! \cdots r_m! s_m!}.$$

Since $H(x, y) \in \mathbb{Q}L$ by Theorem 9.11, we can apply the Dynkin operator δ to the homogeneous components $H_n(x, y)$, to obtain an explicit expression for $H(x, y)$ as a linear combination of commutators in x, y. To wit, $H_n(x, y) = \frac{1}{n}\delta(H_n(x, y))$ by Lemma 9.9, so that

$$H(x, y)$$

$$= \sum_{m \in \mathbb{N}} \sum_{\substack{r_i, s_i \in \mathbb{N} \cup \{0\} \\ r_i + s_i > 0}} \frac{(-1)^{m-1}}{m \sum_{i=1}^{m}(r_i + s_i)} \frac{[\overbrace{x, \ldots, x}^{r_1}, \overbrace{y, \ldots, y}^{s_1}, \ldots, \overbrace{y, \ldots, y}^{s_m}]}{r_1! s_1! \cdots r_m! s_m!}. \qquad (9.16)$$

Here, in A, the sums are taken over all monomials of degrees $\leq c$.

Remark 9.17. The commutators on the right of (9.16) may well be linearly dependent, but this explicit formula already gives important information about $H(x, y)$. For example, note that no prime divisors of the denominators exceed c, the nilpotency class of A.

§ 9.3. Nilpotent Q-powered groups

By Corollary 9.14, $e^{\mathbb{Q}L}$ is a nilpotent group of class c. Moreover, $e^{\mathbb{Q}L}$ is a $\mathbb{Q}$-powered group, that is, torsion-free and divisible: $(e^u)^k = e^{ku} = 1 \Rightarrow u = 0 \Rightarrow e^u = 1$ and $(e^{\frac{1}{k}u})^k = e^u$, for any $e^u \in e^{\mathbb{Q}L}$ and $k \in \mathbb{N}$. Recall that $\mathbb{Q}$-powered groups can be regarded as algebraic systems that are groups with additional unary operations of taking powers in $\mathbb{Q}$, see Example 1.40. We shall prove that the set of all roots of the elements of $F = \langle e^{x_i} \rangle$ is a free nilpotent $\mathbb{Q}$-powered group (and coincides with $e^{\mathbb{Q}L}$). This will enable us to embed every torsion-free nilpotent group in its $\mathbb{Q}$-powered hull (often called its *Mal'cev completion*).

For a subgroup H of a nilpotent group G, we denote the set of all roots of all elements from H by $\sqrt{H} = \{g \in G \mid g^n \in H \text{ for some } n = n(g) \in \mathbb{N}\}$. Note that, obviously, $\sqrt{\sqrt{H}} = \sqrt{H}$.

Theorem 9.18. *If H is a subgroup of a nilpotent group G, then $\sqrt{H}$ is a subgroup. If $A \trianglelefteq B$ for some subgroups $A, B \leq G$, then $\sqrt{A} \trianglelefteq \sqrt{B}$.*

Proof. In proving that $\sqrt{H}$ is a subgroup, we may clearly assume that $G = \langle \sqrt{H} \rangle$; note that this condition is inherited by every factor-group of G with respect to the image of H. Then we need to prove that $G = \sqrt{H}$. Induction on the nilpotency class c of G. For G abelian, the result is obvious: if $a_i^{k_i} \in H$, then, for example, $(a_1 \cdots a_m)^{k_1 \cdots k_m} \in H$. Now let $c > 1$. By 3.6 and 6.13, $\gamma_c(G)$ is generated by the commutators $[g_1, \dots, g_c]$ of weight c in the generators $g_i \in \sqrt{H}$. For each g_i there is n_i such that $g_i^{n_i} \in H$. Using Lemma 6.13, we get

$$H \ni [g_1^{n_1}, \dots, g_c^{n_c}] = [g_1, \dots, g_c]^{n_1 \cdots n_c}.$$

Thus, $\gamma_c(G)$ is generated by the roots of elements from $H \cap \gamma_c(G)$, and therefore, as an abelian group, $\gamma_c(G)$ is contained in $\sqrt{H}$, as shown above. By the induction hypothesis applied to $G/\gamma_c(G)$, for every $g \in G$ there is $n = n(g) \in \mathbb{N}$ such that $g^n \in H\gamma_c(G)$, that is, $g^n = hz$ for some $h \in H$ and $z \in \gamma_c(G)$. But $z^m \in H$ for some $m \in \mathbb{N}$, as shown above. Then $g^{nm} = (hz)^m = h^m z^m \in H$, since h and z commute, so that $g \in \sqrt{H}$, as required.

Now let $A \trianglelefteq B$; we need to show that $[a, b] \in \sqrt{A}$ for any $a \in \sqrt{A}$ and $b \in \sqrt{B}$. We use reverse induction on the weight of a commutator $\varkappa = \varkappa(a, b)$ in a and b to show that $\varkappa(a, b) \in \sqrt{A}$. Let $a^s \in A$ and $b^t \in B$. If the weight of $\varkappa$ is c, with weight c_1 in a and c_2 in b, then $\varkappa(a^s, b^t) = \varkappa(a, b)^{s^{c_1} t^{c_2}}$ by Lemma 6.13. Since $A \trianglelefteq B$, the left-hand side is in A; hence $\varkappa(a, b) \in \sqrt{A}$, as required. Now for $\varkappa$ of any weight $k < c$, of weight k_1 in a and k_2 in b, we have, by Lemma 6.13, $\varkappa(a^s, b^t) = \varkappa(a, b)^{s^{k_1} t^{k_2}} \lambda$, where λ is a product of commutators in $a^{\pm 1}$ and $b^{\pm 1}$ of weight $\geq k + 1$. Again, the left-hand side is in A. By the induction hypothesis and since $\sqrt{A}$ is a subgroup, $\lambda \in \sqrt{A}$, which implies $\varkappa(a, b)^{s^{k_1} t^{k_2}} \in \sqrt{A}$, whence $\varkappa(a, b) \in \sqrt{A}$. Finally, we shall arrive at $[a, b] \in \sqrt{A}$. $\square$

Recall that a $\mathbb{Q}$-powered subgroup of a $\mathbb{Q}$-powered group is a subgroup closed under taking all roots (powers in $\mathbb{Q}$). Saying that a $\mathbb{Q}$-powered (sub-)group is generated by a set, we shall always mean generation as a $\mathbb{Q}$-powered (sub)group. To avoid confusion, we shall speak of "abstract" (sub)groups (generated by a set), meaning just subgroups, regardless of additional operations.

Corollary 9.19. (a) *Suppose that G is a $\mathbb{Q}$-powered group and H is an abstract subgroup of G. Then $\sqrt{H}$ is a $\mathbb{Q}$-powered subgroup generated by H.*

(b) *If G is a $\mathbb{Q}$-powered group generated by a set M, then $G = \sqrt{K}$, where K is an abstract subgroup generated by M.*

Proof. (a) By Theorem 9.18, the set $\sqrt{H}$ is an abstract subgroup. It remains to note that, obviously, all roots of elements of $\sqrt{H}$ are in $\sqrt{H}$.

(b) By (a), $\sqrt{K}$ is a $\mathbb{Q}$-powered subgroup generated by K and hence by M; hence $\sqrt{K} = G$. $\square$

We return to our main construction. Since F is a subgroup of the $\mathbb{Q}$-power-ed group $e^{\mathbb{Q}L}$, we can form $\sqrt{F}$, which here takes the form $\sqrt{F} = \{e^{ru} \mid e^u \in F, \ r \in \mathbb{Q}\}$. By Corollary 9.19, $\sqrt{F}$ is a $\mathbb{Q}$-powered group generated by the e^{x_i} and $\sqrt{F}$ is nilpotent of class c, since so is $e^{\mathbb{Q}L}$ by Corollary 9.14. (In fact, $\sqrt{F} = e^{\mathbb{Q}L}$, as we shall see later.) Now we use $\sqrt{F}$ to prove existence and uniqueness of the Mal'cev completions; as a by-product, we shall prove that $\sqrt{F}$ is a free nilpotent $\mathbb{Q}$-powered group.

Theorem 9.20. (a) *Every torsion-free nilpotent group G of class c can be embedded as a subgroup in a nilpotent $\mathbb{Q}$-powered group $\widehat{G}$ of the same nilpotency class c such that $\widehat{G} = \sqrt{G}$.*

(b) *The group $\widehat{G}$ is unique up to isomorphism; moreover, every isomorphism $\varphi : G \to G'$ extends to an isomorphism of $\widehat{G}$ onto $\widehat{G'}$. In particular, every automorphism of G extends to an automorphism of $\widehat{G}$.*

Proof. (a) We fix some set of generators for $G = \langle g_i \mid i \in I \rangle$ and consider the free nilpotent group F of class c on the corresponding free generators $\{e^{x_i} \mid i \in I\}$ constructed as in §9.1. The mapping $e^{x_i} \to g_i$ extends to a homomorphism of F onto G; let N be the kernel of this homomorphism. Then we may identify G with F/N. By Theorem 9.18 and Corollary 9.19, $\sqrt{N}$ is a normal $\mathbb{Q}$-powered subgroup of $\sqrt{F}$. Then $\sqrt{F}/\sqrt{N}$ is also a $\mathbb{Q}$-powered group. We have $\sqrt{N} \cap F = N$, since if $a \in (\sqrt{N} \cap F) \setminus N$, then the image of a in $G = F/N$ is non-trivial and has finite order, contrary to the fact that G is torsion-free. Hence G can be identified with $F\sqrt{N}/\sqrt{N}$. We can put $\widehat{G} = \sqrt{F}/\sqrt{N}$. Indeed, this is a nilpotent $\mathbb{Q}$-powered group of class c which coincides with $\sqrt{G} = \sqrt{F\sqrt{N}/\sqrt{N}}$, since some power of every element in $\sqrt{F}$ belongs to F.

(b) Recall that for $\mathbb{Q}$-powered groups, abstract isomorphisms are automatically isomorphisms of $\mathbb{Q}$-powered groups (Example 1.40). So we need only construct an abstract isomorphism of $H_1 = \sqrt{G}$ onto $H_2 = \sqrt{G'}$. Consider the direct product $H_1 \times H_2$ (identifying H_i with the factors) and the subgroup $D = \{(g, g^\varphi) \mid g \in G\}$ in $H_1 \times H_2$, which is obviously isomorphic to G. Let $\sqrt{D}$ be the $\mathbb{Q}$-powered subgroup of $H_1 \times H_2$ generated by D. We have

$$\sqrt{D}\, H_i = H_1 \times H_2 \qquad \text{and} \qquad \sqrt{D} \cap H_i = 1 \tag{9.21}$$

for each $i = 1, 2$. Indeed, for any $(h_1, h_2) \in H_1 \times H_2$, there is $k \in \mathbb{N}$ such that $h_1^k \in G$. There is $h_2' \in H_2$ such that $h_2'^k = (h_1^k)^\varphi$; then $(h_1, h_2')^k = (h_1^k, h_2'^k) = (h_1^k, (h_1^k)^\varphi) \in D$ so that $(h_1, h_2') \in \sqrt{D}$ and $(h_1, h_2) = (h_1, h_2')(1, h_2'^{-1}h_2) \in \sqrt{D}\, H_2$. As a result, $H_1 \times H_2 = \sqrt{D}\, H_2$. A similar calculation proves that $\sqrt{D}\, H_1 = H_1 \times H_2$. If $(x, 1) \in \sqrt{D} \cap H_1$, then $(x, 1)^k = (x^k, 1) \in D$ for some $k \in \mathbb{N}$, whence $x^k = 1$; this implies that $x = 1$, since H is torsion-free together with G. Thus, $\sqrt{D} \cap H_1 = 1$; similarly $\sqrt{D} \cap H_2 = 1$. It follows from (9.21) that the restriction of the projection $\pi_1 : (h_1, h_2) \to h_1$ to $\sqrt{D}$ is an isomorphism σ of $\sqrt{D}$ onto H_1; similarly, the restriction of $\pi_2 : (h_1, h_2) \to h_2$ to $\sqrt{D}$ is an isomorphism τ of $\sqrt{D}$ onto H_2. Then the composition of σ^{-1} and τ is the required isomorphism of H_1 onto H_2, which coincides with φ on G. (Note that for each $h_1 \in H_1$ there is a unique $h_2 \in H_2$ such that $(h_1, h_2) \in \sqrt{D}$; this follows from (9.21) and is also encoded in the isomorphisms σ and τ: the composition of σ^{-1} and τ simply maps h_1 to this h_2.) □

Corollary 9.22. $\sqrt{F}$ *is a free nilpotent Q-powered group of class c freely generated by the* e^{x_i}.

Proof. Let U be a free nilpotent Q-powered group of class c on free generators h_i, $i \in I$. The mapping $h_i \to e^{x_i}$ extends to a Q-powered homomorphism ϑ of U onto $\sqrt{F}$. By Corollary 9.19, $U = \sqrt{H}$, where H is generated by the h_i as an abstract group. Since F is a free nilpotent group of class c, the mapping $e^{x_i} \to h_i$ extends to an abstract homomorphism of F onto H. As we saw in the proof of Theorem 9.20, the latter extends to a Q-powered homomorphism φ of $\sqrt{F}$ onto $U' = \sqrt{H}$, a Q-powered hull of H, constructed as $\sqrt{F}/\sqrt{N}$, while F/N is identified with H. By Theorem 9.20(b), there is a Q-powered isomorphism ψ of U' onto U which extends the abstract isomorphism of F/N onto H. The composition of ϑ, φ and ψ is the identity mapping of U since it is identical on the set of its Q-powered generators:

$$h_i \xrightarrow{\vartheta} e^{x_i} \xrightarrow{\varphi} e^{x_i} N \xrightarrow{\psi} h_i.$$

Hence ϑ is an isomorphism of U onto $\sqrt{F}$, so that $\sqrt{F}$ is a free nilpotent Q-powered group of class c on the free generators e^{x_i}. □

Exercises 9

1. Check that $\log(e^a) = a$, $e^{\log(1+a)} = 1 + a$, and $(e^a)^k = e^{ka}$ for any $a \in A$, $k \in \mathbb{Z}$.

2. Prove that $\gamma_i(F) = \zeta_{c-i+1}(F)$ in the free nilpotent group F of class c.

3. Prove that all elements of finite order in a nilpotent group form a subgroup.

4. Let G be a nilpotent Q-powered group. Prove that $\sqrt{\zeta_i(G)} = \zeta_i(\sqrt{G})$ and $\zeta_i(G) = \zeta_i(\sqrt{G}) \cap G$ for all i.

5. Prove that $e^{-x} e^y e^x = e^{y + [y,x] + [y,x,x]/2! + [y,x,x,x]/3! + \cdots}$.

6. Prove Remarks 9.4, 9.5.

Chapter 10

The correspondences
of A. I. Mal'cev and M. Lazard

This chapter is an immediate continuation of Chapter 9. The Baker–Hausdorff Formula and its inverses establish the Mal'cev Correspondence between nilpotent $\mathbb{Q}$-powered groups and nilpotent Lie $\mathbb{Q}$-algebras. Unlike the associated Lie ring, this construction cannot be applied to every nilpotent group, but it provides a much better correspondence, a so-called "equivalence of categories". As an application, we derive a corollary of Kreknin's Theorem for torsion-free nilpotent groups with regular automorphisms. Surprisingly, we shall also be able to apply this technique to finite p-groups with "almost regular" p-automorphisms in Chapter 12. The Lazard Correspondence between nilpotent p-groups and Lie rings of class $\leq p-1$ is also based on the Baker–Hausdorff Formula; this gadget will be used in Chapters 13 and 14.

§ 10.1. The Mal'cev Correspondence

We continue to study the groups F, $e^{\mathbb{Q}L}$, the Lie ring L and the Lie algebra $\mathbb{Q}L$ constructed in § 9.1 using the free nilpotent associative $\mathbb{Q}$-algebra A of class c. First, we prove a technical lemma, where $e^{\mathbb{Q}L_c} = \{e^a \mid a \in \mathbb{Q}L_c = A_c \cap \mathbb{Q}L\}$, which obviously equals $1 + \mathbb{Q}L_c = \{1 + a \mid a \in \mathbb{Q}L_c\}$.

Lemma 10.1. *We have* $\sqrt{\gamma_c(F)} = e^{\mathbb{Q}L_c} = 1 + \mathbb{Q}L_c = \gamma_c(e^{\mathbb{Q}L}) = \gamma_c(\sqrt{F})$.

Proof. By Lemma 9.1(b), we have $\gamma_c(F) = 1 + L_c - e^{L_c}$. Then, obviously, $\sqrt{\gamma_c(F)} = e^{\mathbb{Q}L_c} = 1 + \mathbb{Q}L_c$. The group $\gamma_c(e^{\mathbb{Q}L})$ is generated by commutators of the form $[e^{a_1}, \ldots, e^{a_c}]$, $a_i \in \mathbb{Q}L$. Let ϑ be an endomorphism of $A_0 \oplus A$ that extends the mapping mapping $1 \to 1$, $x_i \to a_i$. Then

$$
\begin{aligned}
[e^{a_1}, \ldots, e^{a_c}] &= [e^{x_1}, \ldots, e^{x_c}]^{\vartheta} \\
&= 1 + [x_1, \ldots, x_c]^{\vartheta} \\
&= 1 + [a_1, \ldots, a_c] \in 1 + \mathbb{Q}L_c,
\end{aligned} \tag{10.2}
$$

whence

$$
\gamma_c(\sqrt{F}) \subseteq \gamma_c(e^{\mathbb{Q}L}) \subseteq 1 + \mathbb{Q}L_c. \tag{10.3}
$$

In the other direction, every element of the group $1 + \mathbb{Q}L_c$ has the form

$$1 + \sum_j \alpha_j [x_{j_1}, \dots, x_{j_c}] = \prod_j e^{\alpha_j [x_{j_1}, \dots, x_{j_c}]}, \quad \alpha_j \in \mathbb{Q}.$$

For every factor we have

$$
\begin{aligned}
e^{\alpha_j [x_{j_1}, \dots, x_{j_c}]} &= 1 + \alpha_j [x_{j_1}, x_{j_2}, \dots, x_{j_c}] \\
&= 1 + [\alpha_j x_{j_1}, x_{j_2}, \dots, x_{j_c}] \\
&= [e^{\alpha_j x_{j_1}}, e^{x_{j_2}}, \dots, e^{x_{j_c}}] \in \gamma_c(\sqrt{F}),
\end{aligned}
$$

where we used (10.2) again. As a result, $\sqrt{\gamma_c(F)} = 1 + \mathbb{Q}L_c \subseteq \gamma_c(\sqrt{F})$. Together with (10.3), this yields the required equalities. □

Now we can easily prove the following basic fact.

Theorem 10.4. $\sqrt{F} = e^{\mathbb{Q}L}$.

Proof. Since $\sqrt{F} \leq e^{\mathbb{Q}L}$, we need only prove the inclusion $e^{\mathbb{Q}L} \subseteq \sqrt{F}$. Induction on the nilpotency class c of A. For $c = 1$ the result follows from Lemma 10.1. For $c > 1$, the images of A, F, $\sqrt{F}$ and $\mathbb{Q}L$ in A/A_c are naturally isomorphic to the corresponding objects constructed for the class $c-1$. By the induction hypothesis, for every $a \in \mathbb{Q}L$, we have $e^a = e^u + b = e^u(1+b) = e^u e^b$, for some $e^u \in \sqrt{F}$ and $b \in A_c$. It remains to show that $e^b \in \sqrt{F}$. Since $e^{\mathbb{Q}L}$ is a group by Corollary 9.14, we have $e^b = e^{-u} e^a \in e^{\mathbb{Q}L}$, whence $b \in \mathbb{Q}L \cap A_c = \mathbb{Q}L_c$. Then $e^b \in e^{\mathbb{Q}L_c} = \sqrt{\gamma_c(F)} \leq \sqrt{F}$ by Lemma 10.1, as required. □

As an important corollary, we derive the inversions of the Baker-Hausdorff Formula. By Theorem 10.4, for any $x, y \in \mathbb{Q}L$, both e^{x+y} and $e^{[x,y]}$ are elements of $\sqrt{F}$. Moreover, there are certain universal formulae expressing e^{x+y} and $e^{[x,y]}$ in terms of the $\mathbb{Q}$-powered group generated by e^x and e^y. Indeed, we have such expressions for the free generators of A:

$$e^{x_1+x_2} = h_1(e^{x_1}, e^{x_2}) \qquad \text{and} \qquad e^{[x_1, x_2]} = h_2(e^{x_1}, e^{x_2}), \qquad (10.5)$$

where $h_1(e^{x_1}, e^{x_2})$ and $h_2(e^{x_1}, e^{x_2})$ are some $\mathbb{Q}$-powered group words in e^{x_1} and e^{x_2} (obtained from e^{x_1} and e^{x_2} by taking products and rational powers). We may indeed assume that only elements e^{x_1} and e^{x_2} are involved, by considering the subalgebra of A generated by x_1 and x_2, which is also a free nilpotent algebra of class c. Since x_1 and x_2 are free generators of A, the same formulae hold for any $x, y \in A$, as images of (10.5) under the homomorphism of A extending the mapping $x_1 \to x$, $x_2 \to y$, $1 \to 1$:

$$e^{x+y} = h_1(e^x, e^y) \qquad \text{and} \qquad e^{[x,y]} = h_2(e^x, e^y), \qquad (10.6)$$

where the $\mathbb{Q}$-powered group words h_j, $j = 1$, 2, are the same as in (10.5). We call (10.6) the *Inverse Baker–Hausdorff Formulae*. By Corollary 9.19, we may assume that $h_1(e^{x_1}, e^{x_2})$ and $h_2(e^{x_1}, e^{x_2})$ are roots of some elements of F; it is, however, more convenient to represent the $h_j(e^{x_1}, e^{x_2})$ as products of roots of group commutators in e^{x_1} and e^{x_2}.

Lemma 10.7. *The Inverse Baker–Hausdorff Formulae (10.6) have the following form:*

(a)
$$e^{x_1 + x_2} = h_1(e^{x_1}, e^{x_2}) = e^{x_1} e^{x_2} \prod_j \varkappa_j^{r_j},$$

where $r_j \in \mathbb{Q}$ and the $\varkappa_j$ are group commutators in e^{x_1} and e^{x_2} of weight ≥ 2 taken in the product in some order agreeing with the increase of the weight;

(b)
$$e^{[x_1, x_2]} = h_2(e^{x_1}, e^{x_2}) = [e^{x_1}, e^{x_2}] \prod_j \varkappa_j^{s_j},$$

where $s_j \in \mathbb{Q}$ and the $\varkappa_j$ are group commutators in e^{x_1} and e^{x_2} of weight ≥ 3 taken in the product in some order agreeing with the increase of the weight.

Proof. (a) Induction on the nilpotency class c of A. If $c = 1$, the assertion is trivial: $e^{x_1 + x_2} = e^{x_1} e^{x_2}$. For $c > 1$, the induction hypothesis gives

$$e^{x_1 + x_2} = e^{x_1} e^{x_2} \prod_j \varkappa_j^{r_j} + a = e^{x_1} e^{x_2} \prod_j \varkappa_j^{r_j} \cdot e^a, \qquad (10.8)$$

where $r_j \in \mathbb{Q}$, the $\varkappa_j$ are commutators in e^{x_1} and e^{x_2} of weight ≥ 2, and $a \in A_c$. Since $e^{x_1 + x_2}$, $e^{x_1} e^{x_2} \prod_j \varkappa_j^{r_j} \in e^{\mathbb{Q}L}$ we have

$$e^a = (e^{x_1} e^{x_2} \prod_j \varkappa_j^{r_j})^{-1} \cdot e^{x_1 + x_2} \in e^{\mathbb{Q}L}$$

by 9.14 and 10.4, whence $a \in \mathbb{Q}L_c$ and $e^a \in e^{\mathbb{Q}L_c}$; hence $e^a \in \sqrt{\gamma_c(\langle e^{x_1}, e^{x_2} \rangle)}$ by Lemma 10.1. Then $e^a = \prod_j \lambda_j(e^{x_1}, e^{x_2})^{\alpha_j}$, where $\alpha_j \in \mathbb{Q}$ and the $\lambda_j(e^{x_1}, e^{x_2})$ are some commutators of weight c in e^{x_1}, e^{x_2}. This is the required form for the last factor on the right of (10.8).

(b) Induction on the nilpotency class c. If $c = 1$, then $e^{[x_1, x_2]} = 1 = [e^{x_1}, e^{x_2}]$. For $c = 2$, we have $e^{[x_1, x_2]} = 1 + [x_1, x_2] = [e^{x_1}, e^{x_2}]$ by Lemma 9.1. Let $c > 2$; by the induction hypothesis, we have

$$e^{[x_1, x_2]} = [e^{x_1}, e^{x_2}] \prod_j \varkappa_j^{s_j} + b = [e^{x_1}, e^{x_2}] \prod_j \varkappa_j^{s_j} \cdot e^b, \qquad (10.9)$$

where $s_j \in \mathbb{Q}$, the $\varkappa_j$ are commutators in e^{x_1} and e^{x_2} of weight ≥ 3, and $b \in A_c$. Again, $e^b \in e^{\mathbb{Q}L_c}$ and hence $e^b = \prod_j \mu_j(e^{x_1}, e^{x_2})^{\beta_j}$, where $\beta_j \in \mathbb{Q}$ and the $\mu_j(e^{x_1}, e^{x_2})$ are some commutators of weight c in e^{x_1}, e^{x_2}. This is the required form for the last factor on the right of (10.9). $\qquad \square$

The expressions of Lemma 10.7 for $h_1(e^{x_1}, e^{x_2})$ and $h_2(e^{x_1}, e^{x_2})$ for the nilpotency class $c-1$ can be taken for the initial factors of the corresponding expressions for nilpotency class c. Hence we may assume that *these formulae are "the same" for all c. From now on, we regard the right-hand sides of Lemma 10.7 as the fixed expressions for the Inverse Baker–Hausdorff Formulae* $h_1(e^x, e^y)$ and $h_2(e^x, e^y)$.

Informally, the Baker–Hausdorff Formula $H(x, y)$ transforms the Lie algebra $\mathbb{Q}L$ into the $\mathbb{Q}$-powered group $\sqrt{F} = e^{\mathbb{Q}L}$, while the inverse formulae $h_1(e^x, e^y)$, $h_2(e^x, e^y)$ transform the group $e^{\mathbb{Q}L}$ back into $\mathbb{Q}L$. We shall use the free objects $\mathbb{Q}L$ and $e^{\mathbb{Q}L}$ to extend this correspondence to arbitrary nilpotent $\mathbb{Q}$-powered groups and nilpotent Lie $\mathbb{Q}$-algebras. This is a very good correspondence indeed, an *equivalence of categories*: every statement in terms of a nilpotent Lie $\mathbb{Q}$-algebra can be translated into a statement in terms of the corresponding $\mathbb{Q}$-powered group, and conversely. However, we shall avoid referring to categories and prove independently certain properties of this correspondence (it is not always quite obvious what is the counterpart of what).

Instead of $\mathbb{Q}L$, we may consider the set $e^{\mathbb{Q}L}$ as a Lie algebra isomorphic to $\mathbb{Q}L$, by keeping $\mathbb{Q}L$ "upstairs". That is, we define the new operations on the set $e^{\mathbb{Q}L}$, addition, Lie bracket and multiplying by scalars, as follows:

$$e^a \mathbin{\hat{+}} e^b = e^{a+b} = h_1(e^a, e^b); \qquad \big[e^a, e^b\big]\hat{} = e^{[a,b]} = h_2(e^a, e^b); \left. \vphantom{\begin{array}{c}a\\b\end{array}} \right\} \quad (10.10)$$
$$r \mathbin{\hat{\cdot}} e^a = e^{ra} = (e^a)^r, \quad r \in \mathbb{Q}.$$

(The hats are used to avoid confusion with the operations in $A_0 \oplus A$ and $\mathbb{Q}L$.) We denote by $\widehat{e^{\mathbb{Q}L}}$ the Lie $\mathbb{Q}$-algebra on the set $e^{\mathbb{Q}L}$ with respect to the operations $\hat{+}, \hat{\cdot}$ and $[\hat{\ },\hat{\ }]$, which is obviously isomorphic to $\mathbb{Q}L$ under the isomorphism $e^a \to a$. These new Lie algebra operations in $\widehat{e^{\mathbb{Q}L}}$ are expressed in terms of the operations of the $\mathbb{Q}$-powered group $e^{\mathbb{Q}L}$ by fixed formulae that are $\mathbb{Q}$-powered group words. Let us regard the Lie $\mathbb{Q}$-algebra laws that hold in $\widehat{e^{\mathbb{Q}L}}$ for the elements e^{x_1}, e^{x_2}, e^{x_3} as equalities of some $\mathbb{Q}$-powered group words in $e^{\mathbb{Q}L}$. For example, the commutative law for addition,

$$e^{x_1} \mathbin{\hat{+}} e^{x_2} = e^{x_2} \mathbin{\hat{+}} e^{x_1},$$

and the Jacobi identity,

$$\big[[e^{x_1}, e^{x_2}]\hat{}, e^{x_3}\big]\hat{} \mathbin{\hat{+}} \big[[e^{x_2}, e^{x_3}]\hat{}, e^{x_1}\big]\hat{} \mathbin{\hat{+}} \big[[e^{x_3}, e^{x_1}]\hat{}, e^{x_2}\big]\hat{} = e^0,$$

become equalities of $\mathbb{Q}$-powered group words in e^{x_1}, e^{x_2}, e^{x_3} on expressing the hat-operations by (10.10). Now we can define the structure of a Lie $\mathbb{Q}$-algebra L_G on any $\mathbb{Q}$-powered nilpotent group G of class c. To wit, we define the new operations of addition, bracket multiplication and multiplying by scalars on the set $L_G = G$ in terms of the $\mathbb{Q}$-powered group operations of G by the same formulae:

$$g_1 + g_2 = h_1(g_1, g_2); \qquad [g_1, g_2] = h_2(g_1, g_2); \qquad rg = g^r, \quad r \in \mathbb{Q}.$$

(Here we can switch back to the usual notation for Lie algebra operations, since there is no danger of confusion with any other operations.) To prove that the laws of a Lie $\mathbb{Q}$-algebra hold for any elements $a, b, c \in L_G$ under these new operations, we apply the homomorphism of the free $\mathbb{Q}$-powered group $e^{\mathbb{Q}L}$ into G extending the mapping $e^{x_1} \to a$, $e^{x_2} \to b$, $e^{x_3} \to c$ to the same laws for $e^{x_1}, e^{x_2}, e^{x_3} \in \widehat{e^{\mathbb{Q}L}}$ written as equalities of $\mathbb{Q}$-powered group words in $e^{\mathbb{Q}L}$. The Lie $\mathbb{Q}$-algebra L_G is nilpotent of class $\leq c$, because the nilpotency law of class c that holds on $\widehat{e^{\mathbb{Q}L}} \cong \mathbb{Q}L$ can also be written as a law of the free nilpotent $\mathbb{Q}$-powered group $e^{\mathbb{Q}L}$ and hence this law holds on G.

Conversely, instead of $e^{\mathbb{Q}L}$, we can define the structure of a $\mathbb{Q}$-powered group $\mathbb{Q}L^*$ on the set $\mathbb{Q}L$ with respect to the new multiplication $a * b = H(a, b)$ (the Baker–Hausdorff Formula) and taking $\mathbb{Q}$-powers $a^{*r} = ra$, $r \in \mathbb{Q}$. Then $\mathbb{Q}L^*$ is a $\mathbb{Q}$-powered group isomorphic to $e^{\mathbb{Q}L}$ under the isomorphism $a \to e^a$. The laws of a $\mathbb{Q}$-powered group (associative law for multiplication $*$, etc.) for $x_1, x_2, x_3 \in \mathbb{Q}L^*$ can be expressed as equalities of some elements of $\mathbb{Q}L$, Lie $\mathbb{Q}$-algebra words in x_1, x_2, x_3. Now we can define the structure of a $\mathbb{Q}$-powered group G_M on any nilpotent Lie $\mathbb{Q}$-algebra M of class c, defining the group operations on the set $G_M = M$ by the same formulae:

$$m_1 \cdot m_2 = H(m_1, m_2); \qquad m^r = rm, \quad r \in \mathbb{Q}.$$

The laws of a $\mathbb{Q}$-powered group hold for any $a, b, c \in G_M$ as images of the same laws for $x_1, x_2, x_3 \in \mathbb{Q}L^*$ written as equalities in $\mathbb{Q}L$, under the homomorphism of the Lie $\mathbb{Q}$-algebra $\mathbb{Q}L$ into M that extends the mapping $x_1 \to a$, $x_2 \to b$, $x_3 \to c$. The group G_M is nilpotent of class $\leq c$ because the nilpotency law of class c that holds on $\mathbb{Q}L^* \cong e^{\mathbb{Q}L}$ can also be written as a law of the free nilpotent Lie $\mathbb{Q}$-algebra $\mathbb{Q}L$ and hence holds on M.

If we construct the $\mathbb{Q}$-powered group $G_{\widehat{e^{\mathbb{Q}L}}}$, as described above, from the Lie $\mathbb{Q}$-algebra $\widehat{e^{\mathbb{Q}L}}$ (with the operations $\hat{+}$, $\hat{\cdot}$ and $\hat{[,]}$), then we arrive at the original group $e^{\mathbb{Q}L}$. Indeed, the sets are the same all the time, and we need only to show that the operations are the same, that is, $(e^u)^r = (e^u)^{*r} = r \hat{\cdot} e^u$ and $e^u e^v = e^u * e^v = \hat{H}(e^u, e^v)$, where $\hat{H}$ means the Baker–Hausdorff Formula applied with respect to the operations $\hat{+}$, $\hat{\cdot}$ and $\hat{[,]}$. But $a \to e^a$ is an isomorphism of Lie $\mathbb{Q}$-algebras $\mathbb{Q}L$ and $\widehat{e^{\mathbb{Q}L}}$: hence $r \hat{\cdot} e^u = e^{ru} = (e^u)^r$ and $\hat{H}(e^u, e^v) = e^{H(u,v)} = e^u e^v$, as required. The fact that $G_{\widehat{e^{\mathbb{Q}L}}} = e^{\mathbb{Q}L}$ can itself be written as equalities of $\mathbb{Q}$-powered group words in the free generators of the free nilpotent $\mathbb{Q}$-powered group $e^{\mathbb{Q}L}$. Hence the same equalities hold for any elements of an arbitrary nilpotent $\mathbb{Q}$-powered group G, which means that $G_{L_G} = G$. Similarly, $L_{G_M} = M$ for an arbitrary nilpotent Lie $\mathbb{Q}$-algebra M.

We summarize the above considerations as follows.

Mal'cev Correspondence 10.11. *For every nilpotent $\mathbb{Q}$-powered group G, the corresponding nilpotent Lie $\mathbb{Q}$-algebra L_G is defined on the same underlying set $L_G = G$, with Lie $\mathbb{Q}$-algebra operations $a + b = h_1(a, b)$, $[a, b] = h_2(a, b)$, $ra = a^r$ for $r \in \mathbb{Q}$. Conversely, for every nilpotent Lie $\mathbb{Q}$-algebra M, the corresponding nilpotent $\mathbb{Q}$-powered group G_M is defined on the same underlying set $G_M = M$, with group operations $a \cdot b = H(a, b)$ and $a^r = ra$ for $r \in \mathbb{Q}$. These transformations are inverses of one another: $L_{G_M} = M$ as Lie $\mathbb{Q}$-algebras (that is, not only sets, but all operations coincide), and, similarly, $G_{L_G} = G$ as $\mathbb{Q}$-powered groups.*

Suppose that $G = G_M$ is a nilpotent $\mathbb{Q}$-powered group and $M = L_G$ is a nilpotent Lie $\mathbb{Q}$-algebra that are in the Mal'cev Correspondence with each other. Since the sets coincide, we have to adjust notation: to avoid confusion, we denote by $[a, b]_M$ commutators in the Lie algebra and $[a, b]_G$ those in the group. Addition is, of course, for the additive group of the Lie algebra, and multiplication is for the group. Note that a rational power a^r in the group, $a \in G$, $r \in \mathbb{Q}$, is the rth multiple in the Lie algebra: $G \ni a^r = ra \in M$. Lemmas 9.15 and 10.7 can be restated as follows.

Lemma 10.12. *Suppose that a nilpotent $\mathbb{Q}$-powered group G and a nilpotent Lie $\mathbb{Q}$-algebra M are in the Mal'cev Correspondence with each other: $G = G_M$ and $M = L_G$. Then, for any elements $a, b \in G = M$,*

 (a) $a + b = h_1(a, b) = ab \prod_j \varkappa_j^{r_j}$, *where $r_j \in \mathbb{Q}$ and the $\varkappa_j$ are group commutators in a and b of weight ≥ 2 taken in the product in some order agreeing with the increase of the weight;*

 (b) $[a, b]_M = h_2(a, b) = [a, b]_G \prod_j \varkappa_j^{s_j}$, *where $s_j \in \mathbb{Q}$ and the $\varkappa_j$ are group commutators in a and b of weight ≥ 3 taken in the product in some order agreeing with the increase of the weight;*

 (c) $ab = H(a, b) = a + b + \sum_j t_j \varkappa_j$, *where $t_j \in \mathbb{Q}$ and the $\varkappa_j$ are Lie ring commutators in a and b of weight ≥ 2;*

 (d) $[a, b]_G = K(a, b) = [a, b]_M + \sum_j u_j \varkappa_j$, *where $u_j \in \mathbb{Q}$ and the $\varkappa_j$ are Lie ring commutators in a and b of weight ≥ 3.*   □

Now we prove some of the nice properties of the Mal'cev Correspondence. Recall that the nilpotency class and the derived length of $\mathbb{Q}$-powered groups are defined via the corresponding laws, rather than series with abelian or central factors. (The equivalent definitions based on such series can be more easily produced using the Mal'cev Correspondence — Exercise 10.1.)

Theorem 10.13. *Suppose that a nilpotent $\mathbb{Q}$-powered group G and a nilpotent Lie $\mathbb{Q}$-algebra M are in the Mal'cev Correspondence with each other: $G = G_M$ and $M = L_G$.*

(a) *A subset $H \subseteq G$ is a $\mathbb{Q}$-powered subgroup if and only if H is a Lie $\mathbb{Q}$-subalgebra of M, and then H as $\mathbb{Q}$-powered group is in the Mal'cev Correspondence with H as a Lie $\mathbb{Q}$-algebra.*

(b) *H is a normal $\mathbb{Q}$-powered subgroup of G if and only if H is an ideal of M.*

(c) *H_1 is a normal $\mathbb{Q}$-powered subgroup of a $\mathbb{Q}$-powered subgroup H_2 of G if and only if H_1 is an ideal of H_2 in M, and then the factor-group H_2/H_1 is abelian if and only if the factor-algebra H_2/H_1 is commutative; the factor-group H_2/H_1 is central in G if and only if the factor-algebra H_2/H_1 is central in M.*

(d) *The nilpotency class of G coincides with the nilpotency class of M.*

(e) *The derived length of G coincides with the derived length of M.*

(f) *A mapping of the set $G = M$ is an endomorphism of the $\mathbb{Q}$-powered group G if and only if it is an endomorphism of the Lie $\mathbb{Q}$-algebra M. In particular, the automorphism groups $\operatorname{Aut} G$ of the $\mathbb{Q}$-powered group G and $\operatorname{Aut} M$ of the Lie $\mathbb{Q}$-algebra M coincide as permutation groups of the set $G = M$.*

(g) *The $\mathbb{Q}$-powered group G is free nilpotent of class c on the set of free generators X if and only if the Lie $\mathbb{Q}$-algebra M is free nilpotent of class c on the set of free generators X.*

Proof. (a) A subset H is closed under the operations of the $\mathbb{Q}$-powered group if and only if H is closed under the operations of the Lie $\mathbb{Q}$-algebra, since the operations in one object are expressed as formulae in terms of the operations in the other. If this is the case, H as a $\mathbb{Q}$-powered group is in the Mal'cev Correspondence with H as a Lie $\mathbb{Q}$-algebra by the definition.

(b) If a $\mathbb{Q}$-powered subgroup H is normal in G, then $[h, g] \in H$ for any $h \in H$, $g \in G$, and all commutators of greater weight involving h and g belong to H together with all their $\mathbb{Q}$-powers. By Lemma 10.12(b), we then have $[h, g]_M \in H$ too, so that H is an ideal. Conversely, if H is an ideal, Lemma 10.12(d) shows that H is normal in G.

(c) Since H_2 as a $\mathbb{Q}$-powered group is in the Mal'cev Correspondence with H_2 as a Lie $\mathbb{Q}$-algebra, (a) and (b) imply that H_1 is a normal $\mathbb{Q}$-powered subgroup of H_2 if and only if H_1 is an ideal of H_2. The factor-group H_2/H_1 is abelian if and only if $[h, h']_G \in H_1$ for any $h, h' \in H_2$. By Lemma 10.12 this is equivalent to $[h, h']_M \in H_1$ for any $h, h' \in H_2$, which, in turn, means that H_2/H_1 is a commutative factor-algebra. The factor-group H_2/H_1 is central in G if and only if $[h, g]_G \in H_1$ for any $h \in H_2$ and $g \in G$. By Lemma 10.12

this is equivalent to $[h, g]_M \in H_1$ for any $h \in H_2$ and $g \in M = G$, which means that H_2/H_1 is a central factor-algebra.

(d) Let c_M and c_G be the nilpotency classes of M and G respectively. Then $c_M \leq c_G \leq c_M$ as noted earlier, whence $c_M = c_G$.

(e) Suppose that M is soluble of derived length k; then the $M^{(i)}$ form a series of ideals of M of length k with commutative factor-algebras. By (c), the same sets form a normal series of G with abelian factor-groups, and hence G is soluble of derived length $\leq k$. Now suppose that G is soluble of derived length d, that is, the law $\delta_d(x_1, \ldots, x_{2^d}) = 1$ holds on G. Then G is soluble of derived length d as an abstract group too: $G^{(d)} = 1$, for the abstract dth derived subgroup. We use induction on s to prove that $M^{(s)} \subseteq \sqrt{G^{(s)}}$ for all s. For $s = 0$ we simply have $M = G$. Suppose that $M^{(k)} \subseteq \sqrt{G^{(k)}}$. For any $a, b \in M^{(k)} \subseteq \sqrt{G^{(k)}}$ there are $m, n \in \mathbb{N}$ such that $a^m, b^n \in G^{(k)}$. Then, by Lemma 10.12(b), $mn[a, b]_M = [ma, nb]_M = [a^m, b^n]_M$ is a product of roots of group commutators in a^m and b^n, all these commutators lying in $[G^{(k)}, G^{(k)}] = G^{(k+1)}$. Since $\sqrt{G^{(k+1)}}$ is a subgroup, we obtain that $[a, b]_M^{mn} = mn[a, b]_M \in \sqrt{G^{(k+1)}}$ and hence $[a, b]_M \in \sqrt{G^{(k+1)}}$. It follows that $M^{(k+1)}$, which is equal to ${}_+\langle [a, b]_M \mid a, b \in M^{(k)} \rangle$, is contained in $\sqrt{G^{(k+1)}}$, which is an ideal of M by (b). This completes the induction step; as a result, we have $M^{(d)} \subseteq \sqrt{G^{(d)}} = \sqrt{1} = 1 = 0_M$, since G is torsion-free. Thus, $d_M \leq d_G$, for the derived lengths of M and G respectively. Together with the reverse inequality $d_G \leq d_M$ proved above, this gives $d_M = d_G$.

(f) If α is a homomorphism of G (of M) into itself, then the same mapping α of the set $G = M$ is a homomorphism of M (respectively, of G), since the operations in M (in G) are expressed as fixed formulae in terms of the operations of G (in M) and hence are preserved under α.

(g) If G is a free nilpotent $\mathbb{Q}$-powered group, we can identify G with $e^{\mathbb{Q}L}$ constructed as above, for suitable nilpotency class c and the cardinality of the set of free generators. Then $M = L_G \cong \mathbb{Q}L$ is a free nilpotent Lie $\mathbb{Q}$-algebra. Similarly, if M is a free nilpotent Lie $\mathbb{Q}$-algebra, then M can be identified with a suitable $\mathbb{Q}L$ constructed as above. Then $G = G_M \cong e^{\mathbb{Q}L}$ is a free nilpotent $\mathbb{Q}$-powered group. A set of free generators of one object is a set of free generators of the other, since the x_i correspond to e^{x_i} under those isomorphisms $\mathbb{Q}L \cong L_{e^{\mathbb{Q}L}}$ and $G_{\mathbb{Q}L} \cong e^{\mathbb{Q}L}$. $\square$

As an application of the newly acquired technique, we prove here the following analogue of Kreknin's Theorem for (locally) nilpotent torsion-free groups with regular automorphisms of finite order.

Corollary 10.14. *If a locally nilpotent torsion-free group* H *admits a regular automorphism* φ *of finite order* n, *then* H *is soluble of derived length at most* $k(n)$, *where* k *is Kreknin's function.*

Proof. Since the law of solubility of given derived length involves only finitely many variables, it is sufficient to prove that every finitely generated subgroup K of H is soluble of derived length at most $k(n)$. Replacing K by the φ-invariant subgroup $\langle K, K^\varphi, \ldots, K^{\varphi^{n-1}} \rangle$, which is also finitely generated, we may consider only finitely generated φ-invariant subgroups. So let us assume that H is finitely generated and hence nilpotent from the outset. In accordance with Theorem 9.20, let $G = \sqrt{H}$ be the $\mathbb{Q}$-powered hull of H, and let $\bar\varphi$ denote the extension of φ to an automorphism of G. Then $\bar\varphi$ is a regular automorphism of G: if $g^{\bar\varphi} = g \neq 1$ for $g \in G$, then $(g^k)^\varphi = g^k \in H$ for some $k \in \mathbb{N}$ and $g^k \neq 1$, since G is torsion-free; a contradiction with the hypothesis $C_H(\varphi) = 1$. Let $M = L_G$ be the Lie $\mathbb{Q}$-algebra in the Mal'cev Correspondence with G. By Theorem 10.13(f), the same bijection $\bar\varphi$ of the set $M = G$ is an automorphism of M. This is a regular automorphism since $C_M(\bar\varphi) = C_G(\bar\varphi) = 1 = 0_M$. Of course, $\bar\varphi$ as an element of $\operatorname{Aut} M$ has the same finite order n. By Corollary 7.27, M is soluble of derived length $\leq k(n)$; hence, by Theorem 10.13(e), G is soluble of the same derived length $\leq k(n)$, and so is H. $\qquad\square$

Remarks. 10.15. It can be proved that the Inverse Baker–Hausdorff Formulae h_1, h_2 are unique, for a fixed ordered sequence of the so-called basic commutators.

10.16. Although every statement about nilpotent Lie $\mathbb{Q}$-algebras can be translated into a statement about nilpotent $\mathbb{Q}$-powered groups, using Lie rings in the proof of Corollary 10.14 seems to be a genuine advantage; it is unlikely that this proof could have been invented in terms of nilpotent $\mathbb{Q}$-powered groups.

§ 10.2. The Lazard Correspondence

Let σ be some set of prime numbers. An integer is said to be a *σ-number* if it is a product of powers of primes from σ. A group H is *σ-divisible* if for every σ-number n, every element $h \in H$ has an nth root of h in H, an element $g \in H$ such that $g^n = h$. A group is *σ-torsion-free* if it has no elements whose orders are σ-numbers.

Example 10.17. Every p-group is p'-divisible, where p' denotes the set of all primes distinct from p. All p'-roots of a p-element g can be found as powers of g, since, for a p'-number n, the mapping $h \to h^n$ is an automorphism of the cyclic subgroup $\langle g \rangle$. Clearly, every p-group is p'-torsion-free.

Keeping an eye on the prime divisors involved, one can see that literally the same arguments prove the following analogue of Lemma 3.16.

Lemma 10.18. *If $x^n = y^n$, for a σ-number n and for elements x, y in a σ-torsion-free nilpotent group, then $x = y$.*   □

For a subgroup H of a group G, we denote by $\sqrt[\sigma]{H}$ the set $\{g \in G \mid g^n \in H$ for some σ-number $n = n(g)\}$. The same care taken of the prime divisors in the proof of Theorem 9.18 gives the following analogous result.

Theorem 10.19. *If H is a subgroup of a nilpotent group G, then $\sqrt[\sigma]{H}$ is a subgroup. If $A \trianglelefteq B$ for some subgroups $A, B \leq G$, then $\sqrt[\sigma]{A} \trianglelefteq \sqrt[\sigma]{B}$.*   □

Let $\mathbb{Q}_\sigma$ denote the ring of all rational numbers whose denominators are σ-numbers. By definition, the σ-divisible σ-torsion-free groups are $\mathbb{Q}_\sigma$-*powered groups*, algebraic systems that are groups with additional unary operations of taking powers in $\mathbb{Q}_\sigma$ satisfying the laws $(x^r)^s = x^{rs}$ for all $r, s \in \mathbb{Q}_\pi$. The Mal'cev Completion Theorem 9.20 can be specialized as follows.

Theorem 10.20. (a) *Every σ-torsion-free nilpotent group G of class c can be embedded as a subgroup in a $\mathbb{Q}_\sigma$-powered nilpotent group $\widehat{G}^\sigma$ of the same class c such that $\widehat{G}^\sigma = \sqrt[\sigma]{G}$.*

(b) *The group $\widehat{G}^\sigma$ is unique up to isomorphism; moreover, every isomorphism $\varphi : G \to G'$ extends to an isomorphism of $\widehat{G}^\sigma$ onto $\widehat{G'}^\sigma$. In particular, every automorphism of G extends to an automorphism of $\widehat{G}^\sigma$.*

(c) *If F is a free nilpotent group of class c on the free generators e^{x_i}, then $\sqrt[\sigma]{F}$ is the free nilpotent $\mathbb{Q}_\sigma$-powered group of class c freely generated by the e^{x_i}.*

Proof. Existence is proved along the same lines as for Theorem 9.20, where σ was the set of all primes. That is, we identify G with F/N, where $F = \langle e^{x_i} \rangle$ is a suitable free nilpotent group from § 9.1. Then it is similarly proved that $\sqrt[\sigma]{N} \cap F = N$ so that we can put $\widehat{G}^\sigma = \sqrt[\sigma]{F}/\sqrt[\sigma]{N}$, where $\sqrt[\sigma]{F} = \{e^{rl} \mid e^l \in F, r \in \mathbb{Q}_\sigma\}$ is a $\mathbb{Q}_\sigma$-powered group by Theorem 10.19. Extending isomorphisms is proved similarly to the proof of Theorem 9.20(b). The proof of (c) is similar to that of Corollary 9.22.   □

There is, however, a significant difference with what gave us the Mal'cev Correspondence in § 10.1. In general, there is no good correspondence between $\sqrt[\sigma]{F}$ and its would-be counterpart in $\mathbb{Q}L$, the Lie $\mathbb{Q}_\sigma$-algebra $\mathbb{Q}_\sigma L$. First of all the set $e^{\mathbb{Q}_\sigma L} = \{e^{rl} \mid l \in L, r \in \mathbb{Q}_\sigma\}$ may not be a group, because, when σ is small and the nilpotency class c is large, the Baker–Hausdorff Formula $H(x, y) = \log(e^x e^y)$ may well have denominators of coefficients at commutators in x, y that are not σ-numbers.

But the desired correspondence does exist in some special cases. Given the nilpotency class c of A, we fix for what follows the set $\pi = \pi(c!)$ of all primes not greater than c. By the explicit form of the Baker–Hausdorff Formula (9.16), the denominators of the coefficients of $H(x,y)$ at commutators in x,y are all π-numbers. In other words, $H(x,y)$ is a Lie polynomial over $\mathbb{Q}_\pi$, with coefficients from $\mathbb{Q}_\pi$ at commutators in x, y.

Corollary 10.21. *The set $e^{\mathbb{Q}_\pi L}$ is a $\mathbb{Q}_\pi$-powered group.*

Proof. Indeed, this set is closed under the multiplication $e^u e^v = e^{H(u,v)}$ and, obviously, under taking inverses. It is also clear that $e^{\mathbb{Q}_\pi L}$ is closed under taking π-roots. $\qquad\square$

Since $e^{x_i} \in e^{\mathbb{Q}_\pi L}$ for all i, it follows that $F \le \sqrt[\pi]{F} \le e^{\mathbb{Q}_\pi L}$. For any $\sigma \supseteq \pi$ the same argument shows that $e^{\mathbb{Q}_\sigma L}$ is a $\mathbb{Q}_\sigma$-powered group and $\sqrt[\sigma]{F} \le e^{\mathbb{Q}_\sigma L}$. Moreover, essentially the same argument as in the proof of Theorem 10.4 yields the equality.

Theorem 10.22. *For any $\sigma \supseteq \pi = \pi(c!)$, we have $\sqrt[\sigma]{F} = e^{\mathbb{Q}_\sigma L}$.*

Proof. The inclusion $\sqrt[\sigma]{F} \subseteq e^{\mathbb{Q}_\sigma L}$ is already noticed above. To prove the inclusion $e^{\mathbb{Q}_\sigma L} \subseteq \sqrt[\sigma]{F}$, we follow the proof of Theorem 10.4. By the induction hypothesis, for any $l \in \mathbb{Q}_\pi L$ we have

$$e^l = e^u + a = e^u(1 + a) = e^u e^a,$$

where $e^u \in \sqrt[\sigma]{F}$ and $a \in A_c$. We need only to ensure that $e^a \in \sqrt[\sigma]{F}$. We have $u \in \mathbb{Q}_\sigma L$, since $\sqrt[\sigma]{F} \le e^{\mathbb{Q}_\sigma L}$, and hence $e^a = e^l e^{-u} \in e^{\mathbb{Q}_\sigma L}$, since $e^{\mathbb{Q}_\sigma L}$ is a group. Thus, $a \in \mathbb{Q}_\sigma L \cap A_c = \mathbb{Q}_\sigma L_c$. Then $a = \sum_j \alpha_j \varkappa_j(x_i)$, where $\alpha_j \in \mathbb{Q}_\sigma$ and the $\varkappa_j(x_i)$ are Lie ring commutators of weight c in the x_i. By Lemma 9.1, we have

$$e^a = 1 + a = 1 + \sum_j \alpha_j \varkappa_j(x_i) = \prod_j \varkappa_j(e^{x_i})^{\alpha_j} \in \sqrt[\sigma]{F},$$

where the $\varkappa_j(e^{x_i})$ are group commutators of weight c in the e^{x_i}, as required. $\qquad\square$

Now for any $x, y \in \mathbb{Q}_\pi L$, both e^{x+y} and $e^{[x,y]}$ are elements of $\sqrt[\pi]{F}$. Again, there are universal formulae expressing e^{x+y} and $e^{[x,y]}$ in terms of the operations of the $\mathbb{Q}_\pi$-powered group generated by e^x and e^y (actually, the same Inverse Baker–Hausdorff Formulae (10.6), as may be seen from the proof of Theorem 10.22). First we write such formulae for the free generators of A:

$$e^{x_1+x_2} = h_1(e^{x_1}, e^{x_2}) \qquad \text{and} \qquad e^{[x_1, x_2]} = h_2(e^{x_1}, e^{x_2}), \qquad (10.23)$$

where $h_1(e^{x_1}, e^{x_2})$ and $h_2(e^{x_1}, e^{x_2})$ are some $\mathbb{Q}_\pi$-powered group words in e^{x_1} and e^{x_2}. Then homomorphisms of A into itself show that the same formulae hold for any $x, y \in A$.

Similarly to what was done in § 10.1, we use the free $\mathbb{Q}_\pi$-powered nilpotent group $\sqrt[\pi]{F}$ and the free nilpotent Lie $\mathbb{Q}_\pi$-algebra $\mathbb{Q}_\pi L$ of class c to obtain the *Lazard Correspondence* between arbitrary nilpotent Lie $\mathbb{Q}_\pi$-algebras of class $\leq c$ and nilpotent $\mathbb{Q}_\pi$-powered groups of class $\leq c$ (recall that $\pi = \pi(c!)$). That is, for every nilpotent $\mathbb{Q}_\pi$-powered group G of class $\leq c$, the nilpotent $\mathbb{Q}_\pi$-algebra L_G is defined on the same underlying set $L_G = G$ with Lie $\mathbb{Q}_\pi$-algebra operations $a + b = h_1(a, b)$, $[a, b] = h_2(a, b)$, $ra = a^r$ for $r \in \mathbb{Q}_\pi$. Conversely, for every nilpotent Lie $\mathbb{Q}_\pi$-algebra M of class $\leq c$ a nilpotent $\mathbb{Q}_\pi$-powered group G_M is defined on the same underlying set $G_M = M$ with the group operations $ab = H(a, b)$ and $a^r = ra$, $r \in \mathbb{Q}_\pi$. These transformations are inverses of one another: $L_{G_M} = M$ as Lie $\mathbb{Q}_\pi$-algebras and $G_{L_G} = G$ as $\mathbb{Q}_\pi$-powered groups. Lemma 10.12 continues to hold, with all coefficients and exponents in $\mathbb{Q}_\pi$. Every statement in terms of the Lie $\mathbb{Q}_\pi$-algebra M can be transformed into a statement in terms of the $\mathbb{Q}_\pi$-powered group G_M, and conversely (an equivalence of categories). An analogue of Theorem 10.13 can be proved in exactly the same way.

The following will be our main application of the Lazard Correspondence to finite p-groups. From now on, let $\pi = \pi((p - 1)!)$ be the set of all primes less than the given prime p.

Example 10.24. Suppose that P is a finite p-group of nilpotency class $\leq p - 1$. As shown in Example 10.17, P is π-divisible and π-torsion-free, that is, P is a nilpotent $\mathbb{Q}_\pi$-group. Conversely, every nilpotent Lie ring of class $\leq p - 1$ whose additive group is a p-group can be regarded as a nilpotent Lie $\mathbb{Q}_\pi$-algebra. Thus, as a specialization of the general construction, we have the Lazard Correspondence between nilpotent p-groups of class $\leq p - 1$ and nilpotent Lie rings of class $\leq p - 1$ whose additive group is a p-group. The situation here is even better, since all abstract (normal) subgroups of P are automatically π-divisible and hence are Lie $\mathbb{Q}_\pi$-subalgebras (ideals) of $M = L_P$. For example, we obtain at once that $[M, M] = [P, P]$ as the smallest normal subgroup of P (ideal of M) such that the factor-group (factor-algebra) is abelian.

The precondition of applying the Lazard Correspondence to finite p-groups, the nilpotency class to be $\leq p - 1$, is, of course, quite a restrictive one. Nevertheless, there are situations where these conditions can be met (see Remark 10.29). We shall apply the Lazard Correspondence in Chapters 13 and 14 to perform fast reductions to Lie rings in the proofs of two of the main results on almost regular p-automorphisms of finite p-groups (which also yields a substantial reduction of the length of the arguments!). Here we give only a few examples of applications to p-groups, beginning with a useful formula that will be needed later.

Lemma 10.25. *Let p be a prime number. For any elements a, b in a finite p-group, we have $(ab)^p \equiv a^p b^p c^p \pmod{\gamma_p(G)}$ for some $c \in [G, G]$, where $G = \langle a, b \rangle$.*

Proof. We may assume that $\gamma_p(G) = 1$; then we need to prove that $b^{-p} a^{-p} (ab)^p = c^p$ for some $c \in [G, G]$. Since the nilpotency class of G is $\leq p - 1$, we can apply the Lazard Correspondence: let $M = L_G$ be the corresponding Lie $\mathbb{Q}_\pi$-algebra. We calculate the product $b^{-p} a^{-p} (ab)^p$ in terms of the Lie ring operations. Since here the Baker–Hausdorff Formula is a polynomial over $\mathbb{Q}_\pi$ (that is, the denominators of all coefficients are coprime to p), we have $(ab)^p = pH(a, b) = pa + pb + pc_1$ for some $c_1 \in [M, M]$; then $a^{-p}(ab)^p = H(-pa, pa + pb + pc_1) = pb + pc_2$ for $c_2 \in [M, M]$; then $b^{-p} a^{-p} (ab)^p = H(-pb, pb + pc_2) = pc$ for some $c \in [M, M]$. Since $[M, M] = [P, P]$ (Example 10.24) and $pc = c^p$, the result follows. $\qquad\square$

Example 10.26. *Suppose that P is a finite p-group of nilpotency class $\leq p - 1$ and $\varphi \in \operatorname{Aut} P$ is such that $P/\Phi(P) = \langle \bar{a}_1 \rangle \times \cdots \times \langle \bar{a}_p \rangle$ with $\bar{a}_i^\varphi = \bar{a}_{i+1}$, where $i + 1$ is taken $\bmod\, p$. Then $C_{P/\Phi(P)}(\varphi) = C_P(\varphi)\Phi(P)/\Phi(P)$.*

Proof. Let the a_i denote some preimages of the $\bar{a}_i$. It is easy to see that the product $\bar{a}_1 \cdots \bar{a}_p$ generates $C_{P/\Phi(P)}(\varphi)$, so we need only show that there is an element of $C_P(\varphi)$ in the coset $a_1 \cdots a_p \Phi(P)$. We turn P into a Lie $\mathbb{Q}_\pi$-algebra M by applying the Lazard Correspondence. Then φ is also an automorphism of M, acting on the same set $M = P$ in the same way. The required fixed point is then $a_1 + a_1^\varphi + \cdots + a_1^{\varphi^{p-1}}$. In terms of the group P, its image in $P/\Phi(P)$ is $a_1 \cdot a_1^\varphi \cdots a_1^{\varphi^{p-1}} \Phi(P)$, which coincides with $a_1 \cdots a_p \Phi(P)$, because in the inversion formula $x + y = xyc_1c_2\cdots$ the c_i are (powers of) commutators in x, y, all lying in $\Phi(P)$. $\qquad\square$

The Lazard Correspondence can sometimes help in constructing certain examples, which can be easier for Lie rings.

Example 10.27. *Let p be an odd prime and let e_1, e_2, e_3 be linearly independent generators of the additive group of the Lie ring ($\mathbb{Z}$-algebra) M with structural constants*

$$[e_1, e_2] = pe_3, \qquad [e_2, e_3] = pe_1, \qquad [e_3, e_1] = pe_2.$$

It is easy to see that $\gamma_s(M) = p^{s-1} M$ and $M^{(d)} = p^{2^d - 1} M$. The factor ring $M/p^{p-1} M$ is nilpotent of class $p - 1$ and its additive group is a p-group. Hence we can apply the Lazard Correspondence to obtain a nilpotent p-group P on the same set. It is easy to see that the linear transformations α_1, α_2 defined by

$$\alpha_1 : \quad e_1 \to -e_1, \quad e_2 \to -e_2, \quad e_3 \to e_3;$$

$$\alpha_2 : \quad e_1 \to e_1, \quad e_2 \to -e_2, \quad e_3 \to -e_3$$

are automorphisms of M and hence induce automorphisms of $M/p^{p-1}M$. The group $A = \langle \alpha_1, \alpha_2 \rangle$ is elementary abelian of order 4, and it is easy to see that $C_{M/p^{p-1}M}(A) = 0$. The same group A is an automorphism group of P, with $C_P(A) = 1$. Note that the derived length of P, which is equal to that of $M/p^{p-1}M$, grows to infinity with the growth of p, that is, it is not bounded by any constant ("depending" only on A). This shows that there is no direct analogue of Kreknin's Theorem for non-cyclic regular groups of automorphisms of nilpotent p-groups. Remark 7.35 produced simple Lie algebras with a regular non-cyclic group of automorphisms of order 4. But this example shows that even nilpotency and finiteness of the group do not help (as they do for regular automorphisms, compare with Remark 7.32).

Example 10.28. The same trick with the following Lie algebra over $\mathbb{F}_p$ shows that the derived length of a finite p-group of maximal class can be arbitrarily large (of course, only with the growth of p, in view of Theorem 8.1, say). The basis is $\{e_1, e_2, \dots, e_p\}$, and the structural constants are

$$[e_i, e_j] = \begin{cases} (i-j)e_{i+j}, & \text{if } i+j \leq p, \\ 0, & \text{if } i+j > p. \end{cases}$$

This example was constructed by B. A. Panfërov [1980]; it answered a question raised by C. R. Leedham-Green and S. McKay [1976].

Remarks. 10.29. The exponential map and the Baker–Hausdorff Formula play an outstanding role in the theory of Lie groups. The Mal'cev Correspondence for discrete groups is also widely applied in the theory of torsion-free (locally) nilpotent groups. But applications of the Baker–Hausdorff Formula in the theory of finite groups are rare. As G. Higman remarked in his address at the International Congress of Mathematicians in Edinburgh [1958], the restrictive preconditions of such applications are *"...too severe to be used..., ...the sort of thing one wants in the conclusion of one's theorem, rather than in the hypothesis"*. This makes it even more interesting to see some more examples of applications of the Baker–Hausdorff Formula to finite groups.

- R. Baer [1938] discovered the special case of the Lazard Correspondence in a classification of nilpotent p-groups of class 2 for odd p.

- The special case of Example 10.26 when the group is nilpotent of class 2 is due to J. G. Thompson [1964b], where this result on p-groups was used to prove an important fact on so-called signalizers; however, the proof was based on a lemma of N. Blackburn [1958] on p-groups of maximal class (which, in turn, can be deduced from the result of Example 10.26 for class 2).

- Then H. Bender [1967] extended J. G. Thompson's signalizer theorem by using R. Baer's construction.

- Similarly to Examples 10.27, 10.28, the Lazard Correspondence was used to construct certain subdirect products (as in Exercise 3.17) of unbounded nilpotency class in [E. I. Khukhro, 1982].

- In the very recent work J. Alperin and G. Glauberman [1997] use the Lazard Correspondence to prove certain generalizations of the Thompson–Glauberman replacement theorem for finite p-groups of class $\leq p-1$.

- We shall use the Lazard Correspondence in Chapters 13 and 14, and the Mal'cev Correspondence will be used in Chapter 12.

10.30. Example 2.10 shows implicitly that there can be no correspondence of Lazard type for class p (for $p = 2$); and [A. V. Borovik and E. I. Khukhro, 1976] and [E. I. Khukhro, 1979] contain similar examples for all p (and even for groups of exponent p and of derived length 2).

10.31. It is worth to mention that the Baker–Hausdorff Formula was used in the works of W. Magnus [1950-53] and I. N. Sanov [1952], where they proved that the associated Lie ring of a group of prime exponent p is a $(p-1)$-Engel Lie algebra of characteristic p (although it was neither the Mal'cev, nor the Lazard Correspondence that was used, but some properties of the Baker–Hausdorff Formula itself).

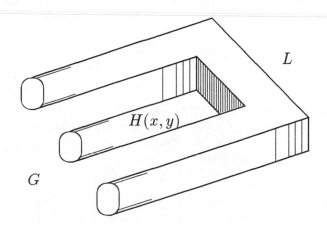

Exercises 10

1. Use the Mal'cev Correspondence to prove that a nilpotent $\mathbb{Q}$-powered group
   (a) has nilpotency class c if and only if it has a central series of $\mathbb{Q}$-powered subgroups of length c;
   (b) has derived length d if and only if it has a (sub)normal series of $\mathbb{Q}$-powered subgroups with abelian factors.

2. Use the Lazard Correspondence to determine all groups of order p^3, where p is an odd prime.

3. Suppose that G is a torsion-free nilpotent group admitting a regular automorphism of order 4. Prove that $\gamma_3(\gamma_2(G)) = 1$. [*Hint:* Use Exercise 7.4.]

4. Let p be an odd prime, P a finite p-group, and $\varphi \in \operatorname{Aut} P$ such that $p \nmid |\varphi|$. Suppose that φ centralizes all elements of order p in P. Prove that $\varphi = 1$. [*Hint:* By induction, φ acts trivially on all φ-invariant subgroups of P; by Exercise 3.22, P is nilpotent of class 2. Apply the Lazard Correspondence and use Exercise 2.5 to obtain the decomposition $L = C_L(\varphi) \oplus U$ for the additive group of the corresponding Lie ring, with φ-invariant U.]

5. [N. Blackburn, 1958] Use Example 10.26 and the fact that every p-group of maximal class contains an element with centralizer of order p^2 to prove that the following group cannot be a proper factor-group of a p-group of maximal class, if $p \neq 2$:

$$G = (\langle \bar{u}_1 \rangle \times \cdots \times \langle \bar{u}_p \rangle) \rtimes \langle \bar{v} \rangle$$

   with $\bar{u}_i^{\bar{v}} = \bar{u}_{i+1}$, where $i+1$ is taken mod p, and $|\bar{u}_i| = |v| = p$ for all i. [*Hint:* If false, $G = P/\langle z \rangle$, where P is of maximal class and $\langle z \rangle = Z(P)$ is of order p. Show that the image of the element v with centralizer of order p^2 in P can be assumed to be $\bar{v}$. Let the u_i denote some preimages of the $\bar{u}_i$. Then the subgroup $\langle u_1, \ldots, u_p \rangle$ has nilpotency class at most 2. Apply Example 10.26 to the inner automorphism φ induced by v to get a fixed point of φ in the coset $u_1 \cdots u_p \langle z \rangle$. Then the centralizer of v has order p^3, since both z and v also belong to $C_P(v)$, a contradiction.]

6. Prove that the unitriangular group $UT_n(\mathbb{Q})$ (see Exercise 2.2) is $\mathbb{Q}$-powered nilpotent. Prove that the Lie $\mathbb{Q}$-algebra that is in the Mal'cev Correspondence with $UT_n(\mathbb{Q})$ is isomorphic to the Lie $\mathbb{Q}$-algebra of null-triangular matrices

$$nt_n(\mathbb{Q}) = \left\{ \begin{pmatrix} 0 & & * \\ & \ddots & \\ 0 & & 0 \end{pmatrix} \,\middle|\, * \in \mathbb{Q} \right\}$$

(where the addition is component-wise and the Lie product of matrices A, B is $[A, B] = AB - BA$). [*Hint:* Compute e^A for $A \in nt_n(\mathbb{Q})$ within the ring of matrices, treating e^A as a matrix polynomial ($A^n = 0$ for $A \in nt_n(\mathbb{Q})$), and show that $A \to e^A$ is an isomorphism of Lie $\mathbb{Q}$-algebras $nt_n(\mathbb{Q})$ and $L_{UT_n(\mathbb{Q})}$.]

7. Check Example 10.28.

8. Let G be a finite p-group of nilpotency class $\leq p - 1$. Prove that $G^p = \{x^p \mid x \in G\}$.

9. Suppose that G is a $\mathbb{Q}$-powered group all of whose 3-generated subgroups are nilpotent. Prove that the Inverse Baker–Hausdorff Formulae can be applied to define a structure of a Lie $\mathbb{Q}$-algebra on the same set G.

Chapter 11
Powerful p-groups

The theory of powerful p-groups was created by A. Lubotzky and A. Mann [1987]; it was also anticipated in an earlier work of M. Lazard [1965]. Powerful p-groups have already found several applications in the theories of finite p-groups, of pro-p-groups, of residually finite groups, of groups with bounded ranks, of groups of given coclass, etc. One can say that the theory of powerful p-groups reflects the properties of the "linear part" of a finite p-group of given rank. Applications to finite p-groups with almost regular p-automorphisms are based on the bounds for the ranks in terms of the number of fixed points and the order of the automorphism (§ 2.2). The exposition in this chapter follows [A. Lubotzky and A. Mann, 1987] and includes some lemmas from [A. Shalev, 1993a] and [J. D. Dixon et al., 1991]. The proofs, however, are here inflated to a more verbose form, to make them accessible for a beginner; some sharper bounds are sacrificed for the same reasons. We shall consider only the case when p is an odd prime; the same results hold for $p = 2$, but the definitions and some proofs are a little different (although not more difficult) and are left as exercises to the reader.

§ 11.1. Definitions and basic properties

Throughout the chapter, p denotes a fixed prime number, which is assumed odd, if not otherwise stated. Recall that $H^p = \langle x^p \mid x \in H \rangle$ is the subgroup generated by the pth powers of all elements of H.

Definition 11.1. A subgroup N of a finite p-group G is *powerfully embedded* in G if $N^p \geq [N, G]$ (for $p = 2$, if $N^4 \geq [N, G]$).

Note that a powerfully embedded subgroup is normal by definition. Since $(N^\varphi)^p = (N^p)^\varphi$ and $[N^\varphi, G^\varphi] = [N, G]^\varphi$ for any homomorphism φ (see 1.7 and (1.14)), the image of a powerfully embedded subgroup is powerfully embedded in the image of the group and, in particular, in any factor-group. We shall freely use this fact in what follows. We begin with two lemmas.

Lemma 11.2. *A normal subgroup K in a finite p-group G is powerfully embedded in G if $[K, G] \leq K^p[K, G, G]$.*

Proof. Taking the commutator subgroup with G, we obtain (using 1.17) that $[K, G, G] \leq [K^p, G][K, G, G, G]$, so that $[K, G] \leq K^p[K, G, G, G]$; and so on. As a result, $[K, G] \leq K^p[K, \underbrace{G, \ldots, G}_{s}]$ for every $s \in \mathbb{N}$, whence $[K, G] \leq K^p$, since G is nilpotent. □

Lemma 11.3. *If M and N are normal subgroups of a finite p-group G such that $[M, N, G, G] = 1$, then $[M^p, N] \leq [M, N]^p$.*

Proof. Applying the standard formula $[ab, c] = [a, c]^b[b, c]$ several times, we have, for any $m \in M$ and $n \in N$,

$$[m^p, n] = [m^{p-1}, n]^m[m, n] = ([m^{p-2}, n]^m[m, n])^m[m, n] = \ldots$$

$$\ldots = (\ldots(((m, n]^m[m, n])^m[m, n])^m \ldots)^m[m, n].$$

Since $a^b = a[a, b]$ and $[m, n, a, b] = 1$ for any $m \in M$, $n \in N$, $a, b \in G$, we obtain

$$[m^p, n] = [m, n]^p[m, n, m]^{p(p-1)/2} \in [M, N]^p$$

(recall that $p \neq 2$). We may assume that $[M, N]^p = 1$; then we need to prove that $[M^p, N] = 1$. We already have $[m^p, n] = 1$ for any $m \in M$, $n \in N$, which means that $m^p \in C_G(N)$. Since $M^p = \langle m^p \mid m \in M \rangle$ and $C_G(N)$ is a subgroup, we have $M^p \leq C_G(N)$, so that $[M^p, N] = 1$, as required. □

Theorem 11.4. *If M and N are powerfully embedded subgroups in a finite p-group G, then*

(a) *$[M, N]$ is powerfully embedded in G;*

(b) *M^p is powerfully embedded in G;*

(c) *MN is powerfully embedded in G.*

Proof. (a) By Lemma 11.2, we may assume $[M, N, G, G] = 1$. By Corollary 3.3, $[M, N, G] \leq [M, G, N][M, [N, G]]$. Since M and N are powerfully embedded, the right-hand side is contained in $[M^p, N][M, N^p]$, which is contained in $[M, N]^p$ by Lemma 11.3.

(b) We need to prove that $[M^p, G] \leq (M^p)^p$. By Lemma 11.2, we may assume that $[M^p, G, G] = 1$. Since $M^p \geq [M, G]$ by the hypothesis, it follows that $[M, G, G, G] = 1$. Then, by Lemma 11.3, $[M^p, G] \leq [M, G]^p$. The subgroup on the right is contained in $(M^p)^p$, since $[M, G] \leq M^p$ by the hypothesis.

(c) We have $[MN, G] \leq [M, G][N, G] \leq M^pN^p \leq (MN)^p$. □

Definition 11.5. A finite p-group is *powerful* if it is powerfully embedded in itself, that is, if $[G, G] \leq G^p$ (for $p = 2$, if $[G, G] \leq G^4$).

Corollary 11.6. *If G is a powerful p-group, then $[G, G]$, G^p, $\Phi(G)$, $G^{(k)}$ and $\gamma_k(G)$ for all $k \in \mathbb{N}$ are powerfully embedded subgroups.* □

We shall need two more lemmas.

Lemma 11.7. *If N is a powerfully embedded subgroup of G, then, for any $h \in G$, the subgroup $H = \langle h \rangle N$ is a powerful p-group and $[H, H] \leq N^p$.*

Proof. We show first that $[H, H] = [H, N]$. Indeed, by Lemma 1.17, $[H, H] = [\langle h \rangle N, H] = [\langle h \rangle, H][N, H]$, since $[\langle h \rangle, H] \trianglelefteq H$, and $[H, \langle h \rangle] = [N\langle h \rangle, \langle h \rangle] = [N, \langle h \rangle][\langle h \rangle, \langle h \rangle] \leq [N, H]$, since $[N, \langle h \rangle] \trianglelefteq N\langle h \rangle$. Now we have $H^p \geq N^p \geq [N, G] \geq [N, H] = [H, H]$. $\qquad \square$

Lemma 11.8. *If a p-group P is nilpotent of class 2 and $p \neq 2$, then $P^p = \{x^p \mid x \in P\}$.*

Proof. By Lemma 6.14, $(ab)^p = a^p b^p [b, a]^{p(p-1)/2}$ for any $a, b \in P$, whence $a^p b^p = (ab)^p [a, b]^{p(p-1)/2} = (ab[a, b]^{(p-1)/2})^p$ (recall that $p > 2$). It follows that every product of pth powers is a pth power as well. $\qquad \square$

Powerful p-groups enjoy many properties of abelian groups, like those in the following three theorems.

Theorem 11.9. *If G is a powerful p-group, then the subgroup G^p coincides with the set $\{x^p \mid x \in G\}$ of pth powers of elements of G.*

Proof. Induction on $|G|$. The factor-group $G/(G^p)^p$ is nilpotent of class 2, since $[G, G, G] \leq [G^p, G] \leq (G^p)^p$ because G^p is powerfully embedded in G by Theorem 11.4. By Lemma 11.8, any product of pth powers in $G/(G^p)^p$ is again a pth power. So, for every $a \in G^p$, there is $b \in G$ such that $a \in b^p(G^p)^p$. Put $H = \langle b, G^p \rangle$; then H is powerful by Lemma 11.7, and $a \in H^p$. If $H \neq G$, then $a \in \{x^p \mid x \in H\}$ by the induction hypothesis. If, however, $H = G$, then $G = \langle b, G^p \rangle = \langle b \rangle$ is cyclic by Theorem 4.7 since $G^p \leq \Phi(G)$, and the theorem obviously holds for cyclic groups. $\qquad \square$

Theorem 11.10. *If G is a powerful p-group, then*

(a) *for every $k \in \mathbb{N}$ the subgroup G^{p^k} coincides with the set $\{x^{p^k} \mid x \in G\}$ of p^kth powers of elements of G; in particular, $(G^{p^i})^{p^j} = G^{p^{i+j}}$ for all $i, j \in \mathbb{N}$;*

(b) *G^{p^k} is powerfully embedded in G for all $k \in \mathbb{N}$;*

(c) *the G^{p^i} form a central series of G; if p^e is the exponent of G, then G is nilpotent of class $\leq e$.*

Proof. We prove (a) and (b) simultaneously by induction on k. For $k = 1$, Theorem 11.9 implies (a) while Theorem 11.4(b) implies (b). For $k > 1$, we have $G^{p^{k-1}} = \{x^{p^{k-1}} \mid x \in G\}$ and $G^{p^{k-1}}$ is powerfully embedded by the induction hypothesis. Then, by Theorem 11.9,

$$G^{p^k} = \left\langle x^{p^k} \mid x \in G \right\rangle = \left\langle y^p \mid y \in G^{p^{k-1}} \right\rangle = \{y^p \mid y \in G^{p^{k-1}}\} = \{x^{p^k} \mid x \in G\},$$

and $G^{p^k} = (G^{p^{k-1}})^p$ is powerfully embedded by Theorem 11.4(b).

(c) We have $[G^{p^i}, G] \le (G^{p^i})^p = G^{p^{i+1}}$ by (a) and (b). $\qquad\square$

Theorem 11.11. *Suppose that a powerful p-group* $G = \langle a_1, \dots, a_k \rangle$ *is generated by elements* $a_1, \dots, a_k$. *Then* $G^p = \langle a_1^p, \dots, a_k^p \rangle$.

Proof. Since $G^{p^2} \le \Phi(G^p)$, we may assume that $G^{p^2} = 1$, by Theorem 4.7. By Theorem 11.10, it follows that G is nilpotent of class 2 and $[G, G]^p \le (G^p)^p = G^{p^2} = 1$. By Lemma 6.14, we then have $(xy)^p = x^p y^p [y, x]^{p(p-1)/2} = x^p y^p$ for all $x, y \in G$. The subgroup G^p is generated by the pth powers of products of the elements a_i, for which

$$(a_{i_1} \cdots a_{i_{k-1}} a_{i_k})^p = (a_{i_1} \cdots a_{i_{k-1}})^p a_{i_k}^p = \dots = a_{i_1}^p \cdots a_{i_{k-1}}^p a_{i_k}^p.$$

Hence G^p is generated by the a_i^p. $\qquad\square$

In subsequent chapters, we shall many times use the following important lemma, due to Shalev [1993a] (the inclusion $[M^{p^i}, N^{p^j}] \subseteq [M, N]^{p^{i+j}}$ was known before).

Interchanging Lemma 11.12. *If* M *and* N *are powerfully embedded subgroups in a finite p-group* P, *then* $[M^{p^i}, N^{p^j}] = [M, N]^{p^{i+j}}$ *for all* $i, j \in \mathbb{N}$.

Proof. By Theorem 11.10, it is sufficient to prove that $[M^p, N] = [M, N]^p$.

Our first objective is to show that $[m^p, n] \equiv [m, n]^p \pmod{[M, N]^{p^2}}$ for every $m \in M$, $n \in N$. Let $K = \langle m, [m, n] \rangle$. By Lemma 10.25, we have

$$(m[m, n])^p \equiv m^p [m, n]^p \pmod{[K, K]^p \gamma_p(K)}.$$

Hence

$$[m, n]^p \equiv m^{-p}(m[m, n])^p = m^{-p}(m^n)^p = m^{-p}(m^p)^n$$

$$= m^{-p} m^p [m^p, n] = [m^p, n] \pmod{[K, K]^p \gamma_p(K)}.$$

Now we show that $[K, K]^p \gamma_p(K) \le [M, N]^{p^2}$. By Lemma 3.6(c), $\gamma_p(K) = \gamma_p(\langle m, [m, n] \rangle)$ is generated by simple commutators of weight $\ge p$ in $m^{\pm 1}$ and $[m, n]^{\pm 1}$. If non-trivial, such a commutator must involve an entry of $[m, n]^{\pm 1}$ in the first or second position. Hence

$$\gamma_p(K) \le [[M, N], \underbrace{M, \dots, M}_{p-1}] \le [M, N, G, G] \le [[M, N]^p, G] \le [M, N]^{p^2},$$

since the ambient subgroups are powerfully embedded. For the same reasons, $[K, K] \le [M, N, M]$, and hence $[K, K]^p \le [M, N, M]^p \le [M, N]^{p^2}$.

Thus, we have $[m^p, n] \equiv [m, n]^p \pmod{[M, N]^{p^2}}$ for any $m \in M$, $n \in N$. The subgroup generated by the elements $[m^p, n]$, $m \in M$, $n \in N$, equals $[M^p, N]$, since $M^p = \{m^p \mid m \in M\}$ by Theorem 11.9. The subgroup generated by the elements $[m, n]^p$, $m \in M$, $n \in N$, equals $[M, N]^p$, since $[M, N]$ is generated by the $[m, n]$ and hence $[M, N]^p$ is generated by the $[m, n]^p$ by Theorem 11.11. The generating sets are the same mod $[M, N]^{p^2}$; hence $[M^p, N][M, N]^{p^2} = [M, N]^p[M, N]^{p^2} = [M, N]^p$. Since $[M, N]^{p^2} \leq \Phi([M, N]^p)$, the required equality $[M^p, N] = [M, N]^p$ follows by Theorem 4.7. □

Interchanging Corollary 11.13. *If G is a powerful p-group, then* $\gamma_i(G^{p^s}) = \gamma_i(G)^{p^{is}}$ *and* $(G^{p^s})^{(i)} = (G^{(i)})^{p^{s2^i}}$ *for all $i, s \in \mathbb{N}$.* □

Now we show that taking pth powers induces homomorphisms of abelian sections in powerful p-groups, similarly to abelian groups.

Lemma 11.14. *Suppose that $B \leq A$ are two powerfully embedded subgroups of a finite p-group P such that A/B is abelian. Then the mapping $x \to x^{p^m}$ induces a homomorphism of A/B onto A^{p^m}/B^{p^m}, for any $m \in \mathbb{N}$.*

Proof. First, we consider the case of $m = 1$. For $x \in A$, the mapping of cosets $\vartheta : xB \to x^p B^p$ is well-defined. Indeed, for any $b \in B$, we have $(xb)^p = x^p b^p z$, where $z \in [\langle x, B \rangle, \langle x, B \rangle] \leq B^p$ by Lemma 11.7. Since $A^p = \{x^p \mid x \in A\}$ by Theorem 11.9, we have $(A/B)^\vartheta = A^p/B^p$. Now we show that ϑ is a homomorphism. For any $x, y \in A$, by Lemma 6.14, we have $(xy)^p = x^p y^p [y, x]^{p(p-1)/2} u$, where $u \in [[A, A], A] \leq [B, A] \leq B^p$ and $[y, x]^{p(p-1)/2} \in [A, A]^p \leq B^p$, since A/B is abelian and B is powerfully embedded. Hence

$$(xB \cdot yB)^\vartheta = (xyB)^\vartheta = (xy)^p B^p = x^p y^p B^p = (xB)^\vartheta \cdot (yB)^\vartheta,$$

as required.

For arbitrary m, the mapping $xB \to x^{p^m} B^{p^m}$ of A/B onto A^{p^m}/B^{p^m} is the composition of the homomorphisms $x^{p^i} B^{p^i} \to x^{p^{i+1}} B^{p^{i+1}}$ of A^{p^i}/B^{p^i} onto $A^{p^{i+1}}/B^{p^{i+1}}$, $i = 0, \ldots, m-1$. □

As a special case, we obtain the following theorem.

Theorem 11.15. *Let G be a powerful p-group. Then*

(a) *the mapping $x \to x^p$ induces a homomorphism of the factor-group $G^{p^i}/G^{p^{i+1}}$ onto $G^{p^{i+1}}/G^{p^{i+2}}$, for every $i \in \mathbb{N}$;*

(b) $|G^{p^i}/G^{p^{i+1}}| \geq |G^{p^{i+1}}/G^{p^{i+2}}|$ *for all $i \in \mathbb{N}$.*

Proof. By Theorem 11.10, all of the G^{p^i} are powerfully embedded subgroups, and all factor-groups $G^{p^i}/G^{p^{i+1}}$ are abelian. Hence the assertion (a) is a special case of Lemma 11.14. Then (b) follows immediately from (a). □

If the equalities hold in 11.15(b), the powerful p-group enjoys properties even more linear.

Definition 11.16. Suppose that p^e is the (minimal) exponent of a powerful p-group G. If $|G^{p^i}/G^{p^{i+1}}| = |G^{p^{i+1}}/G^{p^{i+2}}|$ for all $i \leq e - 2$, then G is *uniformly powerful*.

The following property of such groups makes them similar to abelian homocyclic groups (see § 1.1).

Theorem 11.17. *Suppose that G is a uniformly powerful p-group of (minimal) exponent p^e. Then*

(a) *the mapping $x \to x^p$ induces an isomorphism of the factor-group $G^{p^i}/G^{p^{i+1}}$ onto $G^{p^{i+1}}/G^{p^{i+2}}$, for every $i \leq e - 2$;*

(b) **(Cancellation Property)** *if $x^{p^i} \in G^{p^j}$, for $0 \leq i \leq j \leq e$, then $x \in G^{p^{j-i}}$.*

Proof. Part (a) follows from Theorem 11.15(a) and Definition 11.16. To prove (b), choose s such that $x \in G^{p^s} \setminus G^{p^{s+1}}$. Then, by (a), $x^p \in G^{p^{s+1}} \setminus G^{p^{s+2}}$, and so on: $x^{p^i} \in G^{p^{s+i}} \setminus G^{p^{s+i+1}}$ as long as $s + i < e$. If $s + i < e$, we have both $x^{p^i} \in G^{p^j}$ and $x \notin G^{p^{s+i+1}}$, whence $s + i + 1 > j \Rightarrow s \geq j - i$; then $x \in G^{p^s} \leq G^{p^{j-i}}$, as required. If $s + i \geq e$, then $s + i \geq j$, whence again $s \geq j - i$ and $x \in G^{p^s} \leq G^{p^{j-i}}$, as required. □

§ 11.2. Finite p-groups of bounded rank

A finite p-group is said to have *sectional rank* at most r if each of its abelian sections has rank at most r; the sectional rank is the least such integer. (It is often said that a group has rank r if it has rank $\leq r$.) It is clear that for finite abelian groups, the sectional rank coincides with the rank (as defined in § 1.1). By the Burnside Basis Theorem 4.8, a finite p-group has sectional rank r if and only if all of its subgroups can be generated by r elements. The following theorem shows that powerful p-groups behave like abelian groups with respect to the rank.

Theorem 11.18. *Suppose that G is a powerful p-group generated by d elements. Then every subgroup of G can be generated by d elements (that is, G has sectional rank at most d).*

Proof. By Theorem 11.15, each section $E_i = G^{p^i}/G^{p^{i+1}}$ is an elementary abelian p-group of rank $\leq d$. The E_i can be regarded as vector spaces of dimension $\leq d$ over $\mathbb{F}_p$. For an arbitrary subgroup $H \leq G$, we construct the generators of H inductively as elements in $(H \cap G^{p^i}) \setminus G^{p^{i+1}}$. Let V_i denote the factor-group $(H \cap G^{p^i})G^{p^{i+1}}/G^{p^{i+1}}$ regarded as a vector subspace of E_i.

We choose $h_1, \ldots, h_{n_1}$ in H such that their images in G/G^p form the basis of $V_0 = HG^p/G^p$. Note that $\dim(E_0/V_0) = d - n_1$.

Suppose that we have constructed elements $h_1, h_2, \ldots, h_{n_1} \in H \cap G^{p^0}$, $h_{n_1+1}, \ldots, h_{n_2} \in H \cap G^{p^1}, \ldots, h_{n_{k-1}+1}, \ldots, h_{n_k} \in H \cap G^{p^{k-1}}$ such that the images of $h_1^{p^{k-1}}, \ldots, h_{n_1}^{p^{k-1}}, h_{n_1+1}^{p^{k-2}}, \ldots, h_{n_{k-1}+1}, \ldots, h_{n_k}$ in E_{k-1} span V_{k-1}, and $\dim(E_{k-1}/V_{k-1}) \leq d - n_k$. Let W be the subspace of E_k spanned by the images of the $h_1^{p^k}, \ldots, h_{n_1}^{p^k}, h_{n_1+1}^{p^{k-1}}, \ldots, h_{n_k}^p$ in E_k. Then $\dim(E_k/W) \leq d - n_k$, because taking pth powers induces a homomorphism of E_{k-1}/V_{k-1} onto E_k/W. Choose elements $h_{n_k+1}, \ldots, h_{n_{k+1}} \in H \cap G^{p^k}$ (possibly, an empty set) such that their images form a basis of V_k/W. Then $\dim(E_k/V_k) = \dim(E_k/W) - (n_{k+1} - n_k) \leq d - n_{k+1}$. The definition is complete.

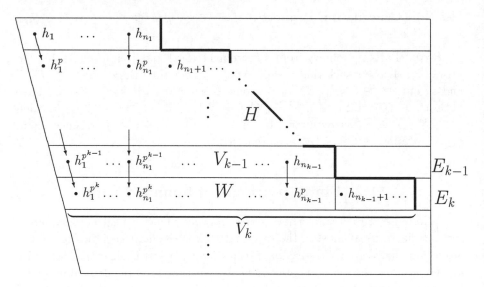

The above construction terminates on reaching the identity subgroup $G^{p^e} = 1$ where p^e is the exponent of G. The last inequality for dimension, $d - n_e \geq \dim(E_{e-1}/V_{e-1}) \geq 0$, implies that $n_e \leq d$, that is, the total number of the elements h_j constructed is at most d. We prove by induction on the exponent of G that the h_j taken together generate H. For $e = 1$ the assertion is obvious. For $e > 1$, we have $H \leq \langle h_1, \ldots, h_{n_e} \rangle G^{p^{e-1}}$ by the induction hypothesis. Since $\langle h_1, \ldots, h_{n_e} \rangle \leq H$, it remains to express an arbitrary element of $H \cap G^{p^{e-1}} = (H \cap G^{p^{e-1}})G^{p^e}/G^{p^e} = V_{e-1}$ in the h_j, which is clearly possible by construction. $\qquad\square$

The following theorem shows that powerful p-groups appear naturally in the theory of groups of given rank; it is sufficient to have only a bound for the ranks of characteristic sections.

Theorem 11.19. *If all characteristic subgroups of a finite p-group G can be generated by r elements, then G contains a characteristic subgroup of (p, r)-bounded index which is a powerful p-group of rank r.*

Proof. Suppose that V is a normal section of G of order p^d. Then G acts on V by conjugation: let $\varphi : G \to \mathbb{S}_V \cong \mathbb{S}_{p^d}$ be the corresponding homomorphism into the symmetric group on p^d symbols. There are upper bounds for the nilpotency class and for the exponent of all p-subgroups of $\mathbb{S}_{p^d}$, some numbers $c = c(d, p)$ and p^e, $e = e(d, p)$, depending only on p and d. Then $\gamma_{c+1}(G)G^{p^e} \leq \operatorname{Ker}\varphi$. Note that $\gamma_{c+1}(G)G^{p^e}$ is a characteristic subgroup of G depending only on d and p. The order of $G/\gamma_{c+1}(G)G^{p^e}$ is bounded in terms of c, p and r by Lemma 6.12(c), so that the index of $\gamma_{c+1}(G)G^{p^e}$ is (p, d, r)-bounded.

We claim that for $d = 2r$ the subgroup $H = \gamma_{c+1}(G)G^{p^e}$ constructed as above is the required powerful subgroup. In proving that H is powerful, we may assume $H^p = 1$; then we need to show that $[H, H] = 1$, that is, $H = Z(H)$. If $H > Z(H)$, then also $\zeta_2(H) > Z(H)$ (see 3.15). The subgroup $\zeta_2(H)$ is characteristic in G and has exponent p (since H has). Hence $|\zeta_2(H)| \leq p^{2r}$, since the ranks of both $Z(H)$ and $\zeta_2(H)/Z(H)$ are at most r by the hypothesis. Put $V = \zeta_2(H)$ in the above construction; then $H \leq \operatorname{Ker}\varphi$, which means that $\zeta_2(H) \leq Z(H)$, a contradiction. Thus, H is a characteristic powerful subgroup. Finally, H is generated by r elements by the hypothesis and hence has rank r by Theorem 11.18. $\qquad\square$

Corollary 11.20. *Suppose that a finite p-group P admits an automorphism φ of order p^n having exactly p^m fixed points. Then P has a characteristic powerful subgroup of (p, n, m)-bounded index.*

Proof. If H is any characteristic subgroup of P, then φ induces on $H/\Phi(H)$ an automorphism of order dividing p^n which has at most p^m fixed points by Lemma 2.12. By Corollary 2.7, the rank of $H/\Phi(H)$ is at most mp^n. The result follows by Theorem 11.19. $\qquad\square$

Corollary 11.21. *If a finite p-group P has sectional rank r and exponent p^e, then the order of P is (r, p, e)-bounded.*

Proof. By Theorem 11.19, we may assume that P is powerful. Then the order of P is the product of the orders of the e elementary abelian factor-groups $P^{p^i}/P^{p^{i+1}}$, $i = 0, \ldots, e - 1$, each of order at most p^r. $\qquad\square$

Remarks. 11.22. If all normal subgroups of a finite p-group P, for odd p, can be generated by r elements, then P has a powerful subgroup of index $\leq p^{r(\log_2 r + 1)}$ (Exercise 11.7). The bound for the index in Theorem 11.19 can be improved to be of the form $p^{f(r)}$.

11.23. There is another way of constructing a Lie ring from a uniformly powerful p-group, which is better in some respects than the associated Lie ring, though it works only for this special class of p-groups. To wit, given the exponent p^e of a uniformly powerful p-group P, we choose $s = [e/4]$ and consider the factor-group $L = P^{p^s}/P^{p^{2s}}$ which is an abelian homocyclic group of exponent p^s. We are going to define Lie ring multiplication on the additive group of L. By Lemma 11.14, the mapping $x \to x^{p^s}$ induces an isomorphism ϑ of L onto $P^{p^{2s}}/P^{p^{3s}}$. For every $x, y \in L$, we choose some preimages $\hat{x}, \hat{y} \in P^{p^s}$ and define the Lie product $[x, y]$ as $([\hat{x}, \hat{y}]P^{p^{3s}})^{\vartheta^{-1}} \in L$, where $[\hat{x}, \hat{y}]P^{p^{3s}}$ is the image of the group commutator $[\hat{x}, \hat{y}]$ in $P^{p^{2s}}/P^{p^{3s}}$. The Jacobi identity for this bracket multiplication can be verified by applying the isomorphism of $P^{p^s}/P^{p^{2s}}$ and $P^{p^{3s}}/P^{p^{4s}}$ induced by $x \to x^{p^{2s}}$ to the Hall–Witt Identity 3.1 written for the elements of P^{p^s} as an equality of some elements in $P^{p^{3s}}$. Similarly, the commutator formulae 1.11 and the isomorphism of $P^{p^s}/P^{p^{2s}}$ and $P^{p^{2s}}/P^{p^{3s}}$ induced by $x \to x^{p^s}$ establish the other Lie ring laws on L. Many important parameters of L as a Lie ring are close to those of the group P. In particular, if d is the derived length of L, then the derived length of G is $\leq d + 3$. This construction stems from the theory of pro-p-groups, where it has important applications (see [J. D. Dixon et al., 1991]); it was also used in the works of A. Shalev [1993a] and Yu. Medvedev [1994a] on almost regular p-automorphisms of finite p-groups.

11.24. T. Weigel [1994] uses the Baker–Hausdorff Formula to construct certain Lie rings from uniformly powerful p-groups for $p > 3$.

Exercises 11

1. Extract an explicit upper bound for the index in Theorem 11.19.

2. Suppose that P is a uniformly powerful p-group P of exponent p^e. Prove that the mapping $x \to x^{p^s}$ induces an injective mapping of P/P^{p^u} onto $P^{p^s}/P^{p^{u+s}}$ for any u and s satisfying $u + s \leq e$ (although it may not be a homomorphism).

3. Suppose that H is a powerful subgroup and N is a powerfully embedded subgroup of a finite p-group G. Prove that NH is powerful.

4. Prove that every finite p-group has a unique maximal powerfully embedded subgroup. [*Hint:* Use Theorem 11.4(c).]

5. A finite p-group P is *regular*, if $(ab)^p = a^p b^p c^p$ for some $c \in [\langle a, b \rangle, \langle a, b \rangle]$, for any $a, b \in P$. Prove that then P^p is a powerful p-group, for odd p.

6. Suppose that N is a normal subgroup of a finite p-group P. Prove that if $[P, N]$ is powerfully embedded in N, then $[N, P]^p = [N^p, P]$.

7. Prove that if all normal subgroups of a finite p-group P, for odd p, can be generated by r elements, then P has a characteristic powerful subgroup of index $\leq p^{r(\log_2 r + 1)}$. [*Hints:* Let H be the intersection of the kernels of all homomorphisms of P into $UT_r(\mathbb{F}_p)$ (see Exercise 4.1). Show that H acts trivially by conjugation on every normal elementary abelian section of P. Prove that $UT_r(\mathbb{F}_p)$ has a normal series of length $\leq \log_2 r + 1$ with elementary abelian factors; use Remak's Theorem (Exercise 1.9) to show that P/H has the same kind of series, derive that $|G : H| \leq p^{r(\log_2 r + 1)}$. Prove that H is powerful: assume $H^p = 1$, and if $H \neq Z(H)$ choose $N \lhd G$ such that $Z(H) < N \leq H$ with $|N : Z(H)| = p$; then N is elementary abelian by Exercise 1.14, whence $[H, N] = 1$, a contradiction.]

8. Prove all results of the chapter for $p = 2$ (with the adjustments for $p = 2$ in Definitions 11.1 and 11.5 and with a few possible adjustments in the statements).

9. Prove that a finite p-group P, for odd p, is powerful if and only if P is a product of r cyclic subgroups, where $|P/\Phi(P)| = p^r$.

10. Prove that a 2-generated finite p-group P, for p odd, is powerful if and only if P has a cyclic normal subgroup with cyclic factor-group.

11. By Corollary 11.20, in the proof of Theorem 8.1 the group P can be assumed powerful from the outset. Examine the proof in order to simplify the argument and/or improve the bound for the index of the subgroup of class $h(p)$.

12. Check that the definition of the Lie ring L in Remark 11.23 is correct. Prove that if the derived length of L is d, then the derived length of G is $\leq d + 3$. Prove that if c is the nilpotency class of L, then the nilpotency class of G is $\leq 4c + 3$.

13. [Yu. Medvedev, 1994a] Use Kreknin's Theorem 7.19(a) and the Lie ring construction of Remark 11.23 (and 12) to perform a reduction of the following conjecture to an analogous conjecture on Lie rings whose additive group is a p-group. Conjecture: if a finite p-group P admits an automorphism of order p^n with exactly p^m fixed points, then P has a subgroup of (p, m, n)-bounded index which is soluble of m-bounded derived length.

Chapter 12

Almost regular automorphism of order p^n: almost solubility of p^n-bounded derived length

The second of the main results on almost regular p-automorphisms of finite p-groups is a match to Kreknin's Theorem on regular automorphisms of Lie rings. If a finite p-group P admits an automorphism φ of order p^n with exactly p^m fixed points, then P contains a subgroup of (p, m, n)-bounded index which is soluble of (p, n)-bounded derived length (that is, of derived length bounded in terms of the order of the automorphism only). Kreknin's Theorem is used twice in the proof. First it is applied to the associated Lie ring $L(P)$, in the case where P is uniformly powerful, to prove that P is an extension of a group of (p, m, n)-bounded nilpotency class by a group of (p, n)-bounded derived length (this already gives a "weak" bound, in terms of p, m and n, for the derived length of P in the general case). Then free nilpotent $\mathbb{Q}$-powered groups and the Mal'cev Correspondence are used to derive a consequence of Kreknin's Theorem, with a kind of a "weak" conclusion that depends on the nilpotency class. Rather miraculously, a combination of two "weak" results yields the desired "strong" bound, in terms of p^n only, for the derived length of a subgroup of (p, m, n)-bounded index.

By Lemma 2.12 the number of fixed points of φ in all φ-invariant sections of P is at most p^m; by Corollary 2.7 all these sections have rank at most mp^n. This is why powerful p-groups appear naturally in the proofs.

Throughout the chapter we shall freely use the facts that if M is a characteristic subgroup of N which is a characteristic (normal) subgroup of G, then M is a characteristic (normal) subgroup of G, and that if U and V are characteristic (normal) subgroups, then $[U, V]$ and $U^n = \langle u^n \mid u \in U \rangle$ are also characteristic (normal) subgroups (see § 1.1 and § 1.3). The automorphisms of φ-invariant sections induced by φ will be denoted by the same letter. We shall also use without reference Theorem 11.4 stating that taking pth powers and commutator subgroups produces powerfully embedded subgroups from powerfully embedded subgroups, and Theorem 11.10 stating that H^{p^j} coincides with the set of p^jth powers of elements of a powerful p-group H.

§ 12.1. Uniformly powerful case

Here Kreknin's Theorem in its combinatorial form is applied to the associated Lie ring of a certain subgroup to prove that the group is an extension of a group of (p, m, n)-bounded nilpotency class by a group of (p, n)-bounded derived length. Remarks at the end of the section explain how this gives a "weak" bound, in terms of p, m and n, for the derived length in the general case. Recall that $k(s)$ is Kreknin's function from § 7.1.

Theorem 12.1. *Suppose that a uniformly powerful p-group P admits an automorphism φ of order p^n with exactly p^m fixed points. Then the $k(p^n)$th derived subgroup $P^{(k(p^n))}$ is nilpotent of (p, m, n)-bounded class.*

Proof. We fix the notation $k = k(p^n)$ for the value of Kreknin's function. Fixing some $s \in \mathbb{N}$, we consider the φ-invariant subgroup P^{p^s}, which is also a uniformly powerful p-group by Theorem 11.15. We denote by the same letter φ the induced automorphism of the associated Lie ring $L = L(P^{p^s})$. An application of Kreknin's Theorem in the combinatorial form 7.19(a) gives $(p^n L)^{(k)} \subseteq {}_{\text{id}}\langle C_L(\varphi)\rangle$. By Lemma 2.12 and by Lagrange's Theorem, we have $p^m C_L(\varphi) = 0$, and hence $p^m {}_{\text{id}}\langle C_L(\varphi)\rangle = {}_{\text{id}}\langle p^m C_L(\varphi)\rangle = 0$ (just as in the proof of Theorem 8.1). It follows that

$$p^{m+n2^k} L^{(k)} = p^m (p^n L)^{(k)} \subseteq p^m {}_{\text{id}}\langle C_L(\varphi)\rangle = 0.$$

In the language of the group P^{p^s} this implies that

$$\left((P^{p^s})^{(k)}\right)^{p^{m+n2^k}} \leq \gamma_{2^k+1}(P^{p^s}).$$

Indeed, $(P^{p^s})^{(k)} \leq \gamma_{2^k}(P^{p^s})$ and the image of $(P^{p^s})^{(k)}$ in $\gamma_{2^k}(P^{p^s})/\gamma_{2^k+1}(P^{p^s})$ is equal to $L^{(k)} \cap \gamma_{2^k}(P^{p^s})/\gamma_{2^k+1}(P^{p^s})$ by Lemma 6.7(a). By the Interchanging Corollary 11.13 for powerfully embedded subgroups, the above inclusion can be rewritten as

$$(P^{(k)})^{p^{s2^k+m+n2^k}} \leq \gamma_{2^k+1}(P)^{p^{s2^k+s}} \leq P^{p^{s2^k+s}}. \qquad (12.2)$$

The idea is to use the extra summand s in the exponent on the right. If p^e is the exponent of P, we shall choose s so that the ratio e/s is (p, m, n)-bounded. Using the Cancellation Property 11.17(b), we shall "cancel" the summand $s2^k$ in the exponents: then $P^{(k)}$ will be "almost" contained in P^{p^s}. By the Interchanging Corollary 11.13, $\gamma_t(P^{p^s}) = \gamma_t(P)^{p^{ts}} \leq P^{p^{ts}}$; so the result will follow since e/s is (p, m, n)-bounded. We have yet to make this plan work.

So now we choose s. Let p^e be the (minimal) exponent of P. We put $s = [e/(2^k + 1)]$, that is, s is the maximal integer satisfying $s2^k + s \leq e$. To be able to apply the Cancellation Property to (12.2) we must have $s2^k + m + n2^k \leq s2^k + s \leq e$. The second inequality holds by the choice of s. If $s2^k + m + n2^k > s2^k + s$, then $m + n2^k > s = [e/(2^k + 1)]$, which implies that e is bounded in terms of p, m and n (recall that $k = k(p^n)$). Then the nilpotency class of P, which is at most e by Theorem 11.10(c), is also (p, m, n)-bounded and the result follows. Thus, we may assume that $s2^k + m + n2^k \leq s2^k + s$. Then we can apply the Cancellation Property 11.17(b) to (12.2) to get rid of the summand $s2^k$:

$$(P^{(k)})^{p^{m+n2^k}} \leq P^{p^s}. \qquad (12.3)$$

We may also assume that $s(2^k + 2) \geq (s + 1)(2^k + 1) \Leftrightarrow [e/(2^k + 1)] = s \geq 2^k + 1$. Otherwise e and hence the nilpotency class of P are (p, m, n)-bounded, and the result follows. Thus, we have $s(2^k + 2) \geq (s+1)(2^k + 1) \geq e$ by the choice of s. Now we take γ_{2^k+2} of both sides of (12.3) and apply the Interchanging Corollary 11.13. On the right we shall have

$$\gamma_{2^k+2}(P^{p^s}) = \gamma_{2^k+2}(P)^{p^{s(2^k+2)}} \leq P^{p^e} = 1,$$

since $s(2^k + 2) \geq e$. Hence the left-hand side will be trivial as well:

$$\left(\gamma_{2^k+2}(P^{(k)})\right)^{p^{(m+n2^k)(2^k+2)}} = \gamma_{2^k+2}\left((P^{(k)})^{p^{m+n2^k}}\right) = 1.$$

This means that $\gamma_{2^k+2}(P^{(k)})$ has (p, m, n)-bounded exponent (recall that $k = k(p^n)$). Hence the order of this powerful subgroup generated by mp^n elements is also (p, m, n)-bounded by Corollary 11.21. By Corollary 3.15, we have strict inequalities $\gamma_{2^k+2+t+1}(P^{(k)}) < \gamma_{2^k+2+t}(P^{(k)})$ unless $\gamma_{2^k+2+t}(P^{(k)}) = 1$. As a result, $\gamma_{2^k+2+r}(P^{(k)}) = 1$ for some (p, m, n)-bounded number r. Thus, $P^{(k)}$ is nilpotent of (p, m, n)-bounded class. □

Remark 12.4. Theorem 12.1 already gives a weak bound, in terms of p, m and n, for the derived length of an arbitrary finite p-group P admitting an automorphism of order p^n with exactly p^m fixed points. Indeed, by Corollary 11.20, P can be assumed to be powerful. Consider the inequalities

$$|P/P^p| \geq \ldots \geq |P^{p^i}/P^{p^{i+1}}| \geq |P^{p^{i+1}}/P^{p^{i+2}}| \geq \ldots,$$

which hold according to Theorem 11.15. Since all factor-groups $P^{p^i}/P^{p^{i+1}}$ are abelian of exponent p and the ranks are (p, m, n)-bounded, there can only be a (p, m, n)-bounded number of strict inequalities in this chain. Every segment with equalities corresponds to a uniformly powerful section, and all of them are soluble of (p, m, n)-bounded derived length by Theorem 12.1. Hence P is soluble of (p, m, n)-bounded derived length.

Such a (p, m, n)-bound for the derived length was obtained by A. Shalev [1993a]. In his work, Kreknin's Theorem was applied to the Lie ring constructed from a uniformly powerful p-group in another way outlined in Remark 11.23. Although we used here the usual associated Lie ring, many ideas of the proof stem from A. Shalev's paper. Since every element of a group acts as an inner automorphism, the following interesting fact follows.

Corollary 12.5. [A. Shalev, 1993a] *The derived length of a finite p-group is bounded in terms of the minimal order of the centralizer of its element.* □

In fact, the proof of Theorem 12.1 above also gives a strong bound, in terms of p^n only, for the derived length of a subgroup of (p, m, n)-bounded index in this special case of a uniformly powerful p-group (Exercise 12.2). But the number of uniformly powerful factors in the general situation may not be (p, n)-bounded.

§ 12.2. Application of the Mal'cev Correspondence

We use the Mal'cev Correspondence to derive a consequence of Kreknin's Theorem for arbitrary nilpotent groups with an automorphism of finite order [E. I. Khukhro, 1993a]. Recall that $k(n)$ denotes Kreknin's function.

Theorem 12.6. *Suppose that a nilpotent group G of class c admits an automorphism φ of finite order t. Then, for some (c, t)-bounded number $N = N(c, t)$, the $k(t)$th derived subgroup of the subgroup generated by all Nth powers is contained in the normal closure of the centralizer of φ in G, that is,*
$$(G^N)^{(k(t))} \leq \langle C_G(\varphi)^G \rangle.$$

Proof. We fix the notation $k = k(t)$ for the value of Kreknin's function.

Recall that the groups admitting the action of $\langle \varphi \rangle$ as a group of automorphisms (not necessarily faithful) can be regarded as algebraic systems with additional unary operator φ, the so-called $\langle \varphi \rangle$-groups (Example 1.39). We start with a free nilpotent $\langle \varphi \rangle$-group F of class c freely generated by $x_1, \ldots, x_{2^k}$ as a $\langle \varphi \rangle$-group. As an abstract group F is free nilpotent of class c with free generators $y_{ij} = x_i^{\varphi^j}$, $i = 1, \ldots, 2^k$, $j = 0, 1, \ldots, t-1$, and φ is an automorphism of F which permutes the free generators in a natural way: $y_{ij}^\varphi = (x_i^{\varphi^j})^\varphi = x_i^{\varphi^{j+1}} = y_{i\,j+1}$, where $j+1$ is taken mod t.

In accordance with Theorem 9.20, we form $\sqrt{F}$, the $\mathbb{Q}$-powered hull of F (a divisible torsion-free nilpotent group of class c consisting of the roots of elements of F), and extend φ to an automorphism of $\sqrt{F}$ (denoted by the same letter). Note that $\sqrt{F}$ is a free nilpotent $\mathbb{Q}$-powered group on free generators y_{ij}. Let $L = L_{\sqrt{F}}$ be the nilpotent Lie $\mathbb{Q}$-algebra which is in the Mal'cev Correspondence 10.11 with $\sqrt{F}$, with the same underlying set $L = \sqrt{F}$. By Theorem 10.13, φ can be regarded as an automorphism of L acting on the same set in the same way.

By Kreknin's Theorem in the combinatorial form 7.19(a), we have

$$L^{(k)} \subseteq \text{ id} \langle C_L(\varphi) \rangle \tag{12.7}$$

(here $|\varphi|L = L$ since L is a $\mathbb{Q}$-algebra). Using Theorem 10.13, we translate this inclusion into the group language. Since $L/L^{(k)}$ is soluble of derived length k, the same subset $L^{(k)}$ is a normal $\mathbb{Q}$-powered subgroup of $\sqrt{F}$ with soluble factor-group of derived length k; hence $\sqrt{F}^{(k)} \subseteq L^{(k)}$. On the other side, we have, of course, $C_{\sqrt{F}}(\varphi) = C_L(\varphi)$ and $_{\mathrm{id}}\langle C_L(\varphi) \rangle = \sqrt{\langle C_{\sqrt{F}}(\varphi)^{\sqrt{F}} \rangle}$, since this is the smallest ideal of L (respectively, the smallest normal $\mathbb{Q}$-powered subgroup of $\sqrt{F}$) containing $C_L(\varphi) = C_{\sqrt{F}}(\varphi)$. By Lemma 3.17, $\sqrt{\langle C_{\sqrt{F}}(\varphi)^{\sqrt{F}} \rangle} = \langle C_{\sqrt{F}}(\varphi)^{\sqrt{F}} \rangle$ (the abstract normal closure), since $C_{\sqrt{F}}(\varphi)$ is a divisible subgroup (being equal to the $\mathbb{Q}$-subalgebra $C_L(\varphi)$). As a result, (12.7) implies

$$F^{(k)} \le \sqrt{F}^{(k)} \subseteq L^{(k)} \subseteq {}_{\mathrm{id}}\langle C_L(\varphi) \rangle = \langle C_{\sqrt{F}}(\varphi)^{\sqrt{F}} \rangle .$$

However, the right-hand side is still larger than desired. In order to qualify for an embedding into $\langle C_F(\varphi)^F \rangle$, we shall take the kth derived subgroup of some smaller subgroup F^N rather than of F.

One of the consequences of the inclusion obtained is the following equality:

$$\delta_k(x_1, \dots, x_{2^k}) = c_1^{g_1} \cdots c_r^{g_r}, \tag{12.8}$$

where on the left is the value of the identity of solubility of derived length k on the $x_i = y_{i0}$, and on the right $c_\alpha \in C_{\sqrt{F}}(\varphi)$ and $g_\alpha \in \sqrt{F}$ for all α. Fixing one such equation and the expressions for the c_α and g_α as $\mathbb{Q}$-powered group words in the $y_{ij} = x_i^{\varphi^j}$, we may regard (12.8) as a law in the free nilpotent $\mathbb{Q}$-powered group $\sqrt{F}$, a law which depends only on t and c.

To make a proper use of powers, we need the following fact.

Lemma 12.9. *For any $w = w(y_{ij}) \in \sqrt{F}$, regarded as a $\mathbb{Q}$-powered group word in the y_{ij}, there is $s \in \mathbb{N}$ such that $\tilde{w} = w(y_{ij}^s)$, the value of the same word on the y_{ij}^s, belongs to F. If $w \in C_{\sqrt{F}}(\varphi)$, then $\tilde{w} \in C_F(\varphi)$. The same inclusions hold for $w(y_{ij}^{sq})$, for any $q \in \mathbb{N}$.*

Proof. First, we prove the existence of s such that $\tilde{w} \in F$ and all multiples of s have this property. Induction on c, the nilpotency class. If $c = 1$, then $w(y_{ij}^s) = w^s \in F$ for some $s \in \mathbb{N}$ by the definition of $\sqrt{F}$; then also $w(y_{ij}^{sq}) = w^{sq} \in F$ for any $q \in \mathbb{N}$. Let $c > 1$. By Lemma 10.1, $\sqrt{\gamma_c(F)} = \gamma_c(\sqrt{F})$ and $\sqrt{F}/\sqrt{\gamma_c(F)}$ can be identified with $\sqrt{F/\gamma_c(F)}$. By the induction hypothesis, there is $s_1 \in \mathbb{N}$ such that the image of $w(y_{ij}^{s_1})$ belongs to $F\sqrt{\gamma_c(F)}/\sqrt{\gamma_c(F)}$. Hence $w(y_{ij}^{s_1}) = h(y_{ij}) \cdot z(y_{ij})$ where $h(y_{ij}) \in F$ and $z = z(y_{ij}) \in \sqrt{\gamma_c(F)}$. Here $h(y_{ij})$ is an abstract group word in the y_{ij} (with exponents in $\mathbb{Z}$), and $z(y_{ij})$ is a product of $\mathbb{Q}$-powers of commutators of weight c in the y_{ij}. By Lemma 6.13 applied to these commutators, we get $z(y_{ij}^u) = z^{u^c}$ for any $u \in \mathbb{N}$.

Since $z^{v^c} \in F$ for some $v \in \mathbb{N}$, we obtain $w(y_{ij}^{vs_1}) = h(y_{ij}^v) \cdot z^{v^c} \in F$. (If two $\mathbb{Q}$-powered group words in free generators are equal, then the equality remains valid on replacing the free generators by any other elements, such a replacing being a homomorphism of the free group; here we replaced the y_{ij} by the y_{ij}^v.) Thus, $s = vs_1$ is the required number. The same inclusion holds for any multiple of s: $w(y_{ij}^{vqs_1}) = h(y_{ij}^{vq}) \cdot z^{v^c q^c} \in F$.

Now suppose that $w \in C_{\sqrt{F}}(\varphi)$; we use the Mal'cev Correspondence to show that then $\widetilde{w} \in C_F(\varphi)$. By Theorem 10.13(g), the Lie $\mathbb{Q}$-algebra $L = L_{\sqrt{F}}$ that is in the Mal'cev Correspondence with $\sqrt{F}$ is free nilpotent of class c with free generators y_{ij}; hence L is homogeneous with respect to the generators y_{ij} (Corollary 5.40). Since the automorphism φ permutes the generators y_{ij}, the homogeneous components of L are φ-invariant and $C_L(\varphi)$ is a homogeneous subalgebra. If $w \in C_{\sqrt{F}}(\varphi) = C_L(\varphi)$, then $w = w_1 + \cdots + w_c$, where $w_d \in C_L(\varphi)$ is a homogeneous Lie polynomial of weight d in the y_{ij}, for each $d = 1, \ldots, c$. By Theorem 10.13(f), the homomorphism of the $\mathbb{Q}$-powered group $\sqrt{F}$ that extends the mapping $y_{ij} \rightarrow y_{ij}^s$ (for all i, j) is also the homomorphism of L extending the mapping $y_{ij} \rightarrow s y_{ij}$. Therefore

$$\widetilde{w} = w(s y_{ij}) = s w_1 + s^2 w_2 + \cdots + s^c w_c \in C_L(\varphi) = C_{\sqrt{F}}(\varphi),$$

since $s^d w_d \in C_L(\varphi)$ for each d. Thus, if s is such that $\widetilde{w} \in F$, then $\widetilde{w} \in C_{\sqrt{F}}(\varphi) \cap F = C_F(\varphi)$, as required. The same calculation is valid for any multiple of s. □

Note that the least s satisfying Lemma 12.9 depends only on w, c and t.

Returning to the proof of Theorem 12.6, we apply Lemma 12.9 to all elements g_α, c_α in the right-hand side of (12.8) regarded as $\mathbb{Q}$-powered group words in the y_{ij}. Let N_1 be the least common multiple of the corresponding minimal numbers given by Lemma 12.9 for all c_α, g_α. Substituting the $y_{ij}^{N_1}$ in place of the y_{ij} in (12.8), we obtain

$$\delta_k(x_1^{N_1}, \ldots, x_{2^k}^{N_1}) = \widehat{c}_1^{\,\widehat{g}_1} \cdots \widehat{c}_r^{\,\widehat{g}_r}, \tag{12.10}$$

where $\widehat{c}_\alpha \in C_F(\varphi)$ and $\widehat{g}_\alpha \in F$ for all α. This equation may be regarded as a law in the y_{ij} that holds in the free nilpotent group F and depends only on t and c.

The right-hand side of (12.10) is an element of $\langle C_F(\varphi)^F \rangle$. However, although the kth derived subgroup is generated by the values of the commutator δ_k, and F^{N_1} is generated by the N_1th, the subgroup $(F^{N_1})^{(k)}$ may not be generated by the values of δ_k only on these powers (but on all their products). We use Lemma 6.15 of A. I. Mal'cev to overcome this difficulty. Put $N = N_1^c$; by Lemma 6.15, every product of Nth powers of elements in a nilpotent group of class c is an N_1th power.

Now we prove the theorem for a free nilpotent $\langle\varphi\rangle$-group F_1 (with arbitrarily many generators). The subgroup $(F_1^N)^{(k)}$ is generated by the values of δ_k on the products of Nth powers of elements of F_1. Since every such product is an N_1th power, the subgroup $(F_1^N)^{(k)}$ is contained in the subgroup generated by all elements of the form $\delta_k(f_1^{N_1},\ldots,f_{2^k}^{N_1})$, $f_i \in F_1$. For any $f_i \in F_1$, the homomorphism of the $\langle\varphi\rangle$-group F into F_1 which extends the mapping $x_i \to f_i$, $i = 1, 2, \ldots, 2^k$ (or, as a homomorphism of abstract groups, the mapping $y_{ij} \to f_i^{\varphi^j}$), obviously maps $C_F(\varphi)$ into $C_{F_1}(\varphi)$. Applying this homomorphism to (12.10) we see that $\delta_k(f_1^{N_1},\ldots,f_{2^k}^{N_1})$, the image of the left-hand side, is contained in $\left\langle C_{F_1}(\varphi)^{F_1}\right\rangle$. It follows that

$$(F_1^N)^{(k)} \le \left\langle C_{F_1}(\varphi)^{F_1}\right\rangle .$$

Now let G be an arbitrary group satisfying the hypothesis of the theorem. There is a homomorphism ϑ of a suitable free nilpotent $\langle\varphi\rangle$-group F_1 onto G. Again, $\left\langle C_{F_1}(\varphi)^{F_1}\right\rangle^{\vartheta} \le \left\langle C_G(\varphi)^G\right\rangle$. Using the result for F_1, we obtain finally

$$(G^N)^{(k)} = ((F_1^N)^{(k)})^{\vartheta} \le \left\langle C_{F_1}(\varphi)^{F_1}\right\rangle^{\vartheta} \le \left\langle C_G(\varphi)^G\right\rangle ,$$

as required. $\square$

§ 12.3. Almost solubility of p^n-bounded derived length

First, we prove that if P is a powerful p-group admitting an automorphism of order p^n with p^m fixed points, then $P^{(k(p^n))}$ is nilpotent of (p, m, n)-bounded class. The proof is in the multiple application of Theorem 12.1 using the Interchanging Lemma 11.12. Then the result of § 12.2 is applied to $P^{(k(p^n))}$ to prove that P has a subgroup of (p, m, n)-bounded index and of derived length $2k(p^n)$. The main result of the chapter follows, since in the general situation the group contains a powerful subgroup of (p, m, n)-bounded index.

Theorem 12.11. *Suppose that a powerful finite p-group P admits an automorphism φ of order p^n with exactly p^m fixed points. Then the $k(p^n)$th derived subgroup $P^{(k(p^n))}$ is nilpotent of (p, m, n)-bounded class.*

Proof. We fix the notation $k = k(p^n)$ for the value of Kreknin's function. As in Remark 12.4, we consider the inequalities

$$|P/P^p| \ge \ldots \ge |P^{p^i}/P^{p^{i+1}}| \ge |P^{p^{i+1}}/P^{p^{i+2}}| \ge \ldots ,$$

which hold according to Theorem 11.15. Since the factor-groups $P^{p^i}/P^{p^{i+1}}$ are elementary abelian, and the ranks are (p, m, n)-bounded, there can only be a (p, m, n)-bounded number of strict inequalities in this chain. Every segment with equalities corresponds to a uniformly powerful section of P; hence we obtain a series of (p, m, n)-bounded length with uniformly powerful factors:

$$P > P^{p^{i_1}} > P^{p^{i_2}} > \ldots > P^{p^{i_{l-1}}} > 1 \qquad (12.12)$$

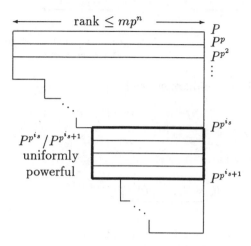

We prove the theorem by induction on the length l of this series. The basis of induction, when P itself is uniformly powerful, is formed by Theorem 12.1. To lighten notation, we denote $r = i_{l-1}$, so that P^{p^r} is the last non-trivial term in (12.12), a uniformly powerful subgroup such that the factor-group P/P^{p^r} has a series of length $l-1$ with uniformly powerful factors.

By Theorem 12.1,

$$[P^{p^r}, \underbrace{(P^{p^r})^{(k)}, \ldots, (P^{p^r})^{(k)}}_{u}] = 1,$$

for some (p, m, n)-bounded number u (here $[P^{p^r}, (P^{p^r})^{(k)}] \le (P^{p^r})^{(k)}$, so $u - 1$ is the nilpotency class of $(P^{p^r})^{(k)}$). A repeated application of the Interchanging Lemma 11.12 transforms this equation to the form

$$[P, \underbrace{P^{(k)}, \ldots, P^{(k)}}_{u}]^{p^{r(1+2^k u)}} = 1.$$

Another application of the Interchanging Lemma 11.12 gives

$$[P^{p^{r(1+2^k u)}}, \underbrace{P^{(k)}, \ldots, P^{(k)}}_{u}] = 1. \qquad (12.13)$$

On the other hand, by the induction hypothesis,

$$[P, \underbrace{P^{(k)}, \ldots, P^{(k)}}_{c}] \le P^{p^r}$$

for some (p, m, n)-bounded number $c = c(p, m, n, l - 1)$ (here $[P, P^{(k)}] \leq P^{(k)}$, so $c - 1$ is the nilpotency class of $(P/P^{p^r})^{(k)}$). Taking repeatedly the mutual commutator subgroup with $P^{(k)}$, we obtain, using the Interchanging Lemma,

$$[P, \underbrace{P^{(k)}, \ldots, P^{(k)}}_{2c}] \quad \leq \quad [P^{p^r}, \underbrace{P^{(k)}, \ldots, P^{(k)}}_{c}]$$

$$= \quad [P, \underbrace{P^{(k)}, \ldots, P^{(k)}}_{c}]^{p^r} \leq P^{p^{2r}},$$

and so on. By an obvious induction, we have

$$[P, \underbrace{P^{(k)}, \ldots, P^{(k)}}_{ic}] \leq P^{p^{ir}}$$

for all $i \in \mathbb{N}$. In particular,

$$[P, \underbrace{P^{(k)}, \ldots, P^{(k)}}_{v}] \leq P^{p^{r(1+2^k u)}} \tag{12.14}$$

for the (p, m, n)-bounded number $v = c(1 + 2^k u)$.

It remains to combine (12.14) with (12.13) to obtain

$$[P, \underbrace{P^{(k)}, \ldots, P^{(k)}}_{v}, \underbrace{P^{(k)}, \ldots, P^{(k)}}_{u}] \leq [P^{p^{r(1+2^k u)}}, \underbrace{P^{(k)}, \ldots, P^{(k)}}_{u}] = 1.$$

This implies that $P^{(k)}$ is nilpotent of (p, m, n)-bounded class $v + u$. □

Now we prove the main result of this chapter, giving a "strong" bound, in terms of the order of the automorphism only, for the derived length of a subgroup of bounded index in the general situation [E. I. Khukhro, 1993a].

Theorem 12.15. *If a finite p-group P admits an automorphism φ of order p^n with exactly p^m fixed points, then P contains a characteristic subgroup of (p, m, n)-bounded index which is soluble of (p, n)-bounded derived length $2k(p^n)$, where $k(p^n)$ is the value of Kreknin's function.*

Proof. Recall the notation fixed, $k = k(p^n)$. By Corollary 11.20, P has a powerful characteristic subgroup of (p, m, n)-bounded index; so we may assume P to be powerful from the outset. By Theorem 12.11, $P^{(k)}$ is nilpotent of (p, m, n)-bounded class s, say. By Theorem 12.6 applied to $P^{(k)}$ and its automorphism φ, we have

$$\left((P^{(k)})^{p^w} \right)^{(k)} \leq \left\langle C_{P^{(k)}}(\varphi)^{P^{(k)}} \right\rangle, \tag{12.16}$$

for some (p, m, n)-bounded number $p^w = N(s, p^n)$.

The subgroup $\left\langle C_{P(k)}(\varphi)^{P^{(k)}} \right\rangle$ is generated by the elements conjugate to elements in $C_{P(k)}(\varphi)$. All elements of $C_{P(k)}(\varphi) \leq C_P(\varphi)$ have order at most p^m by Lagrange's Theorem; hence $\left\langle C_{P(k)}(\varphi)^{P^{(k)}} \right\rangle$ is generated by elements of order dividing p^m. Since $\left\langle C_{P(k)}(\varphi)^{P^{(k)}} \right\rangle \leq P^{(k)}$ is nilpotent of class $\leq s$, the exponent of $\left\langle C_{P(k)}(\varphi)^{P^{(k)}} \right\rangle$ divides the (p, m, n)-bounded number p^{ms} by Lemma 6.12. As a result, it follows from (12.16) that

$$\left(\left((P^{(k)})^{p^w} \right)^{(k)} \right)^{p^{ms}} \leq \left\langle C_{P(k)}(\varphi)^{P^{(k)}} \right\rangle^{p^{ms}} = 1.$$

By the Interchanging Corollary 11.13, this implies that

$$(P^{p^z})^{(2k)} \leq \left(\left((P^{(k)})^{p^w} \right)^{(k)} \right)^{p^{ms}} = 1,$$

for some (p, m, n)-bounded number z. Both the rank and the exponent of the factor-group P/P^{p^z} are (p, m, n)-bounded; hence, by Corollary 11.21, the order of P/P^{p^z} is also (p, m, n)-bounded. Thus, P^{p^z} is the required characteristic subgroup of (p, m, n)-bounded index which is soluble of derived length $\leq 2k(p^n)$. $\qquad\square$

Remarks. 12.17. In the special case of Theorem 12.15 where $|\varphi| = 4$ ($p = 2$, $n = 2$), a stronger result was obtained by N. Yu. Makarenko [1993]: then P has a subgroup of m-bounded index, whose derived subgroup is nilpotent of class at most 3. This matches the stronger Lie ring result for regular automorphism of order 4, see Exercise 7.4

12.18. In the case of $n = 1$, when a finite p-group P admits an automorphism of order p^n with exactly p fixed points, S. McKay [1987] and, independently, I. Kiming [1988], proved that P has a subgroup of (p, n)-bounded index which is nilpotent of class at most 2 (abelian, if $p = 2$). We shall prove this theorem in Chapter 13.

12.19. This result gives rise to the following *Conjecture:* There exists a function $g(m)$, depending only on m, such that a finite p-group admitting an automorphism of order p^n with exactly p^m fixed points has a subgroup of (p, m, n)-bounded index which is soluble of derived length at most $g(m)$. Some evidence in the positive direction appears in Yu. Medvedev [1994a,b], where this conjecture is proved for $n = 1$ (see Chapter 14) and, in the general case, is reduced to a corresponding Lie ring problem using the Lie ring construction outlined in Remark 11.23.

Exercises 12

1. For P as in Theorem 12.1, prove that $P^{(2k+1)}$ has (p, m, n)-bounded order. [*Hint:* Use the fact that $\gamma_{2^k+2}(P^{(k)})$ has (p, m, n)-bounded order.]

2. Use 1 to deduce that, for P as in Theorem 12.1, $C_P(P^{(2k+1)})$ is a subgroup of (p, m, n)-bounded index in P which is soluble of derived length at most $2k + 2$. [*Hint:* The factor-group $P/C_P(P^{(2k+1)})$ embeds in the automorphism group of $P^{(2k+1)}$ and $(C_P(P^{(2k+1)}))^{(2k+1)} \leq C_P(P^{(2k+1)}) \cap P^{(2k+1)} \leq Z(P^{(2k+1)})$.]

3. Prove a version of Theorem 12.6 for $|\varphi| = 4$: if a nilpotent group G of class c admits an automorphism φ of order 4, then $\gamma_3(\gamma_2(G^N)) \leq \langle C_G(\varphi)^G \rangle$, for some c-bounded number $N = N(c)$. [*Hint:* See Exercises 7.3 and 7.4.]

4. Write down an explicit bound for the index of the subgroup in the conclusion of Theorem 12.15.

Chapter 13

p-Automorphisms with p fixed points

In the extreme case, where a p-automorphism of a finite p-group has only p fixed points, the result is extremely strong.

Theorem 13.1. *If a finite p-group P admits an automorphism φ of order p^n with exactly p fixed points, then P has a subgroup of (p,n)-bounded index which is nilpotent of class at most 2 (abelian, if $p = 2$).*

For $|\varphi| = p$ this was proved by C. R. Leedham-Green and S. McKay [1976] and by R. Shepherd [1971]; in the general case it was proved by S. McKay [1987] and by I. Kiming [1988].

We give a proof which is different from the original ones; although with possibly worse bounds for the index of the subgroup, our proof is more Lie ring oriented, making use of Higman's and Kreknin's Theorems from Chapter 7, the theory of powerful p-groups from Chapter 11, and the Lazard Correspondence from Chapter 10. As in Chapters 8 and 12, bounds for the ranks of abelian sections allow us to assume P to be powerful. Using a generalization of Maschke's Theorem, one can show that every φ-invariant abelian section is a kind of "almost one-dimensional" $\mathbb{Z}\langle\varphi\rangle$-module. This information is used in a reduction to the case where P is uniformly powerful, and later in the proof of a Lie ring theorem. An application of Higman's Theorem to a subring of the associated Lie ring $L(P)$ allows us to assume P to be nilpotent of class $h(p)$, the value of Higman's function. This already finishes the proof in the cases of $p = 2$ and $p = 3$, since $h(2) = 1$ and $h(3) = 2$. For $p > 3$, induction on the nilpotency class reduces the proof to the case where P is nilpotent of class 3; then the Lazard Correspondence provides a final reduction to Lie rings. Using Kreknin's and Higman's Theorems, in § 13.3 we prove independently a Lie ring analogue of Theorem 13.1, which is interesting in its own right. There we adopt the approach of Yu. Medvedev [1994b], defining a new "lifted" Lie ring multiplication. (This construction was anticipated in [A. Shalev, 1994] and [A. Shalev and E. I. Zelmanov, 1992], where a new Lie algebra is defined over the polynomial ring $\mathbb{F}_p[\tau]$, with multiplication by τ defined via taking pth powers in the group.)

§ 13.1. Abelian *p*-groups

We shall be able to control the behaviour of certain sections of an arbitrary (powerful) *p*-group via its abelian sections. The information about abelian groups will also be used later for the additive group of a Lie ring with a *p*-automorphism.

Throughout this section, U is an abelian *p*-group and φ is an automorphism of U of order p^n. We use additive notation regarding U as a right $\mathrm{Hom}_{\mathbb{Z}} U$-module and φ as an element of $\mathrm{Hom}_{\mathbb{Z}} U$. The following lemma is a generalization of Maschke's Theorem on complete reducibility of representations; it will also be used in a more general situation in Chapter 14.

Lemma 13.2. *If* $U = V \oplus Y$ *where* V *is a* φ*-invariant subgroup, then there is a* φ*-invariant subgroup* W *such that* $p^n U \leq V + W$ *and* $p^n (V \cap W) = 0$.

Proof. Let π denote the projection of U onto V with respect to Y, that is, $(v + y)\pi = v$ for $v \in V$, $y \in Y$. We define another mapping of U in itself: for $x \in U$ let

$$x\pi^* = \sum_{g \in \langle \varphi \rangle} xg\pi g^{-1}.$$

Clearly, $\pi^* \in \mathrm{Hom}\, U$ as a linear combination of the $g\pi g^{-1} \in \mathrm{Hom}\, U$. Since V is *g*-invariant for any $g \in \langle \varphi \rangle$ and $\pi|_V$ is identical on V, we have $U\pi^* \leq V$ and $vg\pi g^{-1} = v$ for $v \in V$, so that $\pi^*|_V = p^n 1_V$ (recall that $p^n = |\langle \varphi \rangle|$). We have $x\pi^*\pi^* = p^n x\pi^*$ for any $x \in U$ since $x\pi^* \in V$. This implies that $p^n U \leq V + \mathrm{Ker}\, \pi^*$, since $(p^n x - x\pi^*)\pi^* = 0$, that is, $p^n x - x\pi^* \in \mathrm{Ker}\, \pi^*$ for every $x \in U$. If $x \in V \cap \mathrm{Ker}\, \pi^*$, then both $x\pi^* = p^n x$ and $x\pi^* = 0$, so that $p^n(V \cap \mathrm{Ker}\, \pi^*) = 0$. We claim that $W = \mathrm{Ker}\, \pi^*$ is the required subgroup of U. It remains only to show that W is φ-invariant. This follows from the fact that π^* was designed to commute with φ: for any $x \in U$ we have

$$x\varphi\pi^* = \sum_{g \in \langle \varphi \rangle} x\varphi g\pi g^{-1} = \left(\sum_{g \in \langle \varphi \rangle} x\varphi g\pi g^{-1}\varphi^{-1} \right) \varphi = \left(\sum_{h \in \langle \varphi \rangle} xh\pi h^{-1} \right) \varphi = x\pi^*\varphi,$$

where $h = \varphi g$ runs over $\langle \varphi \rangle$ together with g. Hence if $x\pi^* = 0$, then $x\varphi\pi^* = x\pi^*\varphi = 0$ too.   □

From now on, U *is a homocyclic abelian group of exponent* p^e *(that is,* U *is a direct sum of cyclic groups of order* p^e*) and* $|C_U(\varphi)| = p$. By Lemma 2.12, the number of fixed points of φ is at most p in all φ-invariant sections, and by Corollary 2.7 the rank of U (and hence of all of its sections) is at most p^n. We begin to describe in what sense U is essentially a one-dimensional $\mathbb{Z}\langle \varphi \rangle$-module.

Lemma 13.3. (a) *Suppose that V is a φ-invariant homocyclic subgroup of U of the same exponent p^e. If $e \geq 2n + 1$, then $V = U$.*

(b) *Suppose that T is a φ-invariant subgroup of U. Then there is an integer $s \geq 0$ such that $p^{s+2np^n}U \leq T \leq p^s U$.*

Proof. (a) Suppose that $V \neq U$. By Lemma 1.6(a), V is a direct summand of U. By Lemma 13.2, there is a φ-invariant subgroup W such that $p^n U \leq V + W$ and $p^n(V \cap W) = 0$. Let a bar denote the image in $U/(V \cap W)$; then $(V + W)/(V \cap W) = \overline{V} \oplus \overline{W}$ (see 1.4). Since the exponent of U and V is at least p^{2n+1} by the hypothesis, the exponent of W is at least p^{n+1} (for otherwise $p^{2n}U \leq p^n V + p^n W = p^n V$ which is impossible if $V \neq U$). Therefore both $\overline{V}$ and $\overline{W}$ are non-trivial groups. Each of the non-trivial φ-invariant p-subgroups $\overline{V}$ and $\overline{W}$ has at least p fixed points of φ (see 2.8), whence φ has at least p^2 fixed points on $\overline{V} \oplus \overline{W}$. This is a contradiction to the fact that φ has at most p fixed points on every section of U by Lemma 2.12.

(b) Taking pth powers induces isomorphisms of the factor-groups $p^i U/p^{i+1}U$, and pth powers of elements in T are again in T. Therefore we have the following inequalities:

$$|T/(T \cap pU)| \leq |(T \cap pU)/(T \cap p^2 U)| \leq \ldots$$

$$\ldots \leq |(T \cap p^{e-1}U)/(T \cap p^e U)| = |T \cap p^{e-1}U|.$$

Since the ranks are at most p^n, there can be at most p^n strict inequalities in this chain. If there are no steps of height $> p^{2n}$ in this ladder before $T \cap p^i U = p^i U$ for the first time, that is, if each segment of equalities has length $\leq 2n - 1$, with a possible exception for the last segment of equalities satisfying $T \cap p^i U = p^i U$, then the result follows, with s maximal such that $T \leq p^s U$.

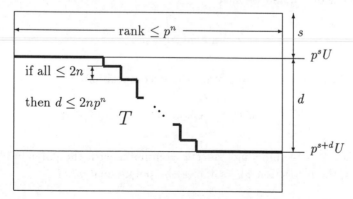

Suppose the opposite; then there is a section $\overline{U} = p^u U/p^{u+v}U$ of exponent $p^v \geq p^{2n+1}$ such that $\overline{T \cap p^u U}$ is a proper homocyclic subgroup of $\overline{U}$ of the same exponent p^v, where the bar denotes the image in $p^u U / p^{u+v} U$. This, however, contradicts (a). $\square$

Now we show that φ as a linear transformation of U "almost" satisfies an irreducible cyclotomic polynomial. It will follow that some power of φ is "almost" an automorphism of order p with (p, n)-bounded number of fixed points.

As a linear transformation of U, φ satisfies the polynomial $x^{p^n} - 1$ since $\varphi^{p^n} = 1$. The polynomial $x^{p^n} - 1$ decomposes as the product of irreducible (cyclotomic) polynomials in $\mathbb{Z}[x]$:

$$x^{p^n} - 1 = g_0(x) g_1(x) \cdots g_n(x),$$

where $g_0(x) = x - 1$ and $\deg g_i = p^i - p^{i-1}$ for $i \geq 1$; the $g_i(x)$ are also irreducible in $\mathbb{Q}[x]$ by Gauss's Lemma. In fact, $g_i(x) = x^{p^i - p^{i-1}} + \cdots + x^{2p^{i-1}} + x^{p^{i-1}} + 1$ for $i \geq 1$. The roots of $g_i(x)$ are precisely all primitive p^ith roots of 1, and $x^{p^l} - 1 = g_0(x) g_1(x) \cdots g_l(x)$ for each $l \in \mathbb{N}$. (See, for example, [B. L. van der Waerden, 1970].)

The greatest common divisor in $\mathbb{Q}[x]$ of the $n + 1$ integral polynomials $\hat{g}_i(x) = (x^{p^n} - 1)/g_i(x)$, $i = 0, 1, \ldots, n$, is 1. Hence there exist polynomials $u_i(x) \in \mathbb{Q}[x]$ such that

$$\sum_{i=0}^n u_i(x) \hat{g}_i(x) = 1.$$

Choosing the $u_i(x)$ so that the greatest power p^a of p dividing the denominators of all coefficients of all of the $u_i(x)$ is minimal possible, we see that a depends only on p and n and hence is a (p, n)-bounded number. Then the $\tilde{u}_i(x) = p^a u_i(x)$ can be regarded as polynomials over $\mathbb{Z}/p^e \mathbb{Z}$ (recall that p^e is the exponent of U), and we have

$$\sum_{i=0}^n \tilde{u}_i(\varphi) \hat{g}_i(\varphi) = p^a 1_U.$$

Applying both sides to U, we obtain the decomposition

$$\sum_{i=0}^n U_i = p^a U, \qquad \text{where } U_i = U \tilde{u}_i(\varphi) \hat{g}_i(\varphi). \tag{13.4}$$

Each U_i is a $\mathbb{Z}\langle \varphi \rangle$-submodule, since φ commutes with the polynomials in φ. For each i, the restriction $\varphi|_{U_i}$ satisfies the polynomial $g_i(x)$:

$$U_i g_i(\varphi) = U \tilde{u}_i(\varphi) \hat{g}_i(\varphi) g_i(\varphi) = U \tilde{u}_i(\varphi)(\varphi^{p^n} - 1_U) = 0. \tag{13.5}$$

In particular, $U_0(\varphi - 1) = 0$, that is, $U_0 \subseteq C_U(\varphi)$.

By Lemma 13.3(b), each U_i satisfies $p^{s_i+2np^n} U \le U_i \le p^{s_i} U$ for some s_i. In view of (13.4), we must have $p^a U \le p^{s_j} U$ for some j. If the exponent p^e of U is large enough, that is, if $a \le e$, then $a \ge s_j$ so that

$$p^{a+2np^n} U \le U_j. \tag{13.6}$$

(It can be shown that such j is unique if e is large enough, but we do not need this fact.) For large e, we must have $j \ne 0$, since otherwise $p^{a+2np^n+1} U \le pU_0 \le pC_U(\varphi) = 0$ which implies that $e \le a + 2np^n + 1$. We assume that $e \ge a + 2np^n + 2$ for what follows. Thus, $j \ne 0$, and we can define $\psi = \varphi^{p^{j-1}}$ which acts on U_j as an automorphism of order p. Indeed, by (13.5), $\varphi|_{U_j}$ satisfies the polynomial $g_j(x)$ which divides $x^{p^j} - 1$, so that $\psi^p = \varphi^{p^j}$ acts trivially on U_j.

Lemma 13.7. *The number of fixed points of ψ on U is (p,n)-bounded.*

Proof. The restriction $\varphi|_{C_U(\psi)}$ satisfies the polynomial

$$x^{p^{j-1}} - 1 = g_0(x)g_1(x) \cdots g_{j-1}(x).$$

Since this polynomial is coprime with $g_j(x)$, there exist polynomials $u(x)$, $v(x) \in \mathbb{Q}[x]$ such that $1 = (x^{p^{j-1}} - 1)u(x) + g_j(x)v(x)$. As above, there is a (p,n)-bounded number b such that p does not divide the denominators of all coefficients of $\tilde{u}(x) = p^b u(x)$ and $\tilde{v}(x) = p^b v(x)$, which may therefore be regarded as polynomials over $\mathbb{Z}/p^e\mathbb{Z}$. Then

$$p^b 1_U = (\varphi^{p^{j-1}} - 1_U)\tilde{u}(\varphi) + g_j(\varphi)\tilde{v}(\varphi).$$

We apply both sides to $p^{a+2np^n} C_U(\psi)$, which is contained in U_j by (13.6), and use (13.5):

$$p^b p^{a+2np^n} C_U(\psi)$$

$$= p^{a+2np^n} C_U(\psi)(\varphi^{p^{j-1}} - 1_U)\tilde{u}(\varphi) + p^{a+2np^n} C_U(\psi)g_j(\varphi)\tilde{v}(\varphi)$$

$$\le C_U(\psi)(\psi - 1|_U)\tilde{u}(\varphi) + U_j g_j(\varphi)\tilde{v}(\varphi) = 0.$$

So the exponent of $C_U(\psi)$ is (p,n)-bounded, whence the order of $C_U(\psi)$ is (p,n)-bounded too, since the rank is at most p^n. $\qquad\square$

§ 13.2. Reduction to Lie rings

Now we begin the reductions in the proof of Theorem 13.1. For the rest of the section, we fix the notation P and φ for a finite p-group and its automorphism of order p^n with exactly p fixed points. By Theorem 12.15, P has a characteristic subgroup of (p,n)-bounded index which is soluble of (p,n)-bounded derived length. Thus, *we may assume that P is soluble of (p,n)-bounded derived length from the outset*:

$$P^{(g)} = 1, \qquad\qquad (13.8)$$

for some (p,n)-bounded number g. By Corollary 11.20, P has a powerful characteristic subgroup of (p,n)-bounded index and of rank $\leq p^n$. Thus, *we may assume P to be powerful from the outset*. Then every section of P has rank $\leq p^n$, and if the exponent of such a section is (p,n)-bounded, then its order is (p,n)-bounded too (Corollary 11.21). For example, if s is a (p,n)-bounded number, then P^{p^s} is a subgroup of (p,n)-bounded index in P. Let p^e be the (minimal) exponent of P. *We may assume that e is large enough*, for if e is (p,n)-bounded, the order of P is (p,n)-bounded too.

We shall use without reference Theorem 11.4 stating that taking pth powers and commutator subgroups produces powerfully embedded subgroups from powerfully embedded subgroups, Theorem 11.10 stating that H^{p^j} coincides with the set of p^jth powers of elements of a powerful p-group H, and the fact that φ has at most p fixed points on every φ-invariant section of P (by Lemma 2.12).

Our next reduction is to uniformly powerful groups. As in the proof of Theorem 12.11, we consider the chain of inequalities (which hold by Theorem 11.15)

$$|P/P^p| \geq |P^p/P^{p^2}| \geq \ldots \geq |P^{p^i}/P^{p^{i+1}}| \geq \ldots ,$$

where each factor-group $P^{p^i}/P^{p^{i+1}}$ is elementary abelian. Since the ranks are at most p^n, there can be at most p^n strict inequalities in this chain. Every segment with equalities gives rise to a uniformly powerful section of P; hence we obtain a series of P of length at most p^n with uniformly powerful factors:

$$P > P^{p^{i_1}} > P^{p^{i_2}} > \ldots > P^{p^{i_l}} > 1. \qquad\qquad (13.9)$$

We shall prove that there is at most one factor in (13.9) of sufficiently large exponent; then all other factors are of (p,n)-bounded exponent and hence of (p,n)-bounded order. Naturally, the single large factor will determine the structure of "most of P". Abelian sections will be used for that, basing on the following simple lemma. Recall that $[a/b]$ denotes the integral part of the ratio a/b; note that $b \cdot [a/b] \geq a - b + 1$ for $a,b \in \mathbb{N}$.

Lemma 13.10. *For every powerful p-group H of exponent p^s, the subgroup $H^{p^{[s/2]}}$ is abelian.*

Proof. Indeed, $[H^{p^{[s/2]}}, H^{p^{[s/2]}}] \leq [H,H]^{p^{s-1}} \leq H^{p^s} = 1$ by the Interchanging Lemma 11.12 and by Theorem 11.10. □

The proof of the following lemma relies on Lemma 13.3 on abelian groups; we take the abelian "lower halves" of the sections.

Lemma 13.11. *There is at most one factor of exponent $\geq p^{4n+1}$ in the series (13.9).*

Proof. Suppose that the two factors $P^{p^{i_r}}/P^{p^{i_r+1}}$ and $P^{p^{i_s}}/P^{p^{i_s+1}}$ in (13.9) have exponents $p^u, p^t \geq p^{4n+1}$ respectively, for some $r < s$. Then the subgroups consisting of all elements of order $\leq p^{2n+1}$ in these factors, namely $U = (P^{p^{i_r}}/P^{p^{i_r+1}})^{p^{u-2n-1}}$ and $T = (P^{p^{i_s}}/P^{p^{i_s+1}})^{p^{t-2n-1}}$, are abelian by Lemma 13.10 and homocyclic by Theorem 11.17(b). By Lemma 11.14, the mapping $x \to x^{p^{i_s+t-i_r-u}}$ induces a homomorphism ϑ of U onto T. The kernel $V = \operatorname{Ker} \vartheta$ is non-trivial since the rank of U is larger than the rank of T. By Lemma 1.6(b), V is a proper direct summand of U. Since φ commutes with taking powers, V is φ-invariant as the kernel of ϑ. This, however, contradicts Lemma 13.3(a). $\qquad\square$

Since the length of the series (13.9) is (p, n)-bounded, Lemma 13.11 means that there are (p, n)-bounded numbers a and b such that $U = P^{p^a}/P^{p^{e-b}}$ is a uniformly powerful section (p^e is the exponent of P). Suppose that U has a subgroup of (p, n)-bounded index which is nilpotent of class at most 2. Then $\gamma_3(U^{p^c}) = 1$ for some (p, n)-bounded number c so that $\gamma_3(P^{p^{a+c}})^{p^b} = 1$. By the Interchanging Corollary 11.13, we then have

$$\gamma_3(P^{p^{a+c+[b/3]+1}}) \leq \gamma_3(P^{p^{a+c}})^{p^b} = 1,$$

so that $P^{p^{a+c+[b/3]+1}}$ is a subgroup of (p, n)-bounded index which is nilpotent of class at most 2. (Analogously, if $\gamma_2(U^{p^c}) = 1$ for some (p, n)-bounded number c, then $\gamma_2(P^{p^{a+c+[b/2]+1}}) = 1$, so that $P^{p^{a+c+[b/2]+1}}$ is an abelian subgroup of (p, n)-bounded index; this will be needed in the case $p = 2$.) Therefore, *we may assume P to be uniformly powerful from the outset.* (The order of φ might have changed, only to become less; the number of fixed points remains p. We renew the same notation, including p^e for the exponent of P.)

Our next step is to produce an automorphism of the associated Lie ring $L(P)$, an automorphism "almost" of order p with fixed-point subring of (p, n)-bounded exponent (as an additive subgroup). But first we use Lemma 13.3(a) to squeeze an arbitrary φ-invariant subgroup into a (p, n)-bounded layer.

Lemma 13.12. *Suppose that T is a φ-invariant subgroup of P. Then there is an integer $s \geq 0$ such that $P^{s+4np^n} \leq T \leq P^{p^s}$.*

Proof. Taking pth powers induces isomorphisms of factor-groups $P^{p^i}/P^{p^{i+1}}$, and pth powers of elements in T are again in T. Therefore we have the following inequalities:

$$|T/(T \cap P^p)| \leq |(T \cap P^p)/(T \cap P^{p^2})| \leq \ldots \leq |(T \cap P^{p^{e-1}})/(T \cap P^{p^e})| = |T \cap P^{p^{e-1}}|.$$

Since the ranks are at most p^n, there can be at most p^n strict inequalities in this chain. If each segment of equalities has length at most $4n - 1$ (with a possible exception for the last segment of equalities satisfying $T \cap P^{p^i} = P^{p^i}$), then the result follows. Suppose the opposite; then there is a segment of equalities within a section $\bar{P}^{p^u} = P^{p^u}/P^{p^{u+v}}$ of exponent $p^v \geq p^{4n+1}$ such that $\bar{T} \cap \bar{P}^{p^u} \neq \bar{P}^{p^u}$, where the bar denotes the image in $P/P^{p^{u+v}}$. The section $\bar{P}^{p^{u+[v/2]}}$ is abelian by Lemma 13.10 and hence homocyclic of exponent $p^{v-[v/2]} \geq p^{2n+1}$. Then $\bar{T} \cap \bar{P}^{p^{u+[v/2]}}$ is a proper homocyclic subgroup of $\bar{P}^{p^{u+[v/2]}}$ of the same exponent $p^{v-[v/2]}$. This, however, contradicts Lemma 13.3(a).   □

To lighten notation, put $d = 4np^n$. Applying Lemma 13.12 to $[P, P]$, we choose an integer s such that

$$P^{p^{s+d}} \leq [P, P] \leq P^{p^s}.$$

(Here s is an analogue of the "degree of commutativity" from the works on p-groups of maximal class.) Then, by the Interchanging Lemma 11.12, we also have

$$P^{p^{c(s+d)}} \leq \gamma_{c+1}(P) \leq P^{p^{cs}} \quad \text{and} \quad P^{p^{(2^l-1)(s+d)}} \leq P^{(l)} \tag{13.13}$$

for any $c, l \in \mathbb{N}$. Since P is soluble of (p, n)-bounded derived length g (see (13.8)), we obtain from (13.13) that $P^{p^{(2^g-1)(s+d)}} \leq P^{(g)} = 1$, whence

$$(2^g - 1)(s + d) \geq e, \tag{13.14}$$

since P has exponent p^e. Now, if s is (p, n)-bounded, so is e; hence *we may assume s to be large enough along with e.*

We apply the considerations of §13.1 to the abelian homocyclic factor-group $A = P/P^s$ and its automorphism φ, switching to the multiplicative notation. Since s is large enough, there are an integer j satisfying $1 \leq j \leq n$ and a (p, n)-bounded number a such that

- $A^{p^a} \leq A_j$ for the φ-invariant subgroup $A_j \leq A$,

- $\psi = \varphi^{p^{j-1}}$ acts on A_j as an automorphism of order p, and

- the number of fixed points of ψ on A is (p, n)-bounded.

Since a is (p, n)-bounded, we may assume $s > a$.

Lemma 13.15. *The number of fixed points of ψ on P is (p, n)-bounded.*

Proof. By Lemma 11.14, the mapping $x \to x^{p^{e-s}}$ induces an isomorphism of the abelian section $A = P/P^{p^s}$ onto $P^{p^{e-s}}/P^{p^e} = P^{p^{e-s}}$, which is an isomorphism of $\mathbb{Z}\langle\psi\rangle$-modules, since ψ commutes with taking powers. Hence the number of fixed points of ψ on $P^{p^{e-s}}$ is the same as on A and is (p, n)-bounded. If there is any fixed point of ψ outside $P^{p^{e-s}}$, then its powers provide at least

p^s fixed points in $P^{p^{e-s}}$ (recall that P is uniformly powerful). In this case s is (p, n)-bounded, whence, by (13.14), the exponent of P is (p, n)-bounded and hence the order of P is (p, n)-bounded. $\square$

Corollary 13.16. *For the induced automorphism ψ of the associated Lie ring $L(P)$, we have $p^b C_{L(P)}(\psi) = 0$ for some (p, n)-bounded number b.*

Proof. By Lemmas 13.15 and 2.12, all of the $|C_{\gamma_i(P)/\gamma_{i+1}(P)}(\psi)|$ are (p, n)-bounded, and $C_{L(P)}(\psi) = \bigoplus_i C_{\gamma_i(P)/\gamma_{i+1}(P)}(\psi)$. $\square$

Proposition 13.17. *The group P has a subgroup of (p, n)-bounded index which is nilpotent of class at most $h(p)$, the value of Higman's function.*

Proof. Recall that $P^{p^{s+d}} \leq [P, P] \leq P^{p^s}$, where $d = 4np^n$, and that that $s > a + d$. By Lemma 11.14, the mapping $x \to x^{p^d}$ induces the homomorphism of $A^{p^a} = P^{p^a}/P^{p^s}$ onto $P^{p^{a+d}}/P^{p^{s+d}}$ which is a homomorphism of $\mathbb{Z}\langle\psi\rangle$-modules. Therefore ψ^p acts trivially on $P^{p^{a+d}}/P^{p^{s+d}}$ and hence so on $P^{p^{a+d}}/[P, P]$. In the additive notation, ψ^p acts trivially on $p^{a+d}P/[P, P]$ regarded as an additive subgroup of the associated Lie ring $L(P)$. Let M denote the Lie subring of $L(P)$ generated by $p^{a+d}P/[P, P]$. Then ψ is an automorphism of M of order dividing p. We apply Higman's Theorem 7.19(b) to obtain that

$$p^{h+1}\gamma_{h+1}(M) = \gamma_{h+1}(pM) \leq {}_{\mathrm{id}}\langle C_M(\psi)\rangle,$$

where $h = h(p)$ is the value of Higman's function. Then, by Corollary 13.16,

$$p^{h+1+b}\gamma_{h+1}(M) \leq p^b{}_{\mathrm{id}}\langle C_M(\psi)\rangle = {}_{\mathrm{id}}\langle p^b C_M(\psi)\rangle = 0,$$

where b is a (p, n)-bounded number. In particular, for the homogeneous component of weight $h + 1$, we have

$$p^{b+(h+1)(a+d+1)} \big[\underbrace{P/[P, P], \ldots, P/[P, P]}_{h+1} \big]$$

$$= p^{h+1+b} \big[\underbrace{p^{a+d}P/[P, P], \ldots, p^{a+d}P/[P, P]}_{h+1} \big] = 0.$$

In terms of the group P, this implies that $\gamma_{h+1}(P)^{p^f} \leq \gamma_{h+2}(P)$ for the (p, n)-bounded number $f = b + (h + 1)(a + d + 1)$. Using (13.13), we obtain

$$P^{p^{h(s+d)+f}} \leq \gamma_{h+1}(P)^{p^f} \leq \gamma_{h+2}(P) \leq P^{p^{(h+1)s}}.$$

If $P^{p^{(h+1)s}} = 1$, then $\gamma_{h+1}(P)^{p^f} = 1$ and we can take $P^{p^{[f/(h+1)]+1}}$ for a subgroup of (p, n)-bounded index which is nilpotent of class at most h since, then

$$\gamma_{h+1}(P^{p^{[f/(h+1)]+1}}) \leq \gamma_{h+1}(P)^{p^f} = 1$$

by the Interchanging Corollary 11.13. If, however, $P^{p^{(h+1)s}} \neq 1$, the inclusion $P^{p^{h(s+d)+f}} \leq P^{p^{(h+1)s}}$ in a uniformly powerful p-group P implies that $h(s+d) + f \geq (h+1)s$, whence $s \leq hd + f$, so that s is (p,n)-bounded. Then by (13.14) the exponent and hence the order of P itself are (p,n)-bounded too. $\square$

Thus, by Proposition 13.17, in proving Theorem 13.1 *we may assume P to be nilpotent of class at most $h = h(p)$.* Since $h(2) = 1$ and $h(3) = 2$ (Lemma 7.11), this already finishes the proof in the cases $p = 2$ and $p = 3$. Hence we may now assume that $p > 3$. It remains to prove the following proposition.

Proposition 13.18. *Let $p > 3$ and suppose that Q is a finite p-group of nilpotency class c admitting an automorphism φ of order p^n with exactly p fixed points. Then Q has a subgroup of (p,n,c)-bounded index which is nilpotent of class at most 2.*

Reduction to Lie rings. To prove this proposition, we use induction on the nilpotency class c. By Lemma 2.12, $|C_{Q/Z(Q)}(\varphi)| \leq p$. By the induction hypothesis, $Q/Z(Q)$ has a subgroup of $(p,n,c-1)$-bounded index which is nilpotent of class 2; its preimage in Q is nilpotent of class 3. Hence we may assume Q to be nilpotent of class 3 from the outset. Since $p > 3$, we can apply the Lazard Correspondence (see Example 10.24). Let L be the Lie ring that is in the Lazard Correspondence with Q; note that the additive group of L is a finite p-group. Then φ can be regarded as an automorphism of the Lie ring L acting in the same way on the set $L = Q$; in particular, $|C_L(\varphi)| = p$. It is sufficient to find a Lie subring of L which is nilpotent of class at most 2 and has (p,n)-bounded index in the additive group of L. Then the same subset is a subgroup of Q which is nilpotent of the same class and has the same index in Q. This provides our final reduction to Lie rings, which are considered in the next section. $\square$

§ 13.3. The Lie ring theorem

We prove independently a result analogous to Theorem 13.1 (although we need only a much more special situation to finish the proof of Proposition 13.18).

Theorem 13.19. *Suppose that L is a Lie ring whose additive group is a finite p-group. If L admits an automorphism φ of order p^n with exactly p fixed points, then L has a nilpotent ideal of class at most 2 (commutative, if $p = 2$) which has (p,n)-bounded index in the additive group of L.*

Proof. For a time, we shall consider the additive group of L (denoted by the same letter) and regard φ as its automorphism. First, we perform a reduction to the case where L is homocyclic, similar to that in § 13.2 for groups. By Corollary 2.7, the rank of L is at most p^n. Hence there can be at most p^n strict inequalities in the chain

$$|L/pL| \geq |pL/p^2L| \geq \ldots \geq |p^iL/p^{i+1}L| \geq \ldots .$$

Every segment with equalities gives rise to a homocyclic section of L, so that we obtain a series of L of length at most p^n with homocyclic factors:

$$L > p^{i_1}L > p^{i_2}L > \ldots > p^{i_t}L > 0. \tag{13.20}$$

Lemma 13.21. *There is at most one factor of exponent $\geq p^{2n+1}$ in the series (13.20).*

Proof. Suppose that both factors $S_1 = p^{i_r}L/p^{i_r+1}L$ and $S_2 = p^{i_s}L/p^{i_s+1}L$ in (13.20) have exponents $p^u, p^t \geq p^{2n+1}$ respectively, for some $r < s$. The mapping $x \to x^{p^{i_s-i_r}}$ induces a homomorphism ϑ of $U = S_1/p^{2n+1}S_1$ onto $T = S_2/p^{2n+1}S_2$. The kernel $V = \mathrm{Ker}\,\vartheta$ is non-trivial since the rank of U is larger than the rank of T. By Lemma 1.6(b), V is a proper direct summand of U. Since φ commutes with taking powers, V is φ-invariant. This contradicts Lemma 13.3(a). □

Since the length of the series (13.20) is (p,n)-bounded, Lemma 13.21 means that there are (p,n)-bounded numbers a and b such that $p^aL/p^{e-b}L$ is a homocyclic section, where p^e is the exponent of L. This section is the additive group of the Lie factor-ring $M = p^aL/p^{e-b}L$. Suppose that M has a nilpotent ideal of class ≤ 2 which has (p,n)-bounded index in the additive group of M. Then $\gamma_3(p^cM) = 0$ for some (p,n)-bounded number c, so that $p^b\gamma_3(p^{a+c}L) = 0$. Then

$$\gamma_3(p^{a+c+[b/3]+1}L) \leq p^b\gamma_3(p^{a+c}L) = 0,$$

and $p^{a+c+[b/3]+1}L$ is a nilpotent ideal of class ≤ 2 which has (p,n)-bounded index in the additive group of L. (Similarly, if $\gamma_2(p^cM) = 0$ for some (p,n)-bounded number c, then $\gamma_2(p^{a+c+[b/2]+1}L) = 0$, and $p^{a+c+[b/2]+1}L$ is a commutative ideal of (p,n)-bounded index in the additive group of L, as required in the case $p = 2$.) Therefore, *we may assume the additive group of L to be homocyclic* from the outset (the order of φ might have changed, but only to become less, and the number of fixed points remains p). Let p^e be the exponent of the additive group of L; we may assume that e is large enough, since if e is (p,n)-bounded, then the order of L is (p,n)-bounded.

Lemma 13.22. *For some (p,n)-bounded number $r = r(p,n)$, the ideal p^rL is nilpotent of class at most $h(p)$, the value of Higman's function.*

Proof. We apply the reasoning of § 13.1 to the additive group of L. Since e is large enough, there are an integer j satisfying $1 \leq j \leq n$ and a (p, n)-bounded number a such that $\psi = \varphi^{p^{j-1}}$ acts as an automorphism of order p on $p^a L$ with (p, n)-bounded number of fixed points. Replacing L by $p^a L$ (whose additive group is homocyclic too), we may assume that ψ is an automorphism of order p with (p, n)-bounded number of fixed points, p^m, say. Let $h = h(p)$ denote the value of Higman's function; by Higman's Theorem 7.19(b), we have $\gamma_{h+1}(pL) \leq {}_{\mathrm{id}}\langle C_L(\psi)\rangle$. Then

$$\gamma_{h+1}(p^{[m/(h+1)]+2}L) \leq p^{h+1+m}\gamma_{h+1}(L)$$
$$= p^m \gamma_{h+1}(pL) \leq p^m{}_{\mathrm{id}}\langle C_L(\psi)\rangle \leq {}_{\mathrm{id}}\langle p^m C_L(\psi)\rangle = 0.$$

Thus, $p^{[m/(h+1)]+2}L$ is the required nilpotent ideal of class at most $h = h(p)$. □

Since $p^r L$ in Lemma 13.22 has (p, n)-bounded index in the additive group of L and remains homocyclic, *we may assume that L is nilpotent of class at most $h(p)$*. This completes the proof of Theorem 13.19 in the case of $p = 2$, since $h(2) = 1$ (see 7.11). We assume $p > 2$ for the rest of the section; then we need to find a nilpotent ideal of class ≤ 2 whose additive group has (p, n)-bounded index in L.

Applying Lemma 13.3(b) to $[L, L]$ as an additive subgroup of L we choose an integer s such that

$$p^{s+2np^n} L \leq [L, L] \leq p^s L.$$

(Here, again, s is an analogue of the "degree of commutativity".) We write, for short, $d = 2np^n$, which is a (p, n)-bounded number. Then

$$p^{c(s+d)}L \leq \gamma_{c+1}(L) \leq p^{cs}L \tag{13.23}$$

for any $c \in \mathbb{N}$. Recall that p^e is the exponent of the additive group of L. If $2s \geq e$, then (13.23) implies that $\gamma_3(L) \leq p^{2s}L \leq p^e L = 0$, so that L itself is nilpotent class ≤ 2. So we may assume $2s < e$. We shall effectively prove that $e - 2s$ is (p, n)-bounded; then the ideal $p^{e-2s}L$ of (p, n)-bounded index will be the required one since it is nilpotent of class ≤ 2: indeed, $\gamma_3(p^{e-2s}L) = p^{3e-6s}\gamma_3(L) \leq p^{3e-6s+2s}L \leq p^e L = 0$, since $3e - 4s > e \Leftrightarrow e > 2s$ by our assumption.

The main idea of the proof is to define new products $[\![a, b]\!] = \frac{1}{p^s}[a, b]$. This operation becomes a Lie multiplication on $\bar{L} = L/p^{e-2s}L$. Since $[L, L]$ (with respect to the old Lie products) "almost equals" $p^s L$, the new Lie ring $\bar{L}$ will be "almost equal" to $[\![\bar{L}, \bar{L}]\!]$. An application of Kreknin's Theorem will then imply that the exponent of the additive group of $\bar{L}$ is (p, n)-bounded, which means exactly that $e - 2s$ is (p, n)-bounded, as desired. We proceed with precise definitions.

Definition 13.24. For any $a, b \in L$, we define the new operation by setting $[\![a, b]\!]$ to be some p^sth root of $[a, b]$ in the additive group of L. In other words, for each pair (a, b), we fix an element $c(a, b)$ such that $p^s c(a, b) = [a, b]$, which is possible since $[a, b] \in p^s L$, and put $[\![a, b]\!] = c(a, b)$.

The new bracket multiplication may not satisfy the laws of Lie rings on L (with the old addition). However, since $p^s [\![a, b]\!] = [a, b]$ for every $a, b \in L$ by the definition, the following hold:

$$p^s [\![a, a]\!] = [a, a] = 0;$$

$$p^s \left([\![a_1 \pm a_2, b]\!] - [\![a_1, b]\!] \mp [\![a_2, b]\!]\right) = [a_1 \pm a_2, b] - [a_1, b] \mp [a_2, b] = 0;$$

$$p^{2s} \left([\![[\![a, b]\!], c]\!] + [\![[\![b, c]\!], a]\!] + [\![[\![c, a]\!], b]\!]\right) = [[a, b], c] + [[b, c], a] + [[c, a], b] = 0.$$

Only the third of these equations is not quite obvious, but it follows from the equalities $p^{2s} [\![[\![u, v]\!], w]\!] = p^s [[\![u, v]\!], w] = [p^s [\![u, v]\!], w] = [[u, v], w]$. Since the additive group of L is homocyclic of exponent $p^e > p^{2s}$, these equalities imply that

$$\left. \begin{array}{rcl} [\![a, a]\!] & \in & p^{e-s} L, \\ [\![a_1 \pm a_2, b]\!] - [\![a_1, b]\!] \mp [\![a_2, b]\!] & \in & p^{e-s} L, \\ [\![[\![a, b]\!], c]\!] + [\![[\![b, c]\!], a]\!] + [\![[\![c, a]\!], b]\!] & \in & p^{e-2s} L \end{array} \right\} \qquad (13.25)$$

for any $a, b, c \in L$.

Proposition 13.26. *The additive factor-group* $\bar{L} = L / p^{e-2s} L$ *endowed with the multiplication* $[\![\bar{x}, \bar{y}]\!] = \overline{[\![x, y]\!]}$, *where the bar denotes images in* $L / p^{e-2s} L$, *is a Lie ring. The automorphism of the additive group of* $\bar{L}$ *induced by* φ *is an automorphism of the Lie ring* $\bar{L}$.

(We use the same new bracket for the images.)

Proof. The anticommutative and distributive laws hold by the inclusions (13.25) and by the definition. For the Jacobi identity, note that $[\![[\![\bar{u}, \bar{v}]\!], \bar{w}]\!] = [\![\overline{[\![u, v]\!]}, \bar{w}]\!] = \overline{[\![[\![u, v]\!], w]\!]}$ by the definition.

We denote the induced automorphism of the additive factor-group $L / p^{e-2s} L$ by the same letter. We need to show that $[\![\bar{x}, \bar{y}]\!]^\varphi = [\![\bar{x}^\varphi, \bar{y}^\varphi]\!]$ for all $\bar{x}, \bar{y} \in \bar{L}$. Since $\bar{a}^\varphi = \overline{a^\varphi}$ for all $x \in L$, the left-hand side is $[\![\bar{x}, \bar{y}]\!]^\varphi = \overline{[\![x, y]\!]}^\varphi = \overline{[\![x, y]\!]^\varphi}$, while the right-hand side is $[\![\bar{x}^\varphi, \bar{y}^\varphi]\!] = \overline{[\![x^\varphi, y^\varphi]\!]}$ by the definition. We have

$$p^s \left([\![x, y]\!]^\varphi\right) = \left(p^s [\![x, y]\!]\right)^\varphi = [x, y]^\varphi = [x^\varphi, y^\varphi] = p^s [\![x^\varphi, y^\varphi]\!],$$

that is, $p^s \left([\![x, y]\!]^\varphi - [\![x^\varphi, y^\varphi]\!]\right) = 0$, which implies that $[\![x, y]\!]^\varphi - [\![x^\varphi, y^\varphi]\!] \in p^{e-s} L \leq p^{e-2s} L$, so that $\overline{[\![x, y]\!]^\varphi} = \overline{[\![x^\varphi, y^\varphi]\!]}$, as required. $\qquad \square$

Completion of the proof of Theorem 13.19. From now on, $\overline{L}$ denotes the Lie ring with additive group $L/p^{e-2s}L$ and multiplication $[\![\,,\,]\!]$, and a bar over elements and subsets of L denotes their images in $\overline{L} = L/p^{e-2s}L$. Let $[\![L, L]\!]$ denote the additive subgroup of L generated by all products $[\![a, b]\!]$, $a, b \in L$; then $\overline{[\![L, L]\!]} = [\![\overline{L}, \overline{L}]\!]$ where the right-hand side is the derived Lie subring of $\overline{L}$. We have $p^s[\![L, L]\!] = [L, L]$ and hence, by (13.23), $p^{s+d}L \leq p^s[\![L, L]\!]$. In the homocyclic additive group of L this implies that $p^d L \leq [\![L, L]\!]$, unless $s + d \geq e$ (Lemma 1.5). If, however, $s + d \geq e$, then together with the inequality $2s < e$ this implies that $e < 2d$, a contradiction with our assumptions that e is large enough. We may therefore assume $p^{s+d} \leq p^e$ so that $p^d L \leq [\![L, L]\!]$ which implies that $p^d \overline{L} \leq \overline{[\![L, L]\!]} = [\![\overline{L}, \overline{L}]\!]$. Then

$$p^{d(2^g-1)}\overline{L} \leq \overline{L}^{(g)} \tag{13.27}$$

for the gth derived subring of $\overline{L}$, for any $g \in \mathbb{N}$.

Note that the order of φ as an automorphism of $\overline{L}$ divides p^n and the number of fixed points of φ on $\overline{L}$ is exactly p. We apply Kreknin's Theorem 7.19(a) to get

$$p^{n2^k}\overline{L}^{(k)} = (p^n\overline{L})^{(k)} \leq {}_{\mathrm{id}}\langle C_{\overline{L}}(\varphi)\rangle, \tag{13.28}$$

where $k = k(p^n)$ is the value of Kreknin's function. Since $p\,{}_{\mathrm{id}}\langle C_{\overline{L}}(\varphi)\rangle = {}_{\mathrm{id}}\langle pC_{\overline{L}}(\varphi)\rangle = 0$, we obtain from (13.27) and (13.28) that

$$p^{1+n2^k+d(2^k-1)}\overline{L} \leq p \cdot p^{n2^k}\overline{L}^{(k)} \leq p\,{}_{\mathrm{id}}\langle C_{\overline{L}}(\varphi)\rangle = 0.$$

Since the exponent of the additive group of $\overline{L}$ is p^{e-2s}, this means that $e - 2s \leq 1 + n2^k + d(2^k - 1)$, so that $e - 2s$ is (p, n)-bounded since d, k and n are (p, n)-bounded numbers.

As a result, the ideal $p^{e-2s}L$ of the original Lie ring L has (p, n)-bounded index in the additive group of L. This is the required nilpotent ideal of class at most 2: indeed, $\gamma_3(p^{e-2s}L) = p^{3e-6s}\gamma_3(L) \leq p^{3e-6s+2s}L \leq p^e L = 0$, since $3e - 4s > e \Leftrightarrow e > 2s$, which is true by our assumption in the case $p \neq 2$. $\quad\square$

As noted at the end of § 13.2, this completes the proof of Theorem 13.1.

Remarks. 13.29. The Lie ring result is, at least formally, more general than for groups, since the Lie ring is not presupposed to be nilpotent or soluble. But of course, the "modular" situation immediately provides reduction to the soluble (or nilpotent) case, see Remark 8.13.

13.30. N. Blackburn [1958] proved that every p-group of maximal class P has a subgroup P_1 of index p and an element $a \in P \setminus P_1$ such that $|C_{P_1}(a)| = p$. Then a induces by conjugation on P_1 an automorphism φ of order p with p fixed points. So we have the following.

Corollary 13.31. ([C. R. Leedham-Green and S. McKay, 1976], [R. Shepherd, 1971]) *Every p-group of maximal class has a subgroup of p-bounded index which is nilpotent of class 2 (abelian if $p = 2$).*

p-Groups of maximal class were studied in a number of papers (by C. R. Leedham-Green, S. McKay and others), where their structure was studied in much detail, approaching even a kind of classification. See also Example 10.28 [B. A. Panfërov, 1980] of p-groups of maximal class of unbounded derived length (for different p, of course).

13.32. A finite p-group P has *coclass* r if $|P| = p^{r+c}$ where c is the nilpotency class of P (so the p-groups of maximal class are precisely the p-groups of coclass 1). The theory of p-groups (and pro-p-groups) of given coclass was developed in the works of S. Donkin, C. R. Leedham-Green, A. Mann, S. McKay, M. Newman, A. Shalev, W. Plesken, E. I. Zelmanov and others. The main result for finite p-groups is similar to that for maximal class: *Every finite p-group of coclass r has a subgroup of (p, r)-bounded index which is nilpotent of class 2 (abelian, if $p = 2$).* In [C. R. Leedham-Green, 1994a,b] even a kind of classification is proposed for p-groups of given coclass. A larger portion of p-groups of given coclass have an element which induces by conjugation an automorphism of order p^n on a subgroup of bounded index. But, unlike p-groups of maximal class, there is no easy reduction of the general situation to this case. Remarkably, in the works of A. Shalev and E. I. Zelmanov [1992] and A. Shalev [1994] theorems of G. Higman and V. A. Kreknin became a new tool in the theory of p-groups and pro-p-groups of given coclass, with effective bounds for the indices. (J. Alperin [1962] was the first to apply Higman's Theorem to studying p-groups of maximal class, see Remark 8.7.)

Exercises 13

1. Extract an explicit upper bound for the index of the nilpotent subgroup of class 2 in Theorem 13.1.

2. Prove that the sum (13.4) is "almost direct": there is a (p, n)-bounded integer t such that if $\sum_{i=0}^n k_i u_i = 0$ for some $u_i \in U_i$, $k_i \in \mathbb{Z}$, then $p^t k_i = 0$ for all i.

3. Use 2 to derive that j satisfying (13.6) is unique if e is large enough.

4. Instead of (13.27), apply Higman's Theorem 7.19(a) to a suitable subring $p^a \overline{L}$ and its automorphism $\psi = \varphi^{p^j}$ of order p with (p, n)-bounded number of fixed points. Examine whether the bound for the index can be improved in this way.

Chapter 14

Automorphism of order p
with p^m fixed points:
almost nilpotency of m-bounded class

Theorem 8.1 states that if a finite p-group P admits an automorphism of order p with p^m fixed points, then P has a subgroup of (p, m)-bounded index which is nilpotent of p-bounded class. In this chapter we prove that the nilpotency class of a subgroup of (p, m)-bounded index can be bounded in terms of m only. The following theorem is due to Yu. Medvedev [1994a,b].

Theorem 14.1. *If a finite p-group P admits an automorphism φ of prime order p with exactly p^m fixed points, then P has a subgroup of (p, m)-bounded index which is nilpotent of m-bounded class.*

Neither Theorem 8.1 nor Theorem 14.1 follows from the other: if p is much less than m, then Theorem 8.1 gives a better result; on the other hand, if m is much less than p, then Theorem 14.1 is better. Theorem 14.1 confirmed the conjecture from [E. I. Khukhro, 1985] (also [Kourovka Notebook, 1986, Problem 10.68]). This conjecture was prompted by the result of C. R. Leedham-Green and S. McKay [1976] and R. Shepherd [1971] on p-groups of maximal class, which amounts to the special case of Theorem 13.1 where $|\varphi| = |C_P(\varphi)| = p$ implies that P has a subgroup of p-bounded index which is nilpotent of class 2.

The proof of Theorem 14.1 is essentially about Lie rings; we use many of the of techniques developed in Chapter 13, including the lifted Lie products from [Yu. Medvedev, 1994b]. The reduction to Lie rings is easier than in Chapter 13, since here we are not constrained by the requirement to obtain such a strong bound for the nilpotency class as 2.

Reduction of the proof of Theorem 14.1 to Lie rings. By Theorem 8.1, P has a subgroup of (p, m)-bounded index which is nilpotent of class at most $h(p)$, the value of Higman's function. Thus, we may assume P to be nilpotent of class at most $h(p)$. Then we may use induction on the nilpotency class: by the induction hypothesis, $P/Z(P)$ has a subgroup of (p, m)-bounded index which is nilpotent of m-bounded class $g(m)$ (since $|C_{P/Z(P)}(\varphi)| \leq p^m$ by Lemma 2.12). Thus, P may be assumed to be nilpotent of class $g(m) + 1$. We may assume that p is large enough, larger than any function of m to be constructed, since otherwise we can simply take $h(p)$ for the bound of the nilpotency class. In particular, we may assume p to be larger than $g(m) + 1$, the nilpotency class of P. Therefore the Lazard Correspondence can be applied (Example 10.24), so that the problem about P is translated into the analogous problem about the corresponding Lie ring.   □

Thus, it suffices to prove the corresponding Lie ring result. In fact, the rest of the chapter is devoted to proving the following theorem on Lie rings of Yu. Medvedev [1994b], which actually asserts more than required for the group-theoretic application and is interesting in its own right.

Theorem 14.2. *Suppose that L is a Lie ring whose additive group is a finite p-group. If L admits an automorphism φ of order p with exactly p^m fixed points, then L has a nilpotent ideal of m-bounded class which has (p, m)-bounded index in the additive group of L.*

§ 14.1. Almost solubility of m-bounded derived length

The first step, however, will be to find a soluble ideal of m-bounded derived length with (p, m)-bounded index in the additive group of L.

Theorem 14.3. *Suppose that L is a Lie ring whose additive group is a finite p-group. If L admits an automorphism φ of order p with exactly p^m fixed points, then L has a soluble ideal of m-bounded derived length which has (p, m)-bounded index in the additive group of L.*

Proof. The rank of the additive group of L is at most pm by Corollary 2.7, so it suffices to find a soluble subring of m-bounded derived length with (p, m)-bounded index in the additive group of L: then, for some (p, m)-bounded number $r(p, m)$, the ideal $p^{r(p,m)}L$ is contained in this subring. Because of the bound for the rank, any section of the additive group of L that has (p, m)-bounded exponent has (p, m)-bounded order.

We shall use notation of a right $\mathbb{Z}\langle \varphi \rangle$-module for L. We may assume that

$$l + l\varphi + l\varphi^2 + \cdots + l\varphi^{p-1} = 0 \qquad (14.4)$$

for all $l \in L$. Indeed, the set $X = \{-x + x\varphi \mid x \in L\}$ is an additive subgroup of L. Since φ acts trivially on the factor-group L/X, we have $|L/X| \leq p^m$ by Lemma 2.12, whence $p^m L \subseteq X$. The ideal $p^m L$ has (p, m)-bounded index in L, and for any $l \in p^m L$ there is $x \in L$ such that $l = -x + x\varphi$, whence

$$l + l\varphi + l\varphi^2 + \cdots + l\varphi^{p-1} = -x + x\varphi - x\varphi + x\varphi^2 - \cdots + x\varphi^p = 0.$$

Replacing L by $p^m L$, from now on *we assume that (14.4) holds for L* (the number of fixed points of φ might become only a divisor of p^m). Note that

(14.4) holds for the induced automorphism φ in every φ-invariant section of L. This equation has a few important consequences.

Lemma 14.5. *It follows from (14.4) that*

    (a) *for any φ-invariant section U of the additive group of L we have* $pC_U(\varphi) = 0$;

    (b) *for any homocyclic φ-invariant section V of the additive group of L we have $|C_V(\varphi)| = |C_{p^i V/p^{i+1} V}(\varphi)|$ whenever $p^i V \neq 0$.*

Proof. (a) If $x\varphi = x$, then (14.4) implies that $px = x + x\varphi + \cdots + x\varphi^{p-1} = 0$.

(b) If p^s is the exponent of V, then $C_V(\varphi) = C_{p^{s-1}V}(\varphi)$ by (a). The mapping $x \to p^{s-1-i}x$ induces an isomorphism of the $\mathbb{Z}\langle\varphi\rangle$-modules $p^i V/p^{i+1} V$ and $p^{s-1}V$. $\qquad\qquad\square$

We shall need later the following information on one-generator $\mathbb{Z}\langle\varphi\rangle$-submodules. Let M be a φ-invariant homocyclic section of the additive group of L. For any $a \in M \setminus pM$, let V_a denote the $\mathbb{Z}\langle\varphi\rangle$-submodule generated by a, that is, $V_a = {}_+\langle a, a\varphi, \dots, a\varphi^{p-1}\rangle$.

Lemma 14.6. *It follows from (14.4) that $V_a \cap pM$ is a homocyclic group of rank $p-1$, and $p^k V_a \geq p^{k+1} M \cap V_a$ for any $k \in \mathbb{N}$.*

Proof. Consider a free abelian group F of the same exponent p^e as M, on free generators $f_1, \dots, f_p$. The cyclic permutation of the generators $f_i \to f_{i+1}$, where $i + 1$ is taken $\bmod\, p$, extends to an automorphism φ of F; in fact, F is a free one-generator abelian $\langle\varphi\rangle$-group of exponent p^e. Put $R = F/D$, where $D = \langle f_1 + \cdots + f_p \rangle$; then R is a free abelian $\langle\varphi\rangle$-group of exponent p^e satisfying the law (14.4). As a cyclic subgroup of exponent p^e, D is a direct summand of F, so that R is a homocyclic group of rank $p - 1$ and of exponent p^e (Lemma 1.6). Direct computation shows that $|C_R(\varphi)| = p$. It follows that given any two φ-invariant subgroups in R, one must contain the other: otherwise, for φ-invariant subgroups $M \neq M \cap N \neq N$, we have $(M + N)/(M \cap N) = \overline{M} \oplus \overline{N}$ with both $\overline{M}$ and $\overline{N}$ non-trivial, so that the number of fixed points of φ on this section is at least p^2, a contradiction, since $|C_{\overline{M} \oplus \overline{N}}(\varphi)| \leq p$ by Lemma 2.12. In particular, every φ-invariant subgroup N of R satisfies $p^k R \geq N \geq p^{k+1} R$ for some k. Since V_a has exponent p^e and satisfies (14.4), the mapping $f_1 \to a$ extends to a $\langle\varphi\rangle$-homomorphism of R onto V_a, with kernel K, say. Since $|a| = p^e$, we have $K \not\geq p^{e-1}R$, whence $p^{e-1}R \geq K$. It follows that $\Omega_{e-1}(V_a) \cong \Omega_{e-1}(R/K)$ is a homocyclic group of exponent p^{e-1} and of rank $p - 1$ (Lemma 1.5(e)). It remains to note that $\Omega_{e-1}(V_a) = V_a \cap \Omega_{e-1}(M) = V_a \cap pM$.

Now, since $V_a \cap pM$ is homocyclic, we have $\Omega_{e-k-1}(V_a \cap pM) = p^k(V_a \cap pM)$.

Then $p^{k+1}M \cap V_a = \Omega_{e-k-1}(V_a) = \Omega_{e-k-1}(V_a \cap pM) = p^k(V_a \cap pM) \leq p^k V_a.$

$\square$

Now we perform a reduction to the case where the additive group of L is homocyclic. We use many of the tools from Chapter 13. The difference is that now there may be not only one "big" homocyclic section, but m, because the dimension of any $\mathbb{Z}\langle\varphi\rangle$-module is essentially $\leq m$, rather than 1. We consider for some time the additive group of L which we denote by the same letter. Since the rank of L is at most mp, there can be at most mp strict inequalities in the chain

$$|L/pL| \geq |pL/p^2L| \geq \ldots \geq |p^iL/p^{i+1}L| \geq \ldots \,.$$

Every segment of equalities gives rise to a homocyclic section of L, so that we obtain a series of L of length at most mp with homocyclic factors:

$$L > p^{i_1}L > p^{i_2}L > \ldots > p^{i_t}L > 0. \tag{14.7}$$

Lemma 14.8. *There are at most m factors of exponent $\geq p^3$ in (14.7).*

Proof. Suppose that both factors $S_1 = p^{i_r}L/p^{i_r+1}L$ and $S_2 = p^{i_s}L/p^{i_s+1}L$ in (14.7) have exponents $\geq p^3$, for some $r < s$. The mapping $x \to x^{p^{i_s-i_r}}$ induces the homomorphism ϑ of $U = S_1/p^3S_1$ onto $T = S_2/p^3S_2$. The kernel $V = \mathrm{Ker}\,\vartheta$ is non-trivial since the rank of U is larger than the rank of T. By Lemma 1.6(b), V is a proper direct summand of U. Since φ commutes with taking powers, V is φ-invariant. By Lemma 13.2, there is a φ-invariant subgroup $W \leq U$ such that $pU \leq V + W$ and $p(V \cap W) = 0$. Let a bar denote the image in $(V + W)/(V \cap W) = \overline{V} \oplus \overline{W}$; then $\overline{V} \neq 0$ and $\overline{W} \neq 0$. Hence $|C_{\overline{W}}(\varphi)| < |C_{\overline{V} \oplus \overline{W}}(\varphi)|$. Applying ϑ to the inclusions $pU \leq V + W \leq U$, we obtain $pT \leq \vartheta(W) \leq T$; hence $|C_{S_2}(\varphi)| = |C_T(\varphi)| = |C_{\vartheta(W)}(\varphi)|$ by Lemma 14.5. Since $\vartheta(W) \cong W/(V \cap W) = \overline{W}$, we have $|C_{S_2}(\varphi)| = |C_{\overline{W}}(\varphi)|$. On the other hand, $|C_{\overline{V} \oplus \overline{W}}(\varphi)| \leq |C_{S_1}(\varphi)|$ by Lemma 2.12. As a result, $|C_{S_2}(\varphi)| = |C_{\overline{W}}(\varphi)| < |C_{\overline{V} \oplus \overline{W}}(\varphi)| \leq |C_{S_1}(\varphi)|$. We see that every successive factor of exponent $\geq p^3$ in (14.7) has a strictly smaller number of fixed points of φ. Since there are at most p^m fixed points at the start (and the number is always a power of p), there can be at most m factors of exponent $\geq p^3$ in (14.7). $\square$

Every factor of the series (14.7) is the additive group of the corresponding Lie ring $p^{i_s}L/p^{i_s+1}L$. Suppose that we proved Theorem 14.3 in the case where the additive group of L is homocyclic. Then there are a (p,m)-bounded number $r = r(p,m)$ and an m-bounded number $g = g(m)$ such that, for every factor H of exponent $\geq p^3$ in (14.7), the subring p^rH is soluble of derived length at most g. The other factors in (14.7), of exponents $\leq p^2$, together with the sections H/p^rH, glue up to at most $m+1$ factor-rings between at most m

sections $p^r H$. Every such piece has (p, m)-bounded exponent dividing p^{2mp+r}, since there are at most mp terms in (14.7). Then

$$\underbrace{p^{2mp+r}(p^{2mp+r}(\ldots(p^{2mp+r}(p^{2mp+r}}_{m+1} L)\underbrace{^{(g)})^{(g)} \ldots)^{(g)})^{(g)}}_{m} = 0,$$

whence, by (5.23), $(p^{u(p,m)}L)^{(mg)} = 0$ for some (p, m)-bounded number $u(p, m)$ and the m-bounded number $mg = mg(m)$, so that Theorem 14.3 would be proved. Thus, *we may assume that the additive group of L is homocyclic* from the outset; we fix the notation p^e for the exponent of the additive group of L.

One of the main ideas of the proof, lifted Lie products, already appeared in Chapter 13. Suppose that b is the minimal and t is the maximal positive integer such that $p^b L \le [L, L] \le p^t L$ ("bottom" and "top"). If $e \le 2t$, then $[L, L, L] \le p^{2t}L \le p^e L = 0$, that is, L is nilpotent of class 2; so we may assume $e > 2t$. We can define the new operation exactly as in Definition 13.24, by putting $[\![x, y]\!]$ to be any of the p^tth roots of $[x, y]$ in the additive group of L. By Proposition 13.26, $\tilde{L} = L/p^{e-2t}L$ is then a Lie ring with respect to the old addition and this new multiplication, and φ becomes an automorphism of this new Lie ring $\tilde{L}$ with at most p^m fixed points. (We use tildes to denote images in $L/p^{e-2t}L$, rather than bars as in § 13.3, because we shall need bars for other purposes here.) Note that (14.4) holds for $\tilde{L}$ too. By Higman's Theorem 7.19(b),

$$\gamma_{h(p)+1}(p^{[m/(h(p)+1)]+2}\tilde{L}) \le p^m \gamma_{h(p)+1}(p\tilde{L}) \le p^m {}_{\mathrm{id}}\big\langle C_{\tilde{L}}(\varphi)\big\rangle = {}_{\mathrm{id}}\big\langle p^m C_{\tilde{L}}(\varphi)\big\rangle = 0.$$

Replacing L by the subring $p^{[m/(h(p)+1)]+2}L$ of (p, m)-bounded index, *we may assume that*

$$\gamma_{h(p)+1}(\tilde{L}) = 0 \tag{14.9}$$

from the outset. Let $[\![L, L]\!]$ denote the additive subgroup of L generated by the $[\![x, y]\!]$, $x, y \in L$; then $\overline{[\![L, L]\!]} = [\![\tilde{L}, \tilde{L}]\!]$. In the homocyclic additive group of L the inclusion $p^b L \le [L, L] = p^t [\![L, L]\!]$ implies that $p^{b-t}L \le [\![L, L]\!]$, unless $b = e$. Then $p^{b-t}\tilde{L} \le [\![\tilde{L}, \tilde{L}]\!]$, including the case of $b = e$, since $p^{e-t}\tilde{L} \le p^{e-2t}\tilde{L} = 0$. By (14.9), it follows that

$$p^{h(p)(b-t)}\tilde{L} \le \gamma_{h(p)+1}(\tilde{L}) = 0. \tag{14.10}$$

Thus, we have the following lemma.

Lemma 14.11. *We have*

    (a) $e - 2t \le h(p)(b - t)$;

    (b) $\gamma_3(p^{e-2t}L) = 0$.

Proof. (a) This follows from (14.10), since p^{e-2t} is the exponent of the additive group of $\tilde{L}$. (b) We have $\gamma_3(p^{e-2t}L) = p^{3e-6t}\gamma_3(L) \le p^{3e-6t+2t}L \le p^e L = 0 \Leftrightarrow 3e - 4t \ge e \Leftrightarrow e \ge 2t$ which is true by our assumption. □

Note that if $e - 2t$ is a (p, m)-bounded number, then $p^{e-2t}L$ is the required subring of (p, m)-bounded index that is nilpotent of class 2 by Lemma 14.11(b), and Theorem 14.3 is proved. In § 13.3 the difference $b - t$ was (p, m)-bounded, and so the result followed as in Lemma 14.11. The main efforts in this chapter will be effectively applied to make the difference $b - t$ small. The new Lie ring with the Lie product $[\![\,,\,]\!]$ will again be used for that. The scheme of the proof is as follows. First, we show that φ may be assumed to have strictly less fixed points on $([L, L] + p^b L)/p^b L$; induction on m then gives that this Lie ring contains an ideal of m-bounded derived length and (p, m)-bounded index. This means that $(p^s L)^{(g)} \le p^b L$ for some m-bounded g and (p, m)-bounded s. The central proposition states that $p^{db+\varepsilon}L$ is nilpotent of class 2 for some m-bounded d and (p, m)-bounded ε. Theorem 14.3 then follows since it is easy to jump from $p^b L$ to $p^{db}L$ simply by taking $\gamma_d(p^b L)$.

In order to lighten notation, we shall denote an elementary abelian section $(H \cap p^x M + p^{x+1}M)/p^{x+1}M$ of an additive group M simply by $\overline{H \cap p^x M}$ and call this section the p^x-*slice* of H (in M).

We consider the inequalities for the orders of the p^i-slices of $[L, L]$:

$$|\overline{[L, L] \cap p^t L}| \le |\overline{[L, L] \cap p^{t+1}L}| \le \dots$$

$$\dots \le |\overline{[L, L] \cap p^{b-1}L}| < |\overline{[L, L] \cap p^b L}| = |p^b L/p^{b+1}L|. \quad (14.12)$$

These inequalities hold since $[L, L]$ is invariant under the mapping $x \to px$ which induces isomorphisms of the factors $p^i L/p^{i+1}L$, $i = t, t + 1, \dots, b$; the last strict inequality and equality hold by the definition of b. The number of strict inequalities in (14.12) is at most pm, since the rank of the additive group of L is at most pm. Let

$$|\overline{[L, L] \cap p^{b_1}L}| = \dots = |\overline{[L, L] \cap p^{b_2-1}L}|$$

be the last "long" segment of equalities of length $b_2 - b_1 - 1 \ge 2$ in (14.12) (that is, with $b_2 - b_1 \ge 3$ equal successive orders). Then $([L, L] \cap p^{b_1}L)/([L, L] \cap p^{b_2}L)$ is a homocyclic group of exponent $\ge p^3$, while $b - b_2$ is (p, m)-bounded, since there are at most pm gaps, each ≤ 2, between b_2 and b in (14.12). If there are no segments of equalities of length ≥ 3 in (14.12), then $b - t \le 2pm$, whence $e - 2t$ is (p, m)-bounded by Lemma 14.11(a), and Theorem 14.3 follows by 14.11(b).

Lemma 14.13. $|C_{([L, L]+p^{b_2}L)/p^{b_2}L}(\varphi)| < p^m$.

Proof. We put $U = p^{b_2-3}L/p^{b_2}L$. By the choice of b_2, the image of $[L,L] \cap p^{b_2-3}L$ in U is a proper homocyclic subgroup V, say, of exponent p^3.

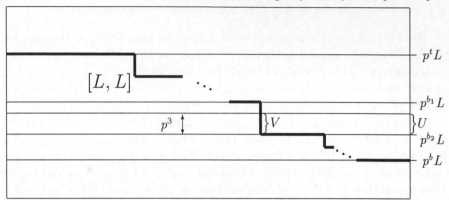

By Lemma 1.6(a), V is a proper direct summand of U. By Lemma 13.2, there is a φ-invariant subgroup $W \le U$ such that $pU \le V + W$ and $p(V \cap W) = 0$. Let a bar denote the image in $(V+W)/(V \cap W) = \overline{V} \oplus \overline{W}$. Then $\overline{W} \ne 0$, whence $|C_{\overline{V}}(\varphi)| < |C_{\overline{V} \oplus \overline{W}}(\varphi)|$. By Lemma 2.12, $|C_{\overline{V} \oplus \overline{W}}(\varphi)| \le |C_L(\varphi)|$. On the other hand, by Lemma 14.5, $|C_{([L,L]+p^{b_2}L)/p^{b_2}L}(\varphi)| = |C_V(\varphi)| = |C_{\overline{V}}(\varphi)|$. Thus,

$$|C_{([L,L]+p^{b_2}L)/p^{b_2}L}(\varphi)| = |C_{\overline{V}}(\varphi)| < |C_{\overline{V} \oplus \overline{W}}(\varphi)| \le |C_L(\varphi)| = p^m,$$

as required.   □

Using induction on m in the proof of Theorem 14.3 we conclude that the factor-ring $([L,L]+p^{b_2}L)/p^{b_2}L$ has a soluble ideal $(p^r[L,L]+p^{b_2}L)/p^{b_2}L$ of m-bounded derived length $f(m)$, for some (p,m)-bounded $r = r(p,m)$. Then

$$p^{b-b_2}(p^r[L,L])^{(f(m))} \le p^{b-b_2}p^{b_2}L = p^bL,$$

whence, for $g = f(m)+1$,

$$(p^sL)^{(g)} \le p^bL \tag{14.14}$$

for some (p,m)-bounded number $s = s(p,m)$, since $b - b_2$ is (p,m,n)-bounded. Theorem 14.3 will easily follow from (14.14) and the following proposition.

Proposition 14.15. *Suppose that L is a Lie ring whose additive group is a homocyclic finite p-group of exponent p^e. Let $b = b(L)$ be the minimal and $t = t(L)$ the maximal positive integer such that $p^bL \le [L,L] \le p^tL$. Suppose that L admits an automorphism φ of order p satisfying (14.4) and having exactly p^m fixed points. Then, for some m-bounded number $d = d(m)$ and a (p,m)-bounded number $\varepsilon = \varepsilon(p,m)$, the subring $p^{d(b-t)+\varepsilon}L$ is nilpotent of class at most 2.*

First we show how Theorem 14.3 follows from Proposition 14.15.

Reduction of Theorem 14.3 to Proposition 14.15. Suppose we have proved Proposition 14.15. For s and g as in (14.14) and for d and ε as in Proposition 14.15, the ideal $H = p^{s+[(\varepsilon+1)/d2^g]}L$ has (p,m)-bounded index in L and

$$
\begin{aligned}
\gamma_3(\gamma_d(H^{(g)})) &\leq \gamma_3(\gamma_d(p^{[(\varepsilon+1)/d2^g]2^g}(p^s L)^{(g)})) \\
&\leq \gamma_3(\gamma_d(p^{b+[(\varepsilon+1)/d2^g]2^g}L)) \\
&\leq \gamma_3(p^{db+\varepsilon}L) \leq \gamma_3(p^{d(b-t)+\varepsilon}L) = 0.
\end{aligned}
$$

Thus, H is the required subring, being soluble of m-bounded derived length $g + \log_2 d + 3 = f(m) + \log_2 d(m) + 4$. $\qquad\qquad\qquad\square$

Proof of Proposition 14.15. First we apply the same arguments as before to L satisfying the hypothesis of Proposition 14.15. We may assume that $e > 2t$, for otherwise $[L, L, L] \leq p^{2t}L \leq p^e L = 0$, so that L itself is nilpotent of class 2. As above, we apply Definition 13.24 and Proposition 13.26 to define the Lie ring $\widetilde{L}$ on the additive group of $L/p^{e-2t}L$ with respect to the new multiplication $[\![\,,\,]\!]$. The same argument as above shows that we may assume that (14.9) holds, and hence (14.10) and Lemma 14.11 hold too.

We proceed by induction on $b - t$. When $b - t$ is (p,m)-bounded, the result follows by Lemma 14.11: then $e - 2t$ is (p,m)-bounded by 14.11(a), and $\gamma_3(p^{e-2t}L) = 0$ by 14.11(b), so that we can take $\varepsilon = e - 2t$ and $d = 0$. (Saying here that $b - t$ is "(p,m)-bounded", we mean that $b - t$ is not large enough to qualify for the subsequent arguments, not larger than, indeed, a certain (p,m)-bounded number to be actually determined in what follows.) Hence we may assume $b - t$ to be large enough. The above calculation also shows that $e - 2t$ may be assumed to be large enough too.

It is sufficient to find a φ-invariant section S of L which is "m-close to L" and has smaller difference of the parameters $b(S)$ and $t(S)$, smaller than $b-t = b(L)-t(L)$, where $b(S)$ and $t(S)$ are, respectively, the minimal and the maximal integer such that $p^{b(S)}S \leq [S,S] \leq p^{t(S)}S$. (By a φ-invariant section we mean a factor-ring of a φ-invariant subring over a φ-invariant ideal of this subring.) More precisely, suppose that we found a φ-invariant section S of L such that, for some m-bounded numbers $d_1 = d_1(m)$ and $d_2 = d_2(m)$,

- $p^{d_1}L/I \leq S \leq L/I$, where I is a φ-invariant ideal such that $p^{d_2}I = 0$;

- the additive group of S is homocyclic;

- $b(S) - t(S) < b - t$.

$$\text{(14.16)}$$

By the induction hypothesis applied to S, we shall have $\gamma_3(p^{d(b-t-1)+\varepsilon}S) = 0$ for some m-bounded number $d = d(m)$ and (p,m)-bounded number $\varepsilon = \varepsilon(p,m)$, which means that $\gamma_3(p^{d(b-t-1)+\varepsilon+d_1}L) \leq I$. Then

$$\gamma_3(p^{d(b-t)+\varepsilon+d_1+[d_2/3]+1-d}L) \leq p^{d_2}\gamma_3(p^{d_1+d(b-t-1)+\varepsilon}L) \leq p^{d_2}I = 0. \quad (14.17)$$

If necessary, we can enlarge d to ensure that $d_1 + [d_2/3] + 1 \leq d$. Since d_1 and d_2 are m-bounded numbers, this enlargement has to be done only once, and the enlarged value of d remains to be m-bounded. Then $\varepsilon + d_1 + [d_2/3] + 1 - d \leq \varepsilon$, and (14.17) implies that $\gamma_3(p^{d(b-t)+\varepsilon}L) = 0$, as required.

Before making our way through the technically difficult and multistage construction of S satisfying (14.16), let us take a more relaxed, informal look at the simpler case of $m = 2$ where we can see more clearly the ideas behind the calculations. Let us use $\approx$ to signify "almost equality", whatever meaning that may have. So let $m = 2$. Recall that $p^bL \leq [L,L] \leq p^tL$ with b minimal and t maximal possible and that p^e is the exponent of the additive group of L. Using Lemma 14.6 and the generalized Maschke's Lemma 13.6, one can show that then $[L,L] \approx p^tV + p^bW$, where both V and W are one-generator $\mathbb{Z}\langle\varphi\rangle$-modules such that $L \approx V + W$. The Lie ring $\tilde{L} = L/p^{e-2t}L$ with respect to the new Lie products $[\![\,,]\!]$ is nilpotent of class $h(p)$ (see (14.9)). Hence we must have $[V,L] \lesssim p^{t+\Delta}L$ for $\Delta \approx (e-2t)/h(p)$: otherwise, since $e-2t$ is large enough, we would have $[V,L] \gtrsim p^{t+\lambda}V \Rightarrow [\tilde{V},\tilde{L}] \geq p^\lambda\tilde{V}$ for some $\lambda < (e-2t)/h(p)$, whence $\gamma_{h(p)+1}(\tilde{L}) \geq p^{h(p)\lambda}\tilde{V} > p^{e-2t}\tilde{V}$ and hence $\gamma_{h(p)+1}(\tilde{L}) \neq 0$, a contradiction. Since $[V,L] \leq [L,L] \approx p^tV + p^bW$, it follows that $[V,L] \lesssim p^{t+\Delta}V + p^bW$.

Now we put $L_1 = V + p^\delta L$ for some δ which is large enough to beat the "$\approx$-fudging", but is much smaller than Δ or $b - t$. Then, on the one hand,

$$
\begin{aligned}
[L_1,L_1] &\approx [V,V] + p^\delta[V,L] + p^{2\delta}[L,L] \lesssim [V,L] + p^{2\delta}[L,L] \\
&\lesssim p^{t+\Delta}V + p^bW + p^{2\delta}[L,L] \\
&\approx p^{t+\Delta}V + p^bW + p^{t+2\delta}V + p^{b+2\delta}W \\
&\lesssim p^{t+2\delta}V + p^{t+3\delta}W,
\end{aligned}
$$

because $\Delta \geq 2\delta$ and $b \geq t + 3\delta$, since $b - t$ is large enough. The right-hand side is exactly $p^{t+2\delta}(V + p^\delta W) = p^{t+2\delta}L_1$. This means that $t(L_1) \gtrsim t + 2\delta$. On the other hand,

$$[L_1,L_1] \geq p^{2\delta}[L,L] \gtrsim p^{t+2\delta}V + p^{b+2\delta}W \gtrsim p^{b+\delta}V + p^{b+2\delta}W,$$

because $t + 2\delta \leq b + \delta$, since $b - t$ is large enough. Here we have exactly $p^{b+\delta}L_1$ on the right. This means that $b(L_1) \lesssim b + \delta$. To make a section with homocyclic additive group, we factor out the bottom putting $S \approx L_1/p^{e-\delta}V$.

The same inequalities hold for S, since S differs from L_1 only at the bottom of V, far from $t + 2\delta$, since $e - t$ is large enough. We see that $t(S) \gtrsim t(L) + 2\delta$ and $b(S) \lesssim b(L) + \delta$. As a result, $b(S) - t(S) \lesssim b(L) - t(L) - \delta$ and hence $b(S) - t(S) < b(L) - t(L)$, so that S is the required section, satisfying (14.16).

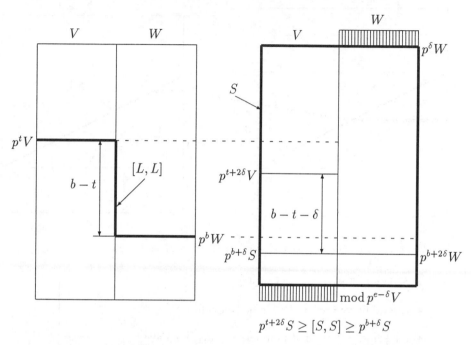

$$p^{t+2\delta} S \geq [S, S] \geq p^{b+\delta} S$$

When m is larger than 2, the picture is more complicated, but the same kind of "skew" choice of S is possible, as a sum of certain one-dimensional $\mathbb{Z}\langle\varphi\rangle$-submodules of L. But, instead of writing an explicit formula for S, we shall approach the desired section S satisfying (14.16) in an m-bounded number of steps. In an inductive construction, in order to diminish the difference $b - t$ we shall "strangle" the top end of the derived subring, tightening the grip by successively subtracting $p - 1$ from the rank of some slice, but moving at each step to a higher slicer, about three times closer to t from below.

To be more precise, in an m-bounded number of steps we shall construct at each step a φ-invariant section S_i which is m-close to S_{i-1}, starting from $S_0 = L$ (so that all S_i will be m-close to L). Let b_k and t_k denote, respectively, the minimal and the maximal integer such that $p^{b_k} S_k \leq [S_k, S_k] \leq p^{t_k} S_k$. At each step either the difference $b_i - t_i$ becomes smaller, in which case $S = S_i$ is the required section satisfying (14.16) and the process terminates, or this difference remains equal to $b - t$ and, for some chosen m-bounded number $\delta = \delta_i$ (which we shall specify later), the rank of the slice $\overline{[S_i, S_i] \cap p^{t_i + \delta} S_i}$ becomes at least $p - 1$ less than the rank of the slice $\overline{[S_{i-1}, S_{i-1}] \cap p^{t_{i-1} + 3\delta + 4} S_{i-1}}$. Since the rank of any section is at most mp, there cannot be more than $2m$ successive subtractions of $p - 1$; this means that after at most $2m$ steps the difference $b_i - t_i$ must become smaller than $b - t$.

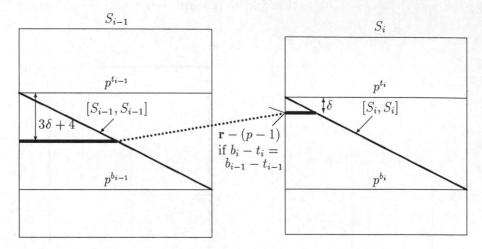

We formalize the construction of S_i from S_{i-1} as follows:

- S_i is a φ-invariant section of S_{i-1};

- the additive group of S_i is homocyclic;

- for some m-bounded numbers d_{i1}, d_{i2}, we have $p^{d_{i1}} S_{i-1}/I_i \leq S_i \leq S_{i-1}/I_i$, where I_i is a φ-invariant ideal of S_{i-1} such that $p^{d_{i2}} I_i = 0$;

- either $b_i - t_i < b - t$, or $b_i - t_i = b - t$ and

$$\mathbf{r}(\overline{[S_i, S_i] \cap p^{t_i+\delta} S_i}) \leq \mathbf{r}(\overline{[S_{i-1}, S_{i-1}] \cap p^{t_{i-1}+3\delta+4} S_{i-1}}) - (p-1)$$

for some m-bounded number $\delta = \delta_i$.

(14.18)

Here $\mathbf{r}U$ denotes the rank of U. Note that at every step we shall have $t_i + 3\delta + 4 < b_i$, because the original difference $b - t$ is large enough and the parameters b_i and t_i change m-boundedly at every step, since S_i is m-close to L. Suppose that this construction can be implemented sufficiently many times. Then, starting with sufficiently large δ, about 3^{2m}, after at most $2m$ steps we must arrive at a situation where the difference $b_i - t_i$ becomes smaller. Otherwise, if the difference $b_i - t_i$ remains constantly equal to $b - t$, there will be $2m$ subtractions of $p - 1$ from the rank, which is impossible, since the rank of L is at most pm. More precisely, we would then have

$$\mathbf{r}\big([\overline{S_{2m}, S_{2m}}] \cap p^{t_{2m}+1} S_{2m}\big)$$

$$\cdots \; \leq \; \mathbf{r}\big([\overline{S_{2m-1}, S_{2m-1}}] \cap p^{t_{2m-1}+7} S_{2m-1}\big) - (p-1) \; \leq \; \cdots$$

$$\cdots \; \leq \; \mathbf{r}\big([\overline{S_{2m-s}, S_{2m-s}}] \cap p^{t_{2m-s}+3^{s+1}-2} S_{2m-s}\big) - s(p-1) \; \leq \; \cdots$$

$$\cdots \; \leq \; \mathbf{r}\big([\overline{L, L}] \cap p^{t+3^{2m+1}-2} L\big) - 2m(p-1), \tag{14.19}$$

where S_1 is constructed from L with $\delta = 3^{2m} - 2$, then S_2 is constructed from S_1 with $\delta = 3^{2m-1} - 2$, and so on, up to S_{2m}, under the assumption that the difference $b_i - t_i$ remains constantly equal to $b - t$. (We used the elementary formula $\underbrace{3(\ldots(3\underbrace{(3+4)+4)+\cdots)+4}_{s}}_{s} = 3^{s+1} - 2$ above.) But the right-hand side of (14.19) is negative, since the rank is at most pm, a contradiction. Thus, there must be the required decrease of the difference $b_{i_0} - t_{i_0} < b - t$ for some $i_0 \leq 2m$, which means that $S = S_{i_0}$ is the required section satisfying (14.16), and Proposition 14.15 is proved.

Each S_i has the same properties as L, and the next section S_{i+1} is constructed based on S_i only. Thus, we need to describe only one step from S_i to S_{i+1} satisfying (14.18) and, moreover, we may consider only the step from L to S_1. The major step in constructing S_1 is the following lemma; we shall see later that an element a satisfying its hypothesis always exists because of the nilpotency of $\widetilde{L}$ (with respect to the new Lie products $[\![\,,\,]\!]$). Again, "m-bounded" here means that the value is bounded by a certain function of m which is determined by the subsequent arguments. Recall that V_a denotes the $\mathbb{Z}\,\langle\varphi\rangle$-submodule generated by a; see Lemma 14.6 for the properties of V_a.

Lemma 14.20. *Suppose that, for some m-bounded number $\delta \in \mathbb{N}$, an element $a \in L \setminus pL$ is such that $p^{t+2\delta+3} V_a \subseteq [L, L]$ and $[V_a, L] \subseteq p^{t+2\delta+4} L$. Put $L_1 = V_a + p^{\delta+2} L$. Then*

(a) L_1 *is a φ-invariant ideal of L such that $p^{b+\delta+2} L_1 \leq [L_1, L_1] \leq p^{t+\delta+2} L_1$, that is, $t(L_1) \geq t + \delta + 2$ and $b(L_1) \leq b + \delta + 2$, so that $b(L_1) - t(L_1) \leq b - t$;*

(b) *if $t(L_1) = t + \delta + 2$, then*

$$\mathbf{r}\big([\overline{L_1, L_1}] \cap p^{t(L_1)+\delta} L_1\big) \; \leq \; \mathbf{r}\big([\overline{L, L}] \cap p^{t+3\delta+4} L\big) - (p-1).$$

Proof. (a) Since $[V_a, L] \subseteq p^{t+2\delta+4} L \leq L_1$, the sum $L_1 = V_a + p^{\delta+2} L$ is a φ-invariant ideal of L. We have

$$\begin{aligned} [L_1, L_1] &= [V_a, V_a] + p^{\delta+2}[V_a, L] + p^{2\delta+4}[L, L] \\ &\leq [V_a, L] + p^{2\delta+4}[L, L] \\ &\leq p^{t+2\delta+4} L \leq p^{t+\delta+2} L_1, \end{aligned} \tag{14.21}$$

so that $t(L_1) \geq t + \delta + 2$, as required. To estimate $b(L_1)$ we again express everything in terms of L rather than L_1; in particular, $p^{b+\delta+2}L_1 = p^{b+\delta+2}V_a + p^{b+2\delta+4}L$. We have

$$[L_1, L_1] \geq p^{2\delta+4}[L, L] \geq p^{b+2\delta+4}L.$$

It remains to show that $p^{2\delta+4}[L, L] \geq p^{b+\delta+2}V_a$. By the hypothesis, $p^{t+2\delta+3}V_a \leq [L, L]$. Since $b - t$ is large enough and δ is m-bounded, we have $b + \delta + 2 \geq t + 4\delta + 7$, whence

$$p^{b+\delta+2}V_a \leq p^{t+4\delta+7}V_a \leq p^{2\delta+4}[L, L].$$

(b) Let now $t(L_1) = t + \delta + 2$. To calculate the rank of the slice in L_1 in question, we express everything in terms of L. As in (14.21) we have $[L_1, L_1] \leq [L, L] \cap p^{t+2\delta+4}L$. Hence the $p^{t(L_1)+\delta}$-slice of $[L_1, L_1]$ in L_1 in question,

$$\overline{[L_1, L_1] \cap p^{t(L_1)+\delta}L_1}$$

$$= ([L_1, L_1] \cap p^{t(L_1)+\delta}L_1 + p^{t+2\delta+3}V_a + p^{t+3\delta+5}L) \big/ (p^{t+2\delta+3}V_a + p^{t+3\delta+5}L),$$

is a subgroup of the factor-group

$$([L, L] \cap p^{t+2\delta+4}L + p^{t+2\delta+3}V_a + p^{t+3\delta+5}L) \big/ (p^{t+2\delta+3}V_a + p^{t+3\delta+5}L).$$

By the Homomorphism Theorems, this factor-group is isomorphic to the factor-group of

$$Q = ([L, L] \cap p^{t+2\delta+4}L + p^{t+3\delta+5}L) \big/ p^{t+3\delta+5}L$$

by R, the image of $([L, L] \cap p^{t+2\delta+4}L + p^{t+3\delta+5}L) \cap p^{t+2\delta+3}V_a$ in $L/p^{t+3\delta+5}L$. We claim that

- the rank of Q is equal to the rank of the $p^{t+3\delta+4}$-slice of $[L, L]$ in L, the other slice in question;

- R contains a direct summand of Q of rank $p - 1$

(then this summand is "cut off" in Q/R, which will finish the proof). Indeed, the rank of Q is equal to the rank of $\overline{[L, L] \cap p^{t+3\delta+4}L}$ because in the homocyclic section $p^{t+2\delta+4}L/p^{t+3\delta+5}L$ the subgroup $p^{t+3\delta+4}L/p^{t+3\delta+5}L$ contains all elements of order p, so that

$$\overline{[L, L] \cap p^{t+3\delta+4}L} = \Omega_1 \left(([L, L] \cap p^{t+2\delta+4}L + p^{t+3\delta+5}L) \big/ p^{t+3\delta+5}L \right) = \Omega_1(Q).$$

Now we consider the second group R. We have

$$([L, L] \cap p^{t+2\delta+4}L + p^{t+3\delta+5}L) \cap p^{t+2\delta+3}V_a \geq [L, L] \cap p^{t+2\delta+4}L \cap p^{t+2\delta+3}V_a.$$

By the choice of a and by the properties of V_a (Lemma 14.6), we have

$$[L, L] \geq p^{t+2\delta+3} V_a \geq p^{t+2\delta+4} L \cap V_a,$$

and $p^{t+2\delta+4} L \cap V_a$ is a homocyclic group of rank $p - 1$. Hence

$$[L, L] \cap p^{t+2\delta+4} L \cap p^{t+2\delta+3} V_a = p^{t+2\delta+4} L \cap V_a \leq [L, L]$$

and the image of $p^{t+2\delta+4} L \cap V_a$ in $L/p^{t+3\delta+5} L$ is a homocyclic subgroup of exponent $p^{\delta+1}$ and of rank $p - 1$ contained in

$$Q = ([L, L] \cap p^{t+2\delta+4} L + p^{t+3\delta+5} L)/p^{t+3\delta+5} L.$$

Since Q also has exponent $p^{\delta+1}$, the image of $p^{t+2\delta+4} L \cap V_a$ in $L/p^{t+3\delta+5} L$ is a direct summand of Q by Lemma 1.6(a). This image is contained in R, and hence $\mathbf{r}(Q/R) \leq \mathbf{r}Q - (p - 1)$. Since the $p^{t(L_1)+\delta}$-slice of $[L_1, L_1]$ in L_1 is a subgroup of Q/R, the result follows. $\square$

Back in the proof of Proposition 14.15, the required section S_1 (satisfying (14.18) with $S_0 = L$) is defined to be $L_1/p^{e-\delta-2} L_1$, where, recall, p^e is the exponent of the additive group of L. The additive group of $L_1/p^{e-\delta-2} L_1$ is homocyclic because that of L is homocyclic of exponent p^e and $L_1 \geq p^{\delta+2} L$ (Lemma 1.5). We have $p^{e-\delta-2} L_1 = p^{e-\delta-2} V_a + p^{e-\delta-2+\delta+2} L = p^{e-\delta-2} V_a$; this is an ideal of L, since $[p^{e-\delta-2} V_a, L] \leq p^{e-\delta-2+t+2\delta+4} L = 0$. The ranks of the ambient slices for S_1 coincide with those for L_1 and $t(S_1) = t(L_1)$ because $e - 2t$ is large enough. (The parameter $b(S_1)$ can only become less than $b(L)$, in which case $b(S_1) - t(S_1) < b - t$ and $S = S_1$ is the required section, satisfying (14.16).)

To finish the proof of Proposition 14.15, and hence of Theorem 14.3, it remains to show that we can always find an element a satisfying the hypothesis of Lemma 14.20 for a given m-bounded δ. Note that $p^s V_a \subseteq [L, L]$ if and only if $p^s a \in [L, L]$, and $[V_a, L] \subseteq p^s L$ if and only if $[a, L] \subseteq p^s L$. Suppose that, for an m-bounded δ, no a exists satisfying the condition of Lemma 14.20. We choose $a_1 \in L \setminus pL$ such that $p^{t+2\delta+3} a_1 \in [L, L]$, which is possible, since $e - 2t$ is large enough. Since $[a_1, L] \not\subseteq p^{t+2\delta+4} L$ by our assumption, there is $a_2 \in L \setminus pL$ such that $p^{t+2\delta+3} a_2 \in [a_1, L]$, and so on. Let $[\![u, L]\!]$ denote the additive subgroup generated by the elements $[\![u, l]\!]$, $l \in L$; then $\widetilde{[\![u, L]\!]} = [\![\tilde{u}, \tilde{L}]\!]$ in the Lie ring $\tilde{L}$ (with new Lie products $[\![\,,]\!]$). Since $e - t \geq b - t$ is large enough, we have the following in the homocyclic additive group of L of exponent p^e:

$$p^{t+2\delta+3} a_1 \in [L, L] = p^t [\![L, L]\!] \;\; \Rightarrow \;\; p^{2\delta+3} a_1 \in [\![L, L]\!];$$

$$p^{t+2\delta+3} a_2 \in [a_1, L] = p^t [\![a_1, L]\!] \;\; \Rightarrow \;\; p^{2\delta+3} a_2 \in [\![a_1, L]\!]$$

$$\Rightarrow \;\; p^{2(2\delta+3)} a_2 \in [\![p^{2\delta+3} a_1, L]\!] \leq [\![[\![L, L]\!], L]\!],$$

and so on. In the Lie ring $\widetilde{L}$, after $h(p)$ steps, we obtain using (14.9)

$$p^{h(p)(2\delta+3)}\widetilde{a}_{h(p)} \in \gamma_{h(p)+1}(\widetilde{L}) = 0.$$

Since $a_{h(p)} \in L \setminus pL$, the order of $\widetilde{a}_{h(p)}$ in the additive group of $\widetilde{L}$ is p^{e-2t}, and hence we must have $e - 2t \leq h(p)(2\delta + 3)$. This, however, contradicts our assumption that $e - 2t$ is large enough. $\qquad\square$

Remarks. 14.22. One can show that (14.4) implies that the rank of L is at most $m(p-1)$, so that, in fact, m steps would suffice in the process of constructing the S_i described, starting with $\delta = 3^{m+1} - 2$.

14.23. In [Yu. Medvedev, 1994b] the part of the proof that deals with decreasing the difference $b - t$ is encoded into essentially one formula; we tried to loosen this tight knot by breaking the calculation into small steps, in the hope of making the idea come through more clearly.

§ 14.2. Almost nilpotency of m-bounded class

In proving Theorem 14.2, we may assume the Lie ring in question to be soluble of m-bounded derived length by Theorem 14.3. Induction on the derived length (using Theorem 5.27) reduces the proof to the main case where the derived length is 2. Many of the steps resemble the proof of Theorem 14.3; a new lifted Lie ring product $[\![\,,]\!]$ is also defined in this section and, again, the main efforts are essentially applied to squeeze the ideal $[L, L, L]$ (instead of $[L, L]$) into a (p, m)-bounded layer of L. It may be left as an exercise to the reader to adjust the reasoning from § 14.1 for the proof of Theorem 14.2 for Lie rings of derived length 2 and then to perform induction on the derived length. So the contents of this section may be regarded as a solution to this exercise.

There is a construction in Lie rings which is analogous to semidirect products for groups: a Lie ring is a *semidirect sum* of the ideal A and a subring B if its additive group is the direct sum $A \oplus B$. The case of derived length 2 can be easily reduced to the case of a semidirect sum, which is dealt with in the following proposition.

Proposition 14.24. *Let $L = A \oplus B$ be a Lie ring whose additive group is a finite p-group, with abelian ideal A and abelian subring B. Suppose that L admits an automorphism φ of order p with exactly p^m fixed points, such that both A and B are φ-invariant. Then $p^{f(p,m)}[A, \underbrace{B, \dots, B}_{g(m)}] = 0$ for a (p, m)-bounded number $f(p, m)$ and an m-bounded number $g(m)$.*

Proof. First, the same argument as in § 14.1 shows that we may assume L

to satisfy (14.4). Next, we perform a reduction to the case, where the additive group of A is homocyclic, similar to that in the proof of Theorem 14.3. Since the rank of A is at most mp, there can be at most mp strict inequalities in the chain

$$|A/pA| \geq |pA/p^2A| \geq \ldots \geq |p^iA/p^{i+1}A| \geq \ldots .$$

Every segment with equalities gives rise to a homocyclic section of A, so that we obtain the following series of length at most mp with homocyclic factors:

$$A > p^{i_1}A > p^{i_2}A > \ldots > p^{i_l}A > 0. \qquad (14.25)$$

Lemma 14.26. *There are at most m factors of exponent $\geq p^3$ in (14.25).*

Proof. Repeat the proof of Lemma 14.8, replacing the additive group of L by that of A. □

For every homocyclic section $H = p^{i_r}A/p^{i_{r+1}}A$ in (14.25) we can form the Lie ring $H \oplus B$ with naturally defined multiplication; this Lie ring admits φ as an automorphism of order dividing p with at most p^m fixed points. Suppose that we proved Proposition 14.24 in the case where the additive group of A is homocyclic. Then there are a (p,m)-bounded number $r = r(p,m)$ and an m-bounded number $u = u(m)$ such that, for every "big" section H in (14.25), of additive exponent p^3, we have $p^r[H, \underbrace{B, \ldots, B}_{u}] = 0$. The small sections in (14.25), of exponents $\leq p^2$, together with the sections H/p^rH glue up to at most $m + 1$ factors between at most m sections p^rH; all these pieces are of (p,m)-bounded exponent dividing p^{2pm+r}. Then

$$p^{(m+1)(2pm+r)}[A, \underbrace{B, \ldots, B}_{mu}] = 0.$$

Thus, *we may assume that the additive group of A is homocyclic* from the outset; we fix the notation p^e for the exponent of the additive group of A.

Suppose that b is the minimal and t is the maximal positive integer such that $p^bA \leq [A, B] \leq p^tA$. If $e - 2t \leq g(p,m)$ for some (p,m)-bounded $g(p,m)$, then $p^{g(p,m)}[A, B, B] \leq p^{g(p,m)+2t}A \leq p^eA = 0$ and the proposition is proved. Thus, *we may assume $e - 2t$ to be large enough.* We define the new operation for any $x \in A$, $y \in B$ by putting $[\![x, y]\!]$ to be any of the p^tth roots of $[x, y]$ in the additive group of A, and $[\![z_1, z_2]\!] = 0$ if either $z_1, z_2 \in A$ or $z_1, z_2 \in B$. Then $\widetilde{L} = (A/p^{e-2t}A) \oplus B$ becomes a Lie ring with respect to the old addition and this new multiplication, and φ becomes an automorphism of the new Lie ring $\widetilde{L}$, with at most p^m fixed points. Note that (14.4) holds for $\widetilde{L}$ too. By Remark 7.20,

$$\gamma_{p+1}(p^{[m/(p+1)]+2}L) \leq p^m\gamma_{p+1}(p\widetilde{L}) \leq p^m\mathrm{id}\,\langle C_L(\varphi)\rangle = \mathrm{id}\,\langle p^mC_L(\varphi)\rangle = 0,$$

for some (p, m) bounded number v. Replacing L by the subring $p^{[m/(p+1)]+2}L$ of (p, m)-bounded index, we may assume that $\gamma_{p+1}(\tilde{L}) = 0$ from the outset. Since $p^b A \leq [A, B] = p^t[A, B]$ (where $[A, B]$ is the additive subgroup of A generated by the $[x, y]$), in the homocyclic additive group of A we obtain $p^{b-t}A \leq [A, B]$, unless $b = e$. Then also $p^{b-t}\tilde{A} \leq [\tilde{A}, \tilde{B}]$, including the case $b = e$, whence

$$p^{p(b-t)}\tilde{A} \leq \gamma_{p+1}(\tilde{L}) = 0. \tag{14.27}$$

Lemma 14.28. *We have*

(a) $e - 2t \leq p(b - t)$;

(b) $p^{e-2t}[A, B, B] = 0$.

Proof. (a) This follows from (14.27) since the exponent of the additive group of $\tilde{A}$ is p^{e-2t}. (b) We have $p^{e-2t}[A, B, B] \leq p^{e-2t+2t}A = 0$. □

As in the proof of Theorem 14.3, we shall effectively try to decrease the difference $b - t$. Recall that $\overline{H \cap p^x M}$ denotes the p^x-slice of H in M, the elementary abelian section $(H \cap p^x M + p^{x+1}M)/p^{x+1}M$ of an additive group M.

We consider the inequalities

$$|\overline{[A, B] \cap p^t A}| \leq |\overline{[A, B] \cap p^{t+1}A}| \leq \ldots$$

$$\ldots \leq |\overline{[A, B] \cap p^{b-1}A}| < |\overline{[A, B] \cap p^b A}| = |p^b A/p^{b+1}A|. \tag{14.29}$$

These inequalities hold since $[A, B]$ is invariant under the mapping $x \to px$ which induces isomorphisms of the factors $p^k A/p^{k+1}A$, $k = t, t+1, \ldots, b$; the last strict inequality and equality hold by the definition of b. The number of strict inequalities in this chain is at most pm, since the rank of the additive group of A is at most pm. Let

$$|\overline{[A, B] \cap p^{b_1}A}| = \ldots = |\overline{[A, B] \cap p^{b_2-1}A}|$$

be the last "long" segment of equalities of length $b_2 - b_1 - 1 \geq 2$ (that is, with $b_2 - b_1 \geq 3$ equal successive orders). Then $([A, B] \cap p^{b_1}A)/([A, B] \cap p^{b_2}A)$ is a homocyclic group of exponent $\geq p^3$, while $b - b_2$ is (p, m)-bounded, since there are at most pm gaps, each ≤ 2, between b_2 and b in (14.29). If there are no segments of equalities of length ≥ 2 in (14.29), then $b - t \leq pm + 1$, whence $e - 2t$ is (p, m)-bounded by Lemma 14.28(a), contrary to our assumption.

Lemma 14.30. $|C_{([A,B]+p^{b_2}A)/p^{b_2}A}(\varphi)| < |C_A(\varphi)|$.

Proof. Repeat the proof of Lemma 14.13, replacing L by A and $[L, L]$ by $[A, B]$. □

Now φ has less than p^m fixed points on the Lie ring $(([A,B]+p^{b_2}A)/p^{b_2}A)\oplus$ B (with naturally defined multiplication and the action of φ). By the induction hypothesis, there are a (p,m)-bounded number $r = r(p,m)$ and an $(m-1)$-bounded number $g = g(m-1)$ such that $p^r[[A,B],\underbrace{B,\dots,B}_{g}] \leq p^{b_2}A$, whence

$$p^{r+b-b_2}[A,\underbrace{B,\dots,B}_{g+1}] \leq p^b A. \tag{14.31}$$

Proposition 14.24 will easily follow from (14.31) and the following proposition.

Proposition 14.32. *Let $L = A \oplus B$ be a Lie ring whose additive group is a finite p-group, with abelian ideal A, whose additive group is homocyclic of exponent p^e, and abelian subring B. Let $b = b(A)$ be the minimal and $t = t(A)$ the maximal positive integer such that $p^bA \leq [A,B] \leq p^tA$. Suppose that L admits an automorphism φ of order p satisfying (14.4) with exactly p^m fixed points, such that both A and B are φ-invariant. Then, for some m-bounded number $d = d(m)$ and a (p,m)-bounded number $\varepsilon = \varepsilon(p,m)$, we have $p^{d(b-t)+\varepsilon}[A,B,B] = 0$.*

First, we show that *Proposition 14.24 follows from Proposition 14.32*. For r and g as in (14.31) and for d and ε as in Proposition 14.32, we have

$$p^{d(r+b-b_2)+\varepsilon}[A,\underbrace{B,\dots,B}_{d(g+1)+2}] \leq p^{db+\varepsilon}[A,B,B] = 0.$$

This is the required result, since $d(r + b - b_2) + \varepsilon$ is (p,m)-bounded, and $g(m) = d(m)(g(m-1)+1)+2$ is m-bounded. $\qquad\square$

Proof of Proposition 14.32. We may assume $e > 2t$, for otherwise $[A,B,B]$ $\leq p^{2t}A = 0$ and the result holds with $d = \varepsilon = 0$. As above, we define the Lie ring $\widetilde{L}$ on the additive group of $(A/p^{e-2t}A) \oplus B$ with respect to the new multiplication $[\![\,,]\!]$. The same argument as above shows that we may assume that (14.27) holds, and hence Lemma 14.28 holds too.

We proceed by induction on $b - t$. When $b - t$ is (p,m)-bounded, the result follows by Lemma 14.28: then $e - 2t$ is (p,m)-bounded by 14.28(a), and $p^{e-2t}[A,B,B] = 0$ by 14.28(b), so that we can take $\varepsilon = e - 2t$ and $d = 0$. Hence we may assume $b - t$ to be large enough. The above calculation also shows that $e - 2t$ may be assumed to be large enough too.

It is sufficient to find a φ-invariant and B-invariant section S of A which is "m-close to A" and has difference $b(S) - t(S)$ smaller than $b-t = b(A)-t(A)$. More precisely, suppose that we found a φ-invariant and B-invariant section

S of A such that, for some m-bounded numbers $d_1 = d_1(m)$ and $d_2 = d_2(m)$,

$\left.\begin{array}{l}\end{array}\right.$

- $p^{d_1} A/I \leq S \leq A/I$, where I is a φ-invariant ideal of L contained in A such that $p^{d_2} I = 0$;

- the additive group of S is homocyclic;

- $b(S) - t(S) < b - t$, where $b(S)$ and $t(S)$ are, respectively, the minimal and the maximal integer such that $p^{b(S)} S \leq [S, B] \leq p^{t(S)} S$.

$$\left.\begin{array}{l}\\\\\\\\\\\\\end{array}\right\} \quad (14.33)$$

By the induction hypothesis applied to $S \oplus B$, we shall have $p^{d(b-t-1)+\varepsilon} [S, B, B] = 0$ for some m-bounded $d = d(m)$ and (p, m)-bounded $\varepsilon = \varepsilon(p, m)$. This means that $p^{d(b-t-1)+\varepsilon} [p^{d_1} A, B, B] \leq I$. Then

$$p^{d(b-t)+\varepsilon+d_1+d_2-d} [A, B, B] = p^{d_2+d(b-t-1)+\varepsilon} [p^{d_1} A, B, B] = 0. \qquad (14.34)$$

We can enlarge d, if necessary, to ensure that $d_1 + d_2 \leq d$. Since d_1 and d_2 are m-bounded numbers, this enlargement has to be done only once, and the enlarged value of d remains m-bounded. Then $\varepsilon + d_1 + d_2 - d \leq \varepsilon$ and (14.34) implies that $p^{d(b-t)+\varepsilon} [A, B, B] = 0$, as required.

We shall approach a section S satisfying (14.33) in an m-bounded number of steps, at each step constructing a φ-invariant section S_i which is m-close to S_{i-1}, starting from $S_0 = A$ (hence all S_i will be m-close to A). Let b_k and t_k denote, respectively, the minimal and the maximal integer such that $p^{b_k} S_k \leq [S, B] \leq p^{t_k} S$. At each step, we shall ensure that either the difference $b_i - t_i$ becomes smaller, in which case $S = S_i$ is the required section and the process terminates, or this difference remains equal to $b - t$ and for some chosen m-bounded number $\delta = \delta_i$ (which we shall specify later) the rank of the slice $\overline{[S_i, B] \cap p^{t_i+\delta} S_i}$ becomes at least $p - 1$ less than the rank of the slice $\overline{[S_{i-1}, B] \cap p^{t_{i-1}+2\delta+2} S_{i-1}}$. At every step, we shall have $t_i + 2\delta + 2 < b_i$, since the original difference $b - t$ is large enough, and the parameters b_i and t_i change m-boundedly at every step, since S_i is m-close to A. We summarize the properties of constructing S_i from S_{i-1} as follows:

- S_i is a φ-invariant and B-invariant section of S_{i-1} with homocyclic additive group;

- for some m-bounded numbers d_{i1}, d_{i2}, we have $p^{d_{i1}} S_{i-1}/I_i \leq S_i \leq S_{i-1}/I_i$, where I_i is a φ-invariant and B-invariant ideal of S_{i-1} such that $p^{d_{i2}} I_i = 0$;

- either $b_i - t_i < b - t$, or $b_i - t_i = b - t$ and

$$\mathbf{r}(\overline{[S_i, B] \cap p^{t_i+\delta} S_i}) \leq \mathbf{r}(\overline{[S_{i-1}, B] \cap p^{t_{i-1}+2\delta+2} S_{i-1}}) - (p-1)$$

for some m-bounded number $\delta = \delta_i$.

$$\left.\begin{array}{l}\\\\\\\\\\\\\\\\\\\end{array}\right\} \quad (14.35)$$

Suppose that this construction can be implemented sufficiently many times. Then, starting with sufficiently large δ, about $3 \cdot 2^{2m}$, after at most $2m$ steps we must arrive at a situation where the difference $b_i - t_i$ becomes smaller. Otherwise, if the difference $b_i - t_i$ remains constantly equal to $b - t$, there will be $2m$ subtractions of $p - 1$ from the rank, which is impossible, since the rank of A is at most pm. More precisely, we would then have

$$\mathbf{r}(\overline{[S_{2m}, B] \cap p^{t_{2m}+1}S_{2m}})$$

$$\ldots \quad \leq \quad \mathbf{r}(\overline{[S_{2m-1}, B] \cap p^{t_{2m-1}+4}S_{2m-1}}) - (p-1) \leq \ldots$$

$$\ldots \quad \leq \quad \mathbf{r}(\overline{[S_{2m-s}, B] \cap p^{t_{2m-s}+3 \cdot 2^s - 2}S_{2m-s}}) - s(p-1) \leq \ldots$$

$$\ldots \quad \leq \quad \mathbf{r}(\overline{[A, B] \cap p^{t+3 \cdot 2^{2m}-2}A}) - 2m(p-1), \qquad (14.36)$$

where S_1 is constructed from A with $\delta = 3 \cdot 2^{2m-1} - 2$, then S_2 is constructed from S_1 with $\delta = 3 \cdot 2^{2m-2} - 2$, and so on, up to S_{2m}, under the assumption that the difference $b_i - t_i$ remains constantly equal to $b - t$. (We used the elementary formula $2(\ldots \underbrace{(2(2 + 2) + 2) + \cdots}_{s}) + \underbrace{2}_{s} = 3 \cdot 2^s - 2$ above.) But the right-hand side of (14.36) is negative, since the rank is at most pm, a contradiction. Thus, there must be the required decrease of the difference $b_{i_0} - t_{i_0} < b - t$ for some $i_0 \leq 2m$, which means that $S = S_{i_0}$ is the required section satisfying (14.33), and Proposition 14.32 is proved.

Each S_i has the same properties as A, and the next section S_{i+1} is constructed based on S_i only. Thus, we need to describe only one step from S_i to S_{i+1} satisfying the conditions of the preceding paragraph, and moreover, we may consider only the step from A to S_1. The major step in constructing S_1 is the following lemma. Recall that V_a denotes the $\mathbb{Z} \langle \varphi \rangle$-submodule generated by a (see Lemma 14.6).

Lemma 14.37. *Suppose that, for some m-bounded number* $\delta \in \mathbb{N}$, *an element* $a \in A \setminus pA$ *is such that* $p^{t+\delta+1}V_a \subseteq [A, B]$ *and* $[V_a, B] \subseteq p^{t+\delta+2}A$. *Put* $A_1 = V_a + p^{\delta+2}A$. *Then*

    (a) A_1 *is a* φ-invariant ideal of L *such that* $p^b A_1 \leq [A_1, B] \leq p^t A_1$, *that is,* $t(A_1) \geq t$ *and* $b(L_1) \leq b$, *so that* $b(A_1) - t(A_1) \leq b - t$;
    (b) *if* $t(A_1) = t$, *then*

$$\mathbf{r}(\overline{[A_1, B] \cap p^{t+\delta}A_1}) \leq \mathbf{r}(\overline{[A, B] \cap p^{t+2\delta+2}A}) - (p-1).$$

Proof. (a) Since $[V_a, B] \leq p^{t+\delta+2}A \leq A_1$, the sum $A_1 = V_a + p^{\delta+2}A$ is a φ-invariant ideal of L. We have

$$[A_1, B] = [V_a, B] + p^{\delta+2}[A, B] \leq p^{t+\delta+2}A \leq p^t A_1, \qquad (14.38)$$

so that $t(A_1) \geq t$, as required. To estimate $b(A_1)$ we again express everything in terms of A rather than A_1; in particular, $p^b A_1 = p^b V_a + p^{b+\delta+2} A$. We have

$$[A_1, B] \geq p^{\delta+2}[A, B] \geq p^{b+\delta+2} A.$$

It remains to show that $p^{\delta+2}[A, B] \geq p^b V_a$. By the hypothesis, $p^{t+\delta+1} V_a \leq [A, B]$. Since $b - t$ is large enough and δ is m-bounded, we have $b \geq t + 2\delta + 3$, whence

$$p^b V_a \leq p^{t+2\delta+3} V_a \leq p^{\delta+2}[A, B].$$

(b) Let now $t(A_1) = t$. To calculate the rank of the slice in A_1 in question, we express everything in terms of A. As in (14.38), we have $[A_1, B] \cap p^{t+\delta} A_1 \leq [A_1, B] \leq [A, B] \cap p^{t+\delta+2} A$. Hence the slice in A_1 in question,

$$\overline{[A_1, B] \cap p^{t+\delta} A_1}$$

$$= \left([A_1, B] \cap p^{t+\delta} A_1 + p^{t+\delta+1} V_a + p^{t+2\delta+3} A\right) \big/ \left(p^{t+\delta+1} V_a + p^{t+2\delta+3} A\right),$$

is a subgroup of the factor-group

$$\left([A, B] \cap p^{t+\delta+2} A\right) + p^{t+\delta+1} V_a + p^{t+2\delta+3} A \big/ \left(p^{t+\delta+1} V_a + p^{t+2\delta+3} A\right).$$

This factor-group is isomorphic to the factor-group of

$$Q = \left([A, B] \cap p^{t+\delta+2} A + p^{t+2\delta+3} A\right) \big/ p^{t+2\delta+3} A$$

by R, the image of $\left([A, B] \cap p^{t+\delta+2} A + p^{t+2\delta+3} A\right) \cap p^{t+\delta+1} V_a$ in $A/p^{t+2\delta+3} A$. We claim that

- the rank of Q is equal to the rank of the $p^{t+2\delta+2}$-slice of $[A, B]$ in A, the other slice in question;

- R contains a direct summand of Q of rank $p - 1$

(then this summand is "cut off" in Q/R, which will finish the proof). Indeed, the rank of Q is equal to the rank of $\overline{[A, B] \cap p^{t+2\delta+2} A}$, since in the homocyclic section $p^{t+\delta+2} A/p^{t+2\delta+3} A$ the subgroup $p^{t+2\delta+2} A/p^{t+2\delta+3} A$ contains all elements of order p, so that

$$\overline{[A, B] \cap p^{t+2\delta+2} A} = \Omega_1 \left(([A, B] \cap p^{t+\delta+2} A + p^{t+2\delta+3} A)/p^{t+2\delta+3} A\right) = \Omega_1(Q).$$

Now we consider the second group R. We have

$$\left([A, B] \cap p^{t+\delta+2} A + p^{t+2\delta+3} A\right) \cap p^{t+\delta+1} V_a \geq [A, B] \cap p^{t+\delta+2} A \cap p^{t+\delta+1} V_a.$$

By the choice of a and by the properties of V_a (Lemma 14.6), we have

$$[A, B] \geq p^{t+\delta+1} V_a \geq p^{t+\delta+2} A \cap V_a,$$

and $p^{t+\delta+2}A \cap V_a$ is a homocyclic group of rank $p-1$. Hence

$$[A,B] \cap p^{t+\delta+2}A \cap p^{t+\delta+1}V_a = p^{t+\delta+2}A \cap V_a \leq [A,B]$$

and the image of $p^{t+\delta+2}A \cap V_a$ in $A/p^{t+2\delta+3}A$ is a homocyclic subgroup of exponent p^δ and of rank $p-1$ contained in

$$Q = \left([A,B] \cap p^{t+\delta+2}A + p^{t+2\delta+3}A\right)/p^{t+2\delta+3}A \ .$$

Since Q also has exponent p^δ, the image of $p^{t+\delta+2}A \cap V_a$ in $A/p^{t+\delta+3}A$ is a direct summand of Q by Lemma 1.6(a). This image is contained in R and hence $\mathbf{r}(Q/R) \leq \mathbf{r}Q - (p-1)$. Since the $p^{t+\delta}$-slice of $[A_1, B]$ in A_1 is a subgroup of Q/R, the result follows. $\qquad\Box$

Back in the proof of Proposition 14.32, the required section S_1 is defined to be $A_1/p^{e-\delta-2}A_1$, where, recall, p^e is the exponent of the additive group of A. In fact, $p^{e-\delta-2}A_1 = p^{e-\delta-2}V_a$; this is an ideal of L, since $[p^{e-\delta-2}V_a, B] \leq p^{e-\delta-2+t+\delta+2}A = 0$. The ranks of the ambient slices for S_1 coincide with those for A_1 and $t(S_1) = t(A_1)$ since $e - 2t$ is large enough. (The parameter $b(S_1)$ can become only smaller, in which case $b(S_1) - t(S_1) < b - t$ and $S = S_1$ is the required section, satisfying (14.33).)

To finish the proof of Propositions 14.32 and 14.24, it remains to show that we can always find an element a satisfying the hypothesis of Lemma 14.37 for a given m-bounded δ. Note that $p^s V_a \subseteq [A,B]$ if and only if $p^s a \in [A,B]$, and $[V_a, B] \subseteq p^s A$ if and only if $[a,B] \subseteq p^s A$. Suppose that, for an m-bounded δ, no a exists satisfying the condition of Lemma 14.37. Choose $a_1 \in A \setminus pA$ such that $p^{t+\delta+1}a_1 \in [A,B]$, which is possible, since $e - t$ is large enough. Since $[a_1, B] \not\subseteq p^{t+\delta+2}A$ by our assumption, there is $a_2 \in A \setminus pA$ such that $p^{t+\delta+1}a_2 \in [a_1, B]$, and so on. Let $[\![u, L]\!]$ denote the additive subgroup generated by the elements $[\![u, l]\!]$, $l \in L$; then $\widetilde{[\![u, L]\!]} = [\![\tilde{u}, \tilde{B}]\!]$ in the Lie ring $\tilde{L} = (A/p^{e-2t}A) \oplus B$ (with new Lie products $[\![\,,]\!]$). Since $e - t \geq b - t$ is large enough, we have the following in the homocyclic additive group of L of exponent p^e:

$$p^{t+\delta+1}a_1 \in [A,B] = p^t[\![A,B]\!] \ \Rightarrow \ p^{\delta+1}a_1 \in [A,B];$$

$$p^{t+\delta+1}a_2 \in [a_1, B] = p^t[\![a_1, B]\!] \ \Rightarrow \ p^{\delta+1}a_2 \in [a_1, B]$$

$$\Rightarrow \ p^{2(\delta+1)}a_2 \in [\![p^{\delta+1}a_1, B]\!] \leq [\![[A,B], B]\!],$$

and so on. In the Lie ring $\tilde{L}$, after p steps, we obtain by (14.27)

$$p^{p(\delta+1)}\tilde{a}_p \in \gamma_{p+1}(\tilde{L}) = 0.$$

Since $a_p \in A \setminus pA$, the order of $\tilde{a}_p$ in the additive group of $\tilde{L}$ is p^{e-2t}, and hence we must have $e - 2t \leq p(\delta + 1)$. This, however, contradicts our assumption that $e - 2t$ is large enough. $\qquad\Box$

Corollary 14.39. *Let L be a soluble Lie ring of derived length 2 whose additive group is a finite p-group. Suppose that L admits an automorphism φ of order p with exactly p^s fixed points. Then $\gamma_{g(s)}(p^{f(p,s)}L) = 0$ for a (p,s)-bounded number $f(p,s)$ and an s-bounded number $g(s)$.*

Proof. Put $A = [L, L]$ and $B = L/[L, L]$; then we can form the semidirect sum $A \oplus B$. To wit, the Lie ring operations within either A or B are the same as in $[L, L]$ and $L/[L, L]$ respectively, and for $a \in A$, $b \in B$, by definition, $[a, b] = [a, y]$, where $b = y + [L, L]$ is the image of $y \in L$ in B. Both A and B are φ-invariant and the number of fixed points of φ on B is at most p^s by Lemma 2.12. Thus, the Lie ring $A \oplus B$ satisfies the hypothesis of Proposition 14.24 with $m \leq 2s$. Hence there are a (p, s)-bounded number $u = u(p, s)$ and an s-bounded number $v = v(s)$ such that

$$p^u[A, \underbrace{B, \dots, B}_{v}] = 0.$$

By the definition of the Lie ring $A \oplus B$, this implies that

$$\gamma_{v+1}(p^{[u/(v+1)]+1}L) \leq p^u[[L, L], \underbrace{L, \dots, L}_{v}] = p^u[A, \underbrace{B, \dots, B}_{v}] = 0.$$

$\square$

Proof of Theorem 14.2. Let L be a Lie ring whose additive group is a finite p-group, and let φ be an automorphism of L of order p with exactly p^m fixed points. By Theorem 14.3 we may assume that L is soluble of m-bounded derived length. Hence the result will follow from the following proposition.

Proposition 14.40. *Suppose that L is a soluble Lie ring of derived length d whose additive group is a finite p-group. If L admits an automorphism φ of order p with exactly p^m fixed points, then L has a nilpotent ideal of (m, d)-bounded class which has (p, m, d)-bounded index in the additive group of L.*

Proof. By Theorem 5.27, if a Lie ring M has nilpotent derived subring $[M, M]$ of class c and the factor-ring $M/M^{(2)}$ by the second derived subring is nilpotent of class k, then M itself is nilpotent of (k, c)-bounded class. Using induction on the derived length d, we shall find the required ideal of L in the form $p^{r(p,m)}L$. By the induction hypothesis $\gamma_v(p^u[L, L]) = 0$ for some $u = u(p, m, d-1)$ and $v = (m, d-1)$. By Corollary 14.39 applied to $L/L^{(2)}$, we have $\gamma_g(p^f L) \leq L^{(2)}$ for some $g = g(m)$ and $f = f(p, m)$. We put $w = \max\{[u/2] + 1, f\}$, which is a (p, m)-bounded integer, and put $M = p^w L$. Then, on the one hand,

$$\gamma_{g+4}(M) \;=\; \gamma_{g+4}(p^w L)$$
$$\leq \; [L^{(2)}, \, p^w L, \, p^w L, \, p^w L, \, p^w L]$$
$$\leq \; p^{4w} L^{(2)}$$
$$= \; (p^w L)^{(2)} = M^{(2)}.$$

On the other hand, $\gamma_v([M, M]) = \gamma_v(p^{2w}[L, L]) \leq \gamma_v(p^u[L, L]) = 0$. By Theorem 5.27, M is nilpotent of class bounded in terms of v and g, that is, of m-bounded class. Thus, M is the required nilpotent ideal of (m, d)-bounded class with (p, m, d)-bounded index in the additive group of L. $\qquad\square$

Remark 14.41. One can show that (14.4) implies that the rank of A is at most $m(p-1)$, so that, in fact, m steps would suffice in the process of constructing the S_i described, starting with $\delta = 3 \cdot 2^m - 2$.

Exercises 14

1. Prove that (14.4) implies that the rank of L is at most $m(p-1)$, where $|C_L(\varphi)| = p^m$.

2. Produce explicit upper bounds for the index, the derived length and the nilpotency class of a subgroup (ideal) in Theorems 14.1, 14.2, 14.3.

3. Follow the instructions on pages 174–175 to complete the proof of Theorem 14.3 in the case of $m = 2$.

4. Extend the argument of the preceding exercise for $m = 2$ to the case of $m = 3$.

5. Expand the proofs of Lemmas 14.26 and 14.30.

Bibliography

Books

Baumslag G. [1971] *Lecture notes on nilpotent groups*, Amer. Math. Soc., Providence, R.I.

Bourbaki N. [1980] *Lie groups and algebras, Chapters 1–3*, Springer, Berlin et al.

Cohn P. M. [1965] *Universal algebra*, Harper and Row, New York.

Dixon J. D., du Sautoy M. P. F., Mann A., Segal D. [1991] *Analytic pro-p-groups* (London Math. Soc. Lecture Note Series, **157**), Cambridge Univ. Press.

Gorenstein D. [1968] *Finite groups*, Harper and Row, New York.

Hall M. [1959] *The theory of groups*, Macmillan, New York.

Huppert B. [1967] *Endliche Gruppen.* I, Springer, Berlin et al.

Huppert B., Blackburn N. [1982] *Finite groups.* II, Springer, Berlin et al.

Kargapolov M. I., Merzlyakov Yu. I. [1982] *Fundamentals of the theory of groups*, 3rd ed., Nauka, Moscow (Russian); English transl. of the 2nd ed., Springer, Berlin et al., 1979.

Khukhro E. I. [1993b] *Nilpotent groups and their automorphisms*, de Gruyter, Berlin.

Kostrikin A. I. [1986] *Around Burnside*, Nauka, Moscow (Russian); English transl., Springer, Berlin et al., 1990.

Kourovka Notebook [1995] *Unsolved problems in group theory: the Kourovka Notebook, 13th ed.*, Institute of Mathematics, Novosibirsk (E. I. Khukhro, V. D. Mazurov, editors).

Mal'cev A. I. [1976] *Selected works. Vol. 1, Classical algebra; Vol. 2, Mathematical logic and the general theory of algebraic systems*, Nauka, Moscow (Russian).

Vaughan-Lee M. R. [1993] *The restricted Burnside problem*, 2nd ed., Clarendon Press, Oxford.

van der Waerden B. L. [1970] *Modern algebra*, vols. 1, 2, Ungar, New York.

Warfield R. B. [1976] *Nilpotent groups* (Lecture Notes in Math., **513**), Springer, Berlin et al.

Survey articles

Hartley B. [1987] Centralizers in locally finite groups, *in: Proc. Conf. Group Theory, Brixen/Bressanone, 1986* (Lecture Notes in Math., **1281**), Springer, Berlin et al., 36–51.

Higman G. [1958] Lie ring methods in the theory of finite nilpotent groups, *in:* *Proc. Intern. Congr. Math. Edinburgh, 1958*, Cambridge Univ. Press, 1960, 307–312.

Shalev A. [1995] Finite p-groups, *in: Proc. Finite and Locally Finite Groups, Istanbul-1994*, Kluwer, Amsterdam, 401–450.

Wall G. E. [1978] Lie methods in group theory, *in: Topics in Algebra (Proc. Eighteenth Summer Res. Inst. Austral. Math. Soc., Canberra, 1978* (Lecture Notes in Math., **697**), Springer, Berlin et al., 137–173.

Zelmanov E. I. [1995] Lie ring methods in the theory of nilpotent groups, *in: Proc. Groups'93/St. Andrews, vol. 2* (London Math. Soc. Lecture Note Ser., **212**), Cambridge Univ. Press, 567–585.

Research papers

Alperin J. L. [1962] Automorphisms of solvable groups, *Proc. Amer. Math. Soc.*, **13**, 175–180.

Alperin J. L., Glauberman G. [1997] Limits of abelian subgroups of finite p-groups, *Preprint*, Univ. of Chicago.

Baer R. [1938] Groups with abelian central quotient groups, *Trans. Amer. Math. Soc.*, **44**, 357–386.

Bauman S. F. [1966] The Klein group as an automorphism group without fixed points, *Pacific J. Math.*, **18**, 9–13.

Bender H. [1967] Über den größten p'-Normalteiler in p-auflösbaren Gruppen, *Arch. Math. (Basel)*, **18**, 15–16.

Berger T. [1973] Nilpotent fixed point free automorphism groups of solvable groups, *Math. Z.*, **131**, 305–312.

Berger T. [1974–78] Hall–Higman type theorems. I, *Canad. J. Math.*, **29** (1974), 513–531; II, *Trans. Amer. Math. Soc.*, **205** (1975), 47–69; III, *Trans. Amer. Math. Soc.*, **228** (1977), 47–83; IV, *Proc. Amer. Math. Soc.*, **37** (1973), 317–325; V, *Pacific J. Math.*, **73** (1977), 1–62; VI, *J. Algebra*, **51** (1978), 416–424; VII, *Proc. London Math. Soc. (3)*, **31** (1975), 21–54.

Blackburn N. [1958] On a special class of p-groups, *Acta Math.*, **100**, 45–92.

Blackburn N. [1965] Conjugacy in nilpotent groups, *Proc. Amer. Math. Soc.*, **16**, 143–148.

Borovik A. V., Khukhro E. I. [1976] Automorphism groups of finite p-groups, *Mat. Zametki*, **19**, 401–418 (Russian); English transl., *Math. Notes*, **19**, 245–255.

Brandl R., Caranti A., Scoppola C. M. [1992] Metabelian thin p-groups, *Quart. J. Math. Oxford Ser. (2)*, **43**, 157–173.

Caranti A., Mattarei S., Newman M. F., Scoppola C. M. [1996] Thin groups of prime-power order and thin Lie algebras, *Quart. J. Math. Oxford Ser. (2)*, **47**, 279–296.

Dade E. C. [1972] Carter subgroups and Fitting heights of finite soluble groups, *Illinois J. Math.*, **13**, 347–369.

Donkin S. [1987] Space groups and groups of prime power order. VIII. Pro-p-groups of finite coclass and p-adic Lie algebras, *J. Algebra*, **111**, 316–342.

Feit W., Thompson J. G. [1963] Solvability of groups of odd order, *Pacific J. Math.*, **13**, 773–1029.

Finken H., Neubüser J., Plesken W. [1980] Space groups and groups of prime power order. II. Classification of space groups by finite factor-groups, *Arch. Math. (Basel)*, **35**, 203–209.

Fong P. [1976] On orders of finite groups and centralizers of p-elements, *Osaka J. Math.*, **13**, 483–489.

Gross F. [1966] Solvable groups admitting a fixed-point-free automorphism of prime power order, *Proc. Amer. Math. Soc.*, **17**, 1440–1446.

Hall P. [1934] A contribution to the theory of groups of prime-power order, *Proc. London Math. Soc. Ser. 2*, **36**, 29–95.

Hall P. [1958] Some sufficient conditions for a group to be nilpotent, *Illinois J. Math.*, **2**, 787–801.

Hall P., Higman G. [1956] The p-length of a p-soluble group and reduction theorems for Burnside's problem, *Proc. London Math. Soc. (3)*, **6**, 1–42.

Hartley B., Isaacs M. [1990] On characters and fixed points of coprime operator groups, *J. Algebra*, **131**, 342–458.

Hartley B., Meixner T. [1980] Periodic groups in which the centralizer of an involution has bounded order, *J. Algebra*, **64**, 285–291.

Hartley B., Meixner T. [1981] Finite soluble groups containing an element of prime order whose centralizer is small, *Arch. Math. (Basel)*, **36**, 211–213.

Hartley B., Turau V. [1987] Finite soluble groups admitting an automorphism of prime power order with few fixed points, *Math. Proc. Cambridge Philos. Soc.*, **102**, 431–441.

Higgins P. J. [1954] Lie rings satisfying the Engel condition, *Math. Proc. Cambridge Philos. Soc.*, **50**, 8–15.

Higman G. [1957] Groups and rings which have automorphisms without nontrivial fixed elements, *J. London Math. Soc. (2)*, **32**, 321–334.

Hughes I. [1985] Groups with fixed-point-free automorphisms, *C. R. Math. Rep. Acad. Sci. Canada*, **7**, 61–66.

Kegel O. H. [1960] Die Nilpotenz der H_p-Gruppen, *Math. Z.*, **75**, 373–376.

Khukhro E. I. [1979] Verbal commutativity and fixed points of p-automorphisms of finite p-groups, *Mat. Zametki*, **25**, 505–512 (Russian); English transl., *Math. Notes*, **25**, 262–265.

Khukhro E. I. [1982] On nilpotent subdirect products, *Sibirskii Mat. Zh.*, **23**, No 6, 178–180 (Russian).

Khukhro E. I. [1985] Finite p-groups admitting an automorphism of order p with a small number of fixed points, *Mat. Zametki*, **38**, 652–657 (Russian); English transl., *Math. Notes.*, **38** (1986), 867–870.

Khukhro E. I. [1989] Groups and Lie rings admitting almost regular automorphisms of prime order, *in: Proc. Int. Conf. Theory of Groups, Brixen/Bressanone, 1989* (Suppl. Rend. Circ. Mat. Palermo, 1990), 183–192.

Khukhro E. I. [1990] Groups and Lie rings admitting an almost regular automorphism of prime order, *Mat. Sbornik*, **181**, 1207–1219 (Russian); English transl., *Math. USSR Sbornik*, **71** (1992), 51–63.

Khukhro E. I. [1993a] Finite p-groups admitting p-automorphisms with few fixed points, *Mat. Sbornik*, **184**, 53–64; English transl., *Russian Acad. Sci. Sbornik Math.*, **80** (1995), 435–444.

Khukhro E. I. [1996] Almost regular automorphisms of finite groups of bounded rank, *Sibirskii Mat. Zh.*, **37**, 1407–1412 (Russian); English transl., *Siberian Math. J.*, **37**, 1237–1241.

Khukhro E. I., L'vov I. V. [1978] Letter to the Editors on problem 5.56, *The Kourovka Notebook, 6th ed.*, Institute of Mathematics, Novosibirsk (Russian).

Khukhro E. I., Makarenko N. Yu. [1996a] On Lie rings admitting an automorphism of order 4 with few fixed points, *Algebra i Logika*, **35**, 41–78 (Russian); English transl., *Algebra and Logic*, **35**, 21–43.

Khukhro E. I., Makarenko N. Yu. [1996b] Nilpotent groups admitting almost regular automorphism of order 4, *Algebra i Logika*, **35**, 314–333 (Russian); English transl., *Algebra and Logic*, **35**, 176–187.

Khukhro E. I., Makarenko N. Yu. [1997] On Lie rings admitting an automorphism of order 4 with few fixed points. II, *to appear in Algebra and Logic.*

Khukhro E. I., Shumyatsky P. [1995] On fixed points of automorphisms of Lie rings and locally finite groups, *Algebra i Logika*, **34**, 706–723 (Russian); English transl., *Algebra and Logic*, **34**, 395–405.

Kiming I. [1988] Structure and derived length of finite p-groups possessing an automorphism of p-power order having exactly p fixed points, *Math. Scand.*, **62**, 153–172.

Kolmakov Yu. A. [1984] Varieties of groups whose elements satisfy some conditions close to solubility, *Mat. Zametki*, **35**, 735–738 (Russian); English transl., *Math. Notes.*, **35** (1985), 389–391.

Kostrikin A. I. [**1959**] The Burnside problem, *Izv. Akad. Nauk SSSR Ser. Mat.*, **23**, 3–34 (Russian); English transl., *Transl. II Ser. Amer. Math. Soc.*, **36** (1964), 63–99.

Kovács L. G. [**1961**] Groups with regular automorphisms of order four, *Math. Z.*, **75**, 277–294.

Kreknin V. A. [**1963**] The solubility of Lie algebras with regular automorphisms of finite period, *Dokl. Akad. Nauk SSSR*, **150**, 467–469 (Russian); English transl., *Math. USSR Doklady*, 4 (1963), 683–685.

Kreknin V. A. [**1967**] The solubility of Lie algebras with a regular automorphism, *Sibirsk. Mat. Zh.*, **8**, 715–716 (Russian); English transl., *Siberian Math. J.*, **8** (1968), 536–537.

Kreknin V. A., Kostrikin A. I. [**1963**] Lie algebras with regular automorphisms, *Dokl. Akad. Nauk SSSR*, **149**, 249–251 (Russian); English transl., *Math. USSR Doklady*, **4**, 355–358.

Kurzweil H. [**1971**] p-Automorphismen von auflösbaren p'-Gruppen, *Math. Z.*, **120**, 326–354.

Lazard M. [**1954**] Sur les groupes nilpotents et les anneaux de Lie, *Ann. Sci. École Norm. Supr.*, **71**, 101–190.

Lazard M. [**1965**] Groupes analytiques p-adiques, *Publ. Math. Inst. Hautes Études Sci.*, **26**, 389–603.

Leedham-Green C. R. [**1994a**] Pro-p-groups of finite coclass, *J. London Math. Soc.*, **50**, 43–48.

Leedham-Green C. R. [**1994b**] The structure of finite p-groups, *J. London Math. Soc.*, **50**, 49–67.

Leedham-Green C. R., McKay S. [**1976**] On p-groups of maximal class. I, *Quart. J. Math. Oxford Ser. (2)*, **27**, 297–311.

Leedham-Green C. R., McKay S. [**1978a**] On p-groups of maximal class. II, *Quart. J. Math. Oxford Ser. (2)*, **29**, 175–186.

Leedham-Green C. R., McKay S. [**1978b**] On p-groups of maximal class. III, *Quart. J. Math. Oxford Ser. (2)*, **29**, 281–299.

Leedham-Green C. R., McKay S., Plesken W. [**1986**] Space groups and groups of prime power order. V. A bound to the dimension of space groups with fixed coclass, *Proc. London Math. Soc.*, **52**, 73–94.

Leedham-Green C. R., Newman M. F. [**1980**] Space groups and groups of prime power order. I, *Arch. Math. (Basel)*, **35**, 193–202.

Lubotzky A., Mann A. [**1987**] Powerful p-groups. I: finite groups, *J. Algebra*, **105**, 484–505; II: p-adic analytic groups, *ibid.*, 506–515.

Magnus W. [**1940**] Über Gruppen und zugeordnete Liesche Ringe, *J. Reine Angew. Math.*, **182**, 142–159.

Magnus W. [1950–53] A connection between the Baker–Hausdorff formula and a problem of Burnside, *Ann. of Math. (2)*, **52** (1950), 111–126; Errata, *Ann. of Math. (2)*, **57** (1953), 606.

Makarenko N. Yu. [1992] Almost regular automorphisms of prime order, *Sibirsk. Mat. Zh.*, **33**, No 5, 206–208 (Russian). English transl., *Siberian Math. J.*, **33**, 932–934.

Makarenko N. Yu. [1993] Finite 2-groups that admit automorphisms of order 4 with a small number of fixed points, *Algebra i Logika*, **32**, 402–427 (Russian); English transl., *Algebra and Logic*, **32**, 215–230.

Makarenko N. Yu. [1994] Nilpotent groups with an almost regular automorphism of prime order, *Sibirsk. Mat. Zh.*, **35**, 630–632 (Russian); English transl., *Siberian Math. J.*, **35**, 564–565.

Mal'cev A. I. [1949] Nilpotent groups without torsion, *Izv. Akad. Nauk SSSR, Ser. Mat.*, **13**, 201–212 (Russian).

Mal'cev A. I. [1958] On homomorphisms onto finite groups, *Učen. Zap. Ivanovo Pedag. Inst.*, **18**, 49–60 (Russian).

Mann A. [1992] Space groups and groups of prime power order. VII. Powerful p-groups and uncovered p-group, *Bull. London Math. Soc.*, **24**, 271–276.

McKay S. [1987] On the structure of a special class of p-groups, *Quart. J. Math. Oxford Ser. (2)*, **38**, 489–502.

McKay S. [1990] On the structure of a special class of p-groups. II, *Quart. J. Math. Oxford Ser. (2)*, **41**, 431–448.

Medvedev Yu. [1994a] p-Groups, Lie p-rings and p-automorphisms, *Preprint*, Univ. of Wisconsin, Madison.

Medvedev Yu. [1994b] p-Divided Lie rings and p-groups, *Preprint*, Univ. of Wisconsin, Madison.

Medvedev Yu. [1994c] Groups and Lie rings with almost regular automorphisms, *J. Algebra*, **164**, 877–885.

Meixner Th. [1979] Über endliche Gruppen mit Automorphismen, deren Fixpunktgruppen beschränkt sind, *Dissertation*, Erlangen–Nürnberg Univ.

Meixner Th. [1986] Solvable groups admitting an automorphism of prime power order whose centralizer is small, *J. Algebra*, **99**, 181–190.

Novikov P. S., Adyan S. I. [1968] Infinite periodic groups. I, II, III, *Izv. Akad. Nauk SSSR Ser. Mat.*, **32**, 212–244, 251–524, 709–731 (Russian); English transl., *Math. USSR Izvestiya*, **2** (1969), 209–236, 241–480, 665–685.

Ol'shanskii A. Yu. [1982] Groups of bounded exponent with subgroups of prime orders, *Algebra i Logika*, **21**, 553–618 (Russian); English transl., *Algebra and Logic*, **21** (1983), 369–418.

Panfërov B. A. [1980] On nilpotent groups with lower central factors of minimal ranks, *Algebra i Logika*, **19**, 701–706 (Russian); English transl., *Algebra and Logic*, **19** (1981), 455–458.

Razmyslov Yu. P. [1971] On Engel Lie algebras, *Algebra i Logika*, **10**, 33–44 (Russian); English transl., *Algebra and Logic*, **10** (1972), 21–29.

Roseblade J. E. [1965] On groups in which every subgroup is subnormal, *J. Algebra*, **2**, 402–412.

Sanov I. N. [1952] Establishment of a connection between periodic groups with period a prime number and Lie rings, *Izv. Akad. Nauk SSSR Ser. Mat.*, **16**, 23–58 (Russian).

Shalev A. [1993a] On almost fixed point free automorphisms, *J. Algebra*, **157**, 271–282.

Shalev A. [1993b] Automorphisms of finite groups of bounded rank, *Israel J. Math.*, **82**, 395–404.

Shalev A. [1994] The structure of finite p-groups: effective proof of the coclass conjectures, *Invent. Math.*, **115**, 315–345.

Shalev A., Zelmanov E. I. [1992] Pro-p-groups of finite coclass, *Math. Proc. Cambridge Philos. Soc.*, **111**, 417–421.

Shepherd R. [1971] p-Groups of maximal class, *Ph. D. Thesis*, Univ. of Chicago.

Shirshov A. I. [1958] On free Lie rings, *Mat. Sbornik*, **45**, 113–122 (Russian).

Shult E. [1965] On groups admitting fixed point free Abelian operator groups, *Illinois J. Math.*, **9**, 701–720.

Shumyatsky P. [1988] Groups with regular elementary abelian 2-groups of automorphisms, *Algebra i Logika*, **27**, 715–730 (Russian); English transl., *Algebra and Logic*, **27**, 447–457.

Stuart A. G. R. [1966] On the class of certain nilpotent groups, *Proc. Roy. Soc., Ser. A*, **292**, 374–379.

Thompson J. G. [1959] Finite groups with fixed-point-free automorphisms of prime order, *Proc. Nat. Acad. Sci. U.S.A.*, **45**, 578–581.

Thompson J. G. [1964a] Automorphisms of solvable groups, *J. Algebra*, **1**, 259–267.

Thompson J. G. [1964b] Fixed points of p-groups acting on p-groups, *Math. Z.*, **86**, 12–13.

Turull A. [1984] Fitting heights of groups and of fixed points, *J. Algebra*, **86**, 555–566.

Turull A. [1991] Groups of automorphisms and centralizers, *Preprint*, Univ. of Miami.

Vaughan-Lee M. R., Wiegold J. [1981] Countable locally nilpotent groups of finite exponent without maximal subgroups, *Bull. London Math. Soc.*, **14**, 45–46.

Weigel T. [1994] Exp and Log functors for the categories of powerful p-central groups and Lie algebras, *Habilitationsschrift*, Freiburg in Breisgau.

Winter D. J. [1968] On groups of automorphisms of Lie algebras, *J. Algebra*, **8**, No 2, 131–142.

Zassenhaus H. [1939a] Liesche Ringe mit Primzahlcharacteristik, *Abhandlungen Math. Sem. Univ. Hamburg*, **13**, 1–100.

Zassenhaus H. [1939b] Ein Verfahren, jeder endlichen p-Gruppe einen Lie-Ring mit der Charakteristik p zuzuordnen, *Abh. Math. Sem. Univ. Hamburg*, **13**, 200–207.

Zelmanov E. I. [1989] On some problems of the theory of groups and Lie algebras, *Mat. Sbornik*, **180**, 159–167 (Russian); English transl., *Math. USSR Sbornik*, **66** (1990), 159–168.

Zelmanov E. I. [1990] A solution of the Restricted Burnside Problem for groups of odd exponent, *Izv. Akad. Nauk SSSR Ser. Mat.*, **54**, 42–59 (Russian); English transl., *Math. USSR Izvestiya*, **36** (1991), 41–60.

Zelmanov E. I. [1991] A solution of the Restricted Burnside Problem for 2-groups, *Mat. Sbornik*, **182**, 568–592 (Russian); English transl., *Math. USSR Sbornik*, **72** (1992), 543–565.

Index of names

Subject index

List of symbols

$\mathbb{N}$	1	τ_g	8
$\mathfrak{N}_c$	39, 63	$UT_n(k)$	34
$N_G(M)$	2	$[u]$	67
$nt_n(k)$	82	$\mathfrak{V}(G)$	21
$\pi(c!)$	122	$\mathbb{Z}$	1
$\mathbb{Q}$	1	$\mathbb{Z}G$	13
Q_8	32	$Z(G)$	3
p'	121	$\zeta_i(G)$	38, 72
$\mathbb{R}$	1	$\Phi(P)$	53
$\mathbf{r}U$	176	$\bar\varphi$	32, 78
$\mathbb{S}_M, \mathbb{S}_n$	1	$\Omega_i(P)$	4
$T_n(k)$	34